# Advances in Agroforestry

Volume 11

**Series editor**
P.K. Ramachandran Nair, Gainesville, USA

**Aims and Scope**

Agroforestry, the purposeful growing of trees and crops in interacting combinations, began to attain prominence in the late 1970s, when the international scientific community embraced its potentials in the tropics and recognized it as a practice in search of science. During the 1990s, the relevance of agroforestry for solving problems related to deterioration of family farms, increased soil erosion, surface and ground water pollution, and decreased biodiversity was recognized in the industrialized nations too. Thus, agroforestry is now receiving increasing attention as a sustainable land-management option the world over because of its ecological, economic, and social attributes. Consequently, the knowledge-base of agroforestry is being expanded at a rapid rate as illustrated by the increasing number and quality of scientific publications of various forms on different aspects of agroforestry.

Making full and efficient use of this upsurge in scientific agroforestry is both a challenge and an opportunity to the agroforestry scientific community. In order to help prepare themselves better for facing the challenge and seizing the opportunity, agroforestry scientists need access to synthesized information on multi-dimensional aspects of scientific agroforesty.

The aim of this new book-series, Advances in Agroforestry, is to offer state-of-the art synthesis of research results and evaluations relating to different aspects of agroforestry. Its scope is broad enough to encompass any and all aspects of agroforestry research and development. Contributions are welcome as well as solicited from competent authors on any aspect of agroforestry. Volumes in the series will consist of reference books, subject-specific monographs, peer-reviewed publications out of conferences, comprehensive evaluations of specific projects, and other book-length compilations of scientific and professional merit and relevance to the science and practice of agroforestry worldwide.

More information about this series at http://www.springer.com/series/6904

Pablo Luis Peri • Francis Dube
Alexandre Varella
Editors

# Silvopastoral Systems in Southern South America

Springer

*Editors*
Pablo Luis Peri
Universidad Nacional de la Patagonia Austral (UNPA)
Instituto Nacional de Tecnología Agropecuaria (INTA), CONICET
Río Gallegos, Argentina

Francis Dube
Department of Silviculture
University of Concepción
VIII - Concepción, Chile

Alexandre Varella
EMBRAPA
Brasilia, Brasília, Brazil

ISSN 1875-1199 ISSN 1875-1202 (electronic)
Advances in Agroforestry
ISBN 978-3-319-24107-4 ISBN 978-3-319-24109-8 (eBook)
DOI 10.1007/978-3-319-24109-8

Library of Congress Control Number: 2016931907

Springer Cham Heidelberg New York Dordrecht London


Printed on acid-free paper

Springer International Publishing AG Switzerland is part of Springer Science+Business Media (www.springer.com)

**Book Abstract**

This multi-authored volume contains peer-reviewed chapters from leading researchers and professionals in silvopastoral systems topic in the subtropical and temperate zones of South America (Argentina, Chile and South Brazil). It is a compendium of original research articles, case studies, and regional overviews and summarizes the current state of knowledge on different components and aspects (pasture production, animal production, trees production, carbon sequestration, conservation) of silvopatoral systems in native forests and tree plantations. The main hypothesis of the book is that farmers have integrated tree and pasture/grassland species in their land use systems to reach higher production per unit of land area, risk avoidance, product diversification, and sustainability. These production systems also impact positively in main ecosystem processes. Management of these productive systems, Policy and Socioeconomic Aspects provide great opportunities and challenges for farmers and policy makers in our region. The book is unique on this subject in Southern South America and constitutes a valuable reference material for graduate students, professors, scientists and extensionists who work with silvopastoral systems.

**Book Keywords**

Animal production • Tree-understorey interaction • Native forest conservation • Silvopastoral systems design • Policy for silvopastoral development

# Foreword

Agroforestry is now well recognized as a land-use activity that is different from sole-crop agriculture and plantation forestry. To most land-use experts, however, agroforestry presents the image of low-input, subsistence, "dirt-poor" tropical farming systems involving an array of some trees and crops and maybe some animals, often in haphazard arrangements in contrast to the uniformly spaced rows of trees or crops they are used to. No wonder the mention of the word also brings out some strange reaction – even disdain – from those who firmly, albeit falsely, believe that the solution to the world's food and agricultural problems rests entirely on commercial and industrialized land management.

The past few decades of significant developments in agroforestry, however, have helped change some such lopsided perceptions. Lately, the role of agroforestry in addressing and mitigating some of the environmental problems created and/or exacerbated by commercial agricultural and forestry production enterprises has become increasingly clear. A wave of belief and enthusiasm, though still supported more by convictions rather than empirical research, is emerging in North America and Europe that in order to meet the society's needs and aspirations for forest-derived goods and services, we must – and can – find ways of deriving some portion of those benefits from agricultural lands through agroforestry. Indeed, in many places, the only opportunity to provide forest-based benefits such as wildlife habitat or forested riparian systems is through the increased adoption of agroforestry on agricultural lands.

While such interesting developments have helped to add North America and Europe to the global agroforestry map, some regions of the world such as the southern parts of South America have seldom been mentioned in agroforestry literature. With limited or no personal exposure to the region, most agroforestry enthusiasts including myself have been ignorant of and had seldom been encouraged to think about the existence or relevance of agroforestry in the regions southward from southern Brazil. So, it was with some trepidation that I entertained Dr. Pablo Peri's desire for a discussion during his visit to the University of Florida in September 2013, to talk about silvopastoral systems in that part of the world and the studies and experience on such systems that he and his colleagues had accumulated. In retrospect, it is rather amazing how Dr. Peri's visit and our discussions have led to this book.

The speed, tenacity, and momentum at which Dr. Peri moved forward with the idea and bring it to fruition are really outstanding. From organizing a team of dedicated coeditors, commissioning a talented group of chapter authors, organizing insightful peer review of the chapters, following up with the publisher, and, above all, pushing me incessantly for doing my part in facilitating the publication, he has spared no stone unturned. I warmly and sincerely congratulate Dr. Peri and his coeditors and all the chapter authors on this outstanding accomplishment.

This publication represents a significant contribution to the growing body of agroforestry literature worldwide. The information presented here will be of immense value to colleagues in other parts of the world where such systems of land use have scope and potential. I hope Dr. Peri and colleagues' commendable effort will provide inspiration to other agroforestry experts to organize such high-quality publications to bring to light the little-reported but valuable results and experiences from other parts of the world.

Editor, Advances in Agroforestry Book-Series P.K. Ramachandran Nair
University of Florida, Gainesville, FL, USA
June 2015

# Contents

# 1 Silvopastoral Systems in the Subtropical and Temperate Zones of South America: An Overview

Pablo Luis Peri, Francis Dube, and Alexandre Costa Varella

**Abstract**

In the subtropical and temperate zones of South America (Argentina, Chile and Southern Brazil), silvopastoral systems have become an economical, ecological and productive alternative. These systems incorporate exotic tree species or managed native forests into farming systems allowing the production of trees and livestock from the same unit of land. This provides diversification of farm income, either directly from the sale of timber and animals, and/or indirectly by the provision of stock shelter and beneficial effects on soil conservation. Thus, these systems can be more biologically productive, profitable and sustainable than forestry or animal production monocultures. However, livestock grazing is one of the most widespread land uses in South America and is arguably the land use that has had the greatest impact on regional biodiversity. This multi-authored book is a compendium of original research articles, case studies, and regional overviews that summarizes the current state of knowledge on different components and aspects (pasture, animal and trees productions, carbon sequestration, and conservation) of silvopastoral systems in native forests and tree plantations.

**Keywords**

Conservation • Livestock • Native forest • Pasture • Policy • Tree plantations

P.L. Peri (✉)
Universidad Nacional de la Patagonia Austral (UNPA), Instituto Nacional de Tecnología Agropecuaria (INTA), CONICET, CC 332, 9400 Río Gallegos, Argentina
e-mail: peri.pablo@inta.gob.ar

F. Dube
Department of Silviculture, Faculty of Forest Sciences, University of Concepción, Victoria 631, Casilla 160-C, VIII - Concepción, Chile

A.C. Varella
Brazilian Agricultural Research Corporation (EMBRAPA), South Livestock Center (CPPSUL), BR 153, Km 603, 242, 96401-970 Bage, RS, Brazil


P.L. Peri et al. (eds.), *Silvopastoral Systems in Southern South America*, Advances in Agroforestry 11, DOI 10.1007/978-3-319-24109-8_1

In many developing countries, trees and shrubs have traditionally been managed and used on agricultural landscapes for shelter, wood, browse, fruit or nuts. Silvopastoral systems, as part of agroforestry practices, are an approach to land-use and technologies where woody perennials (trees, shrubs, palms, etc.) are deliberately used on the same land management unit as livestock with different spatial arrangements or temporal sequences. Thus, these systems incorporate exotic tree species or managed native forests into farming systems allowing the production of trees and livestock from the same unit of land.

In silvopastoral systems, there are both ecological and economic interactions between the different components. Thus, silvopastoral systems are those that have been designed to improve beneficial ecological interactions, which may be demonstrated as improvement in yield per unit area, resource use efficiency and/or enhancement in environmental issues. Silvopastoral advantages can be described as the provision of multiple products (e.g., food, wood, fodder, mulch, medicinal plants) or services (e.g., maintenance of soil fertility, control of erosion, microclimate improvement, biodiversity enhancement, watershed protection, carbon sequestration) by the trees. Silvopastoral constraints mainly deal with potential competition between trees and pasture or grassland for water, light and nutrients as well as for farm resources such as land and labor.

Agroforestry systems have a rich history of development and the concept was initially implemented in tropical regions, within the context of developing nations where land shortages due to rapid population growth of indigenous people demanded efficient production systems for both food and wood resources. A classic example can be found in the managed "conucos" carried out for centuries by native people in Central America, which are gardens where agriculture has been practiced in a traditional manner (Esquivel and Hammer 1988). Also, the swidden cultivation, which is also known as slash-and-burn agriculture, has been widely practiced for centuries in tropical forests of South America. Another example of agroforestry systems is the "chacra" system practiced by native communities in Central and South America, which consists of small-scale shifting cultivation where food crops such as cassava and banana are cultivated under forest tree species, which provide timber and bark. The introduction of trees and/or livestock into agricultural plots appeared as an interesting approach to preserving tropical forests. Worldwide awareness of the value of tropical forests, and the need for conservation through agroforestry practices increased the importance for the research, political, and financial support in the twentieth century. Since the late 1970s, agroforestry has increasingly become accepted as a sustainable and promising land use system on both farms and in forests (Nair 1993).

The range of conditions in South America provides insight about how factors such as geography, climate, culture, and markets affect silvopastoral systems selection and implementation. The area covered in the book includes southern region of Brazil (including the states of Rio Grande do Sul, Santa Catarina and the south-central region of Paraná), and contrasting regions in Argentina and Chile (Fig. 1.1). This region is located between latitudes 24° and 55° S. Climate ranges from subtropical (with frequent frosts in winter and warm summers) to cold temperate (characterized by short, cool summers and long, snowy and frozen winters). Average annual temperatures range from 5 to 20 °C and the annual average rainfall varies from 200 to 2,000 mm. The distribution of natural vegetation is largely determined by climate in the region with presence of rain forests, savannas with tall grasses occupying scattered patches of trees, dry land ecosystems and cold temperate forests. However, anthropogenic influence on land cover change has been more evident in the context of increasing deforestation and agricultural expansion. The UN Population Division predicts that population in Latin America and the Caribbean could rise to 784 million by 2050 (United Nations 2015). Therefore, the forests in the region will very likely continue to be cleared for agriculture and ranching. The conversion of natural environments into managed ones contributed to major environmental problems, such as pollution, land degradation and loss of biodiversity. In this con-

**Fig. 1.1** Location of the area covered in the present book (*dark green*) in South America (Map from Danilo Serra da Rocha, EMBRAPA South Livestock) (Color figure online)

text, sustainable silvopastoral systems are suggested as a key solution to the conflict between expanding agricultural production and conserving natural ecosystems.

In South America, many environmental and social problems have arisen from the embracing of technological agriculture and the green revolution. For example, in Argentina in response to the high international prices of agricultural commodities (particularly soybean) and technological changes (mainly genetically modified soybean cultivars), the country (mainly the Northern provinces) has seen a very rapid expansion of industrial agriculture, a major driver of recent land use changes (Secretaría de Agricultura, Ganadería, Pesca y Alimentación [SAGPyA] 2009). Furthermore, livestock grazing is one of the most widespread land uses in South America and is arguably the land use that has had the greatest impact on regional biodiversity. In Latin America, more than 90 million ha of land is under pasture, mostly as a result of forest conversion to cattle ranching (FAO 1999), where meat and milk consumption probably assume greater political and economic importance than in any other region of the world. However, they are also one of the major causes of environmental degradation. This has led to increased research and practical activity in the development of sustainable farming systems. Governmental and international agencies, universities and research institutes have strongly encouraged many scientific projects on environmentally, socially and economically sustainable development. In this context, in South America in the last two decades, silvopastoral systems became an economical, ecological and social productive alternative. As a result, the benefits that provide silvopastoral systems in some regions or areas are based on the basic premise that these systems can be more biologically productive, profitable and sustainable than forestry or animal production monocultures. The trees in silvopastoral systems enhance nutrient uptake from the soil, which is then passed to the grass through the degradation of organic matter so improving both soil fertility and forage quality. The trees, in addition to enhancing animal welfare due to their moderating effects on climate extremes, can be a source of wood (timber and fuel), bio-diesel, as well as improving environmental aspect (e.g., aesthetics), reducing soil erosion and water loss and increasing biodiversity. Also, silvopastoral systems are capable of fixing significant amounts of carbon (C) in the soil under the improved pastures and in the standing tree biomass (Dube et al. 2012; Peri 2011). Given the vast area of land currently managed as ruminant production systems in Latin America, the potential for climate change mitigation through C sequestration is huge (Dube et al. 2011). Thus, the payment for environmental services for C sequestration in silvopastoral systems can be an important incentive for ensuring their widespread adoption.

In Argentina, while silvopastoral-systems implementation using exotic tree species planted in natural grasslands (or with exotic pasture species) had reached ~140,000 ha (mainly in Argentine northeastern provinces of Misiones and northern Corrientes, Lower Delta of the Paraná River and NW Patagonia) by 2012, native forests under silvopastoral use is ~6,826,000 ha with different degrees of management strategies (Peri 2012). There are also differences in the motivation behind the silvopastoral systems between landowners. For example, while in northeastern Argentina many farmers or companies are seeking to diversify and improve their economic profits, silvopastoral systems in Patagonia can allow ranchers to protect the cattle and sheep in extensive paddocks against strong prevailing winds and to control severe erosion problems. Also, several landowners believe that livestock under silvopastoral use has the advantage to provide annual income under a more certain markets together with the long term wait for timber cash from tree harvest.

In Chile, the Forest Research Institute (INFOR) has been responsible since 2006 for the establishment of silvopastoral systems in small properties. Between 2006 and 2008, 342 ha were established. In 2011, there was an additional 72 ha in the Maule, Biobío, Araucania and Aysén Regions, representing 23 % of the total area under agroforestry for that year. Finally in 2012, new silvopastoral systems were implemented in the O'Higgins, Maule and Los Ríos Regions, totaling 47 ha, i.e. 30 % of the surface with agroforestry (A. Lucero, pers. comm. 2013). The following paragraphs present a brief summary of the main silvopastoral systems established in planted and native forests across the country. *Prosopis tamarugo* is a native and endemic species in northern Chile, which grows mainly in the Tarapacá Region, in extreme soil and climate conditions. It is a multipurpose, drought resistant species. Its main use is for soil protection and carbon sequestration. It is also a source of forage tree, mainly because of its tender shoots and fruits that are used by domestic animals, sheep and goats owned by small farmers, as well as wild animals such as camels (González and Maraboli 1997). *Acacia saligna* and *Atriplex* sp. are other species that have been used in the north and central-south region of the country. Fodder shrubs are valued because of their capacity to be a source of forage for sheep and goats during dry periods (summer-autumn period), when there is no natural source of green fodder. Over 40,000 ha have been planted in northern Chile, mainly in the Coquimbo Region to supplement the shortage of fodder in summer (Perret and Mora 2001). On the other hand, the *Acacia caven* steppes, or "espinal", cover an approximate area of 2 million ha, from the Coquimbo Region in the north to the Biobío Region in the south. They are mostly located in areas where agriculture and livestock are the most important productive activities. Proper management of this species is required, however, because of overgrazing and overexploitation for fuel wood and charcoal production over several decades. It has been shown that with good management systems involving *Acacia* spp. helps increase the productivity of both natural and artificial pastures (Ovalle et al. 1996). The "Agrícola y Forestal El Álamo" Company located near the town of Parral in the Maule Region owns 3,232 ha of *Populus* sp., all under agrosilvopastoral use.

Poplar is mixed with corn during the first 3 years, after which the plantation system is grazed throughout the rotation until harvest. Within the remaining native forest area of temperate Chile, second-growth *Nothofagus* forests play a prominent role, not only for the large covered area but also for its economic and socioeconomic importance (INFOR 2012). A large proportion of this forest belongs to small landowners who use their properties as pasture for cattle grazing and shelter for animal protection. However, the presence of cattle cause stagnation of natural regeneration processes through browsing and leaves the soil compacted by trampling. In other cases the silvicultural system being practiced is "coppice" type, in which selected areas of second-growth forests are clear-cut every 15 years. This practice permits the use of stems and branches for firewood and charcoal, while waiting for the vegetative regeneration of stump shoots, which produces significant exports of nutrients and, in the long term, destroys the capacity of the forest to regenerate on its own. And last, but not the least, there is no governmental regulation regarding the use of second-growth forests for pastures, so this source of degradation is becoming progressively worse over time (Dube et al. 2014). In Chilean Patagonia, the use of silvopastoral systems is a practice that has been studied since 2002 by INFOR and the Agricultural Research Institute (INIA), with the objective of reintroducing the tree in a friendly way to the cattle farming productive systems and with the producers. In this way, it is expected to present a new form of productive management of the soil, to improve their productivity, reduce their level of erosion and protect them from the most important erosive factor in the region, wind (Teuber and Ganderatz 2009; Sotomayor 2010).

In Brazil, the integration of beef cattle with forests date from the mid-eighteenth century (Chang 1985), known as traditional "faxinais" systems, mainly established in the area of occurrence of *Araucaria* native forests (*Araucaria angustifolia*). Usually beef cattle was introduced in conventional tree plantations as a strategy to improve early cash flow, besides getting the benefits of controlling the growth of undesired plants in the understory, reducing the risks of fire inside the forest (Saibro 2001; Porfirio-da-Silva et al. 2009; Varella et al. 2012). More recently, the increase in the costs of agricultural inputs, a decrease of native forest cover, an increase of degraded areas in agricultural activities and livestock (mainly soil erosion), with the phenomenon of rural exodus and concerns about animal welfare were the reasons to research institutions and rural extension service seek for new and more sustainable production systems, especially involving the integration of forestry with agriculture and livestock systems (Dube et al. 2002). The area covered in Brazil in this book is located between latitudes 24° and 33°S, occupying an area of approximately 576.410 km$^2$ (6 % of the national territory). In the southern states of Brazil, agriculture and extensive livestock systems are predominant. Over the last 5 years, forestry investments were maintained stable in this region. Currently, the expansion area for conventional forestry is limited, but there is still land area available for the expansion of forestry-cropping-livestock integrated systems. The agroforestry systems are noticed as a sustainable alternative in Brazil, leading to new challenges from farmer to industry sector. The main issues that match producer interests and needs of this region are related to cattle and sheep welfare, strategic forage supply during periods of extreme weather, soil conservation, self-consumption of wood in rural properties as well as for direct sale and land use efficiency. From these observations, many challenges and opportunities arise for scientific institutions, rural extension agencies and public policies aiming to promote more sustainable and diversified farming systems in this part of Brazil.

Research on silvopastoral systems has also spread widely, and research papers on topics of interest to practitioners have become more prevalent in conference proceedings. In Latin America, a biannual conference series on agroforestry was initiated in Colombia, in 1999, with seven well-attended conferences held to date. In Argentina, since 2009 triennial silvopastoral Congress increased the interest of farmers, scientists and governmental institutions. The proceedings of these conferences are demanded by individuals and groups with an Interest in sustainable animal production, diversified farm economies, biodiversity and animal welfare, among others.

Despite the many promises that silvopastoral systems in South America hold under appropriate conditions, there are also limitations arising from biophysical, socioeconomic and political conditions. For example, depending on political and demographic pressures, it may be necessary to devote highly productive land to intensive food cultivation in order to meet local needs. Other limitations may arise from the time lag until the full benefits of silvopastoral practices become apparent. For example, soil conservation benefits and financial return from tree harvesting may only become apparent after several years (problem similar to those in forestry). However, strategic planning with an appropriate combination of pastures and animals with trees, both in space and time, may help to overcome these problems. Beyond this, silvopastoral system interventions at landscape level are often believed to contribute to several aspects of the social and cultural environment, such as creating social stability (land tenure) or enhancing the aesthetic/recreation values of the landscape. This arises from the combination of production with protection functions give it the potential to meet some, or all, of the society expectations. For example, recent academic interest in managed lands as potential habitat reveals the ecological and conservation value of shaded cropping systems as refuges for wildlife and/or areas capable of providing ecological services (Donald 2004).

This multi-authored book contains peer-reviewed chapters from leading researchers and professionals in silvopastoral systems topic in Southern South America (Argentina, Chile and Southern Brazil). It is a compendium of original research articles, case studies, and regional overviews that summarizes the current state of knowledge on different components and aspects (pasture production, animal production, trees production, carbon sequestration, conservation) of silvopastoral systems in native forests and tree plantations. Some chapters explore the technical details of particular systems, while others take a more holistic perspective. The authors have conducted research in various silvopasture plots and interacted with managers, researchers, landowners and extension agents working in silvopastoral systems in each country. Chapter 2 takes into account silvopastoral systems developed in native forest of the western Chaco (Argentina). In this ecosystem it seems to be an acceptable option for the management of the current vegetation a disturbance regime of medium magnitude where native woody plant communities (secondary forests and shrub thickets) dominate. Management recommendations are shown in order to provide good and services (meat, wood, diversity and wildlife) from Chaco ecosystems in a sustainability framework. Chapter 3 deals with production of poplar (*Populus* sp.) and willow (*Salix* sp.) wood for multiple uses, spontaneous natural pastures, and beef production in the silvopastures on the Lower Delta of the Paraná River, Argentina. Knowledge and management practices of silvopastoral systems in the warm temperate (subtropical) humid climate region of Argentine northeastern provinces of Misiones and Corrientes is the focus of Chap. 4. In general, these highly productive systems integrate a tree component (mainly *Pinus* spp.) planted in natural grasslands, or with exotic pasture species (*Brachiaria* spp. and *Axonopus catarinensis*), and usually cattle cross-breeds of Braford and Brangus. In Chap. 5 results of research conducted over the past 15 years are summarized and management guidelines for the development of silvopastoral systems based on ponderosa pine plantations established on natural grasslands in north west Patagonia, Argentina is also provided. The goals of the studies were to generate knowledge for improving environmental and production efficiency in that region. In Chap. 6, Patagonian experience with silvopastoral systems in native *Nothofagus antarctica* forests is reviewed. Information related to livestock production, productivity and nutritive value of understory grassland, silviculture and wood production, adaptive silvopastoral management and processes (litter decomposition, nutrients dynamic, carbon storage) and criteria and indicator (C&I) to assess ñire forest's sustainability are informed. Silvopastoral systems in arid and semiarid zones of Chile are presented in Chap. 7. It provides valuable information on the *Espinal*, the most important silvopastoral ecosystem in central Chile, where the influence of the leguminous trees on soil characteristics is a fundamental explanatory factor of the tree-grassland association. In addition, it discusses the use of a fodder shrub, *Atriplex nummularia* where sheep are grazed in summer and autumn, and *Prosopis*, which is adapted to extremely dry areas, on soils lacking nitrogen and under conditions of high sun radiation. In Chap. 8, the authors present a novel *Nothofagus obliqua* -based silvopastoral trial that was established in the temperate Andes in order to rejuvenate the over-mature forests, and evaluate the quantity and quality of high nutritional value forage (grasses and herbaceous legumes) sown under different tree coverages. This study provides strategies and advice on how to adapt the results of the study to the particular socioeconomic conditions that are prevalent in the daily lives of rural communities. The chapter also includes technical aspects and economic data of *Pinus radiata*-based silvopastoral systems established in semi-arid to humid zones of Mediterranean Chile. Finally, it summarizes several experimental trials in the last 20 years with exotic species such as Cherry, Walnut, Poplar and Stone pine. Knowledge on silvopastoral systems in Chilean Patagonia is provided in Chap. 9. It discusses the use of *Pinus contorta*, *P. ponderosa* and *Pseudotsuga menziesii* and their effect in reducing wind erosion. Silvopastoral trials associated to *Nothofagus pumilio* forests in Magallanes, where thinning resulted in increased soil temperature and moisture during the growing season. They also allowed for the growth of naturalized species creating an ecotone between pastures and forests of high biodiversity. Chapter 10 includes the main scientific results collected from field experiments in the cold zone of Brazil, including several $C_3$ and $C_4$ pastures (native and exotic) performances under shade, beef cattle and sheep performances and behaviour under tree-pasture systems, potential of trees species as well as a list of future challenges for research and rural extension service to increase the adoption of silvopastoral systems in Southern Brazil.

The main theme of the book is that farmers have integrated tree and pasture/grassland species in their land use systems to reach higher production per unit of land area, risk avoidance, product diversification, and sustainability. These production systems also impact positively in main ecosystem processes. Issues of management of these productive systems and implementation of supportive policy and socioeconomic aspects provide great opportunities and challenges for farmers and policy makers in our region. We believe the book is unique on this subject in Southern South America and constitutes a valuable reference material for undergraduate and graduate students, professors, scientists, landowners and extensionists who work with silvopastoral systems.

## References

Chang MY (1985) Faxinais no Paraná. Instituto Agronômico do Paraná – IAPAR, Londrina, 26p (in Portuguese)

Donald PF (2004) Biodiversity impacts of some agricultural commodity production systems. Conserv Biol 18:17–38

Dube F, Couto L, Silva ML, Leite HG, Garcia R, Arauco GA (2002) A simulation model for evaluating technical and economic aspects of an industrial eucalyptus-based agroforestry system in Minas Gerais, Brazil. Agrofor Syst 55:73–80

Dube F, Thevathasan NV, Zagal E, Gordon AM, Stolpe NB, Espinosa M (2011) Carbon sequestration potential of silvopastoral and other land use systems in the Chilean Patagonia. In: Kumar BM, Nair PKR (eds) Carbon sequestration potential of agroforestry systems, Advances in Agroforestry 8. Springer, Dordrecht, pp 101–127

Dube F, Espinosa M, Stolpe N, Zagal E, Thevathasan NV, Gordon AM (2012) Productivity and carbon storage in silvopastoral systems with *Pinus ponderosa* and *Trifolium* spp., plantations and pasture on an Andisol in Patagonia, Chile. Agrofor Syst 86(2):113–128

Dube F, Müller-Using B, Stolpe NB, Zagal E (2014) Carbon loss and degradation indicators in *Nothofagus* forests of the Andes and Coastal range in the Mediterranean region of Chile. In: Proceedings (CD-ROM) of the XXIV IUFRO 2014 World Congress. Salt Lake City, 5–11 October

Esquivel M, Hammer K (1988) The "conuco"-an important refuge of Cuban plant genetic resources. Kulturpflanzen 36:451–463

FAO (1999) Production yearbook. Food and Agriculture Organization, Rome

González GM, Maraboli FF (1997) Actividades Proyecto 2402 Agroforestería Sector Tambillo. Departamento Desarrollo y Fomento Forestal. CONAF, Región de Antofagasta, 51p

INFOR (2012) Desarrollo de tecnología silvícola productiva como apoyo para la implementación de la Ley de Bosque Nativo. Anexo 6. Elaboración de pautas silvícolas de manejo para renovales. Proyecto INNOVA/CORFO – 09CN14-5854. Valdivia, Chile, 28p

Nair PKR (1993) An introduction to agroforestry. Kluwer Academic Publishers, London, 499p

Ovalle C, Avendaño J, Aronson J, del Pozo A (1996) Espinales of subhumid central Chile. I. Cartography of land occupation. For Ecol Manage 86:129–139

Peri PL (2011) Carbon storage in cold temperate ecosystems in Southern Patagonia, Argentina. In: Atazadeh I (ed) Biomass and remote sensing of biomass. InTech Publisher, Croacia, pp 213–226

Peri PL (2012) Implementación, manejo y producción en SSP: enfoque de escalas en la aplicación del conocimiento aplicado. Proceedings of the 2nd National Congress of Silvopastoral Systems. INTA Editions, Santiago del Estero, pp 8–21

Perret S, Mora F (2001) *Acacia saligna*, leguminosa arbórea de uso múltiple para las zonas áridas y semiáridas de Chile. INFOR, Santiago, 14p

Porfirio-da-Silva V, Medrado MJS, Nicodemo MLF, Dereti RM (2009) Arborização de pastagens com espécies florestais madeireiras: implantação e manejo. Embrapa Florestas, Colombo, 48p (in Portuguese)

Saibro JC (2001) Animal production from tree-pasture association systems in Brazil. In: International Grassland Congress, 19, Brazilian Society of Animal Husbandry, São Pedro, SP. Palestra, pp 637–643

Secretaría de Agricultura, Ganadería, Pesca y Alimentación (2009) Estimaciones agrícolas Buenos Aires, Argentina

Sotomayor A (2010) Sistemas silvopastorales, alternativa de producción integrada para un desarrollo sustentable de la agricultura en Chile. Cienc Investig For 16(1):19–51

Teuber O, Ganderatz S (2009) Características geográficas y edafoclimáticas de la Región de Aysén. In: Teuber O (ed) Sistemas agroforestales para la Región de Aysén: cortinas cortaviento y silvopastoreo. Instituto de Investigaciones Agropecuarias, Coyhaique, pp 9–16

United Nations (2015) World population prospects. The 2015 revision highlights, UN Population Division Department of Economic and Social Affairs, New York

Varella AC, da Silva VP, Ribaski J, Soares AB, de Moraes A, Morais H, de Saibro JC, Barro RS (2012) Estabelecimento de plantas forrageiras em sistemas de integração floresta-pecuária no sul do brasil. In: Fontaneli RS, H P d S, Fontaneli RS (eds) Forrageiras para a integração lavoura-pecuária-floresta na região sul-brasileira, 2nd edn. Embrapa, Brasília, pp 283–303 (in Portuguese)

# Silvopastoral Systems Developed in Misiones and Corrientes, Argentina

# 2

Santiago María Lacorte, Sara Regina Barth, Luis Colcombet, Ernesto Héctor Crechi, Jorge Isaac Esquivel, Hugo E. Fassola, María Cristina Goldfarb, Raúl Pezzuti, Daniel Videla, and Rosa Ángela Winck

**Abstract**

The proper use of the environment in natural or transformed systems is a global issue and the most popular approach to its achievement is the development of diversified and environmentally friendly land-use systems that adopt adequate measures for its protection. In this sense, Silvopastoral Systems (**SPS**) that combine trees and animal production in the same area have an important role in diversifying and improving productivity in an environmentally friendly manner. In Misiones and Corrientes provinces (Argentina) the climate is defined as subtropical with frosts and a regular rainfall pattern; it is characterized by production diversification and yerba mate (*Ilex paraguarensis*) is an iconic crop. Forestry development is the most important in the country and **SPSs** currently take up around 100,000 ha in both provinces. They were particularly promoted in the past two decades following the results obtained in applied research and the transfer of technologies to the productive sector by the INTA (National Institute of Agricultural Technology), the Regional Consortia of Agricultural Experimentation, national and provincial agencies and other organizations of farmers. The information available about **SPSs** is presented, with reference to the production persistence of the forage component – grassland and pastures with an emphasis on the adaptation of $C_4$

S.M. Lacorte (✉) • J.I. Esquivel
Misiones, Argentina
e-mail: santiagomarialacorte@gmail.com; elfacon@outlook.com

S.R. Barth • L. Colcombet • E.H. Crechi
H.E. Fassola • R.Á. Winck
INTA-EEA Montecarlo, Misiones, Argentina
e-mail: barth.sara@inta.gob.ar; colcombet.luis@inta.gob.ar; crechi.ernesto@inta.gob.ar; fassola.enrique@inta.gob.ar; winck.rosa@inta.gob.ar

M.C. Goldfarb
INTA-EEA Corrientes, Corrientes, Argentina
e-mail: goldfarb.maria@inta.gob.ar

R. Pezzuti
Forestal Bosques del Plata S.A, Corrientes, Argentina
e-mail: rpezzutti@cmpc.com.ar

D. Videla
School of Forest Science-National University of Misiones, Misiones, Argentina
e-mail: dangalaret@hotmail.com



P.L. Peri et al. (eds.), *Silvopastoral Systems in Southern South America*, Advances in Agroforestry 11, DOI 10.1007/978-3-319-24109-8_2

forage species, improvement of their nutritive quality, effect of shade on soil fertility and cattle receptivity. The physical and mechanical properties of low density plantations (**SPSs**) were identified as significantly higher than in the intensive forestry system. Various methodologies are presented as applied to the economic and financial analysis of **SPSs**, like planning and management tools. The business limitations to the adoption of **SPSs** in both provinces include the capital investment required and, in the case of Misiones, the small size of the farms. The impact of **SPSs** is promising at a social level (higher income for the farms and genuine job generation), for the environment (sustainability of the resources), and for the economy of the territories (generation of products with a current or future differential value and efficient use of resources).

**Keywords**
Silvopastoral Systems • Properties • Argentinean Mesopotamia provinces

## 2.1 Introduction

The proper use of the environment, in natural or transformed systems, is a global issue and the most popular approach to its achievement is the development of diversified and environmentally friendly systems because they adopt adequate measures for its protection. Throughout the history of mankind, in one period or another, agroforestry was one of the common practices, cultivating trees and agricultural crops in intimate combination with one another (King 1987).

Today, there are different models of agroforestry, and **SPSs** represent a modality for the use of the land; they combine trees and livestock in the same area for the purpose of diversifying and improving productivity in an environmentally friendly manner. The products obtained range from cattle products (meat, milk, wool, hides, etc.) to timber and non-timber forest products. They allow for the diversification of production activities, obtaining income from the sale of timber, animals and/or for family consumption, making a more efficient use of the soil in a sustainable manner (Cameron et al. 1994).

Another benefit of these systems, according to Sotomayor et al. (2009), is the protection provided by the forest to the livestock in adverse climate conditions; to the soil by reducing deterioration (erosion, compacting, strong sunburn, microflora and microfauna), to the water courses and sources and besides, they reduce the risk of fires.

In the Argentinean provinces of Corrientes and Misiones, with an animal husbandry and forestry tradition respectively, **SPSs** were initially adopted by forestry companies that included cattle in a limited time and space. The forest was grazed in the period between 1.5 and 2.5 years and 5–6 years from planting for the sole purpose of eliminating the accumulated forage turned into combustible material and thus reduce the risk of fires. The **SPS** as such was only used around 25 % of the total forestry cycle. More recently, cattle raisers that adopt forestry promote **SPSs** because they integrate both activities until the end of the forestry cycle. At the same time, these farmers demanded technologies for the set up and management of **SPSs**, emphasizing the need to maintain a lower number of plants per hectare in order to know the effect of incident light, obtain a better quality of timber, adapted forage resources and stocking rates for cattle (Ligier 2002). This chapter presents a historical and production description of both provinces and the evolution of these systems, from the first attempts for a partial integration of forestry and grazing, to the technological contribution made by research.

## 2.2 Characteristics of the Region

### 2.2.1 Ecological Characteristics

Misiones and Corrientes are two of the provinces that make up the Argentinean Mesopotamia surrounded by the Paraná river to the NW-W and the Uruguay river to the E. Corrientes includes the wetland known as Iberá; the province has a total area of 9 million hectares (22.2 million acres), and the Iberá covers around 2 million hectares (4.9 million acres) covered by swamps, marshes, lagoons and soils that retain humidity due to deficient drainage.

Both provinces concentrate the largest forest area in the country with exotic species, mainly Pines and Eucalyptus. The latest forest inventory carried out in both provinces indicated a total of 370,000 ha (914,000 acres) and 580,000 ha (1.4 million acres) (SIFIP 2010) in Misiones and Corrientes respectively, with annual increases of 10,000 ha (25,000 acres) in each.

The climate, in the agro-ecological region that includes both provinces, was classified as humid subtropical, with a predominantly regular rainfall pattern. Average annual rainfall is 1,824 ± 435 mm regularly distributed throughout the year (Papadakis 1974; Müller and Bopp, series 1927–2007). Mean annual temperature is 21.5 °C ± 1.3; the average absolute minimum temperature is −0.2 °C for July and the average absolute maximum temperature is 37.2 °C for January (Müller and Bopp, Series 1970–2007). The frost season extends from early May to mid September (Pachas 2010). Other authors also indicate that the rainfall pattern is regular (1,054–2,089 mm per year) (Galeano cit. by Fernández et al. 1996).

The ecological subregions located in the departments in the center and North of Misiones along the border with Paraguay and Brazil, *Alta Misiones* (High Misiones) and *Misiones Centro* (Central Misiones) (Papadakis 1974), are characterized by the forest formation of the *Paranaense Rainforest*, that belongs to the Amazon domain (Cabrera 1976). The subregion known as "Campos Misioneros-Correntinos" (Misiones-Corrientes fields), ranging from the South of Misiones to the NE of Corrientes, mainly presents grasslands and wooded formations in islets associated to water sources, creating a network that flows towards one of the two large surrounding rivers.

The predominant soils in Misiones are dark red, with a clay texture and well drained, from the large groups of Kandihumults and Kandiudalfs (Ligier et al. 1990). The topography is undulating with 3 % slopes (Fernández et al. 1996). Frangi (2008) indicates that the soils, derived from the meteorization of the basaltic rock, are diverse and include mollisols, alfisols and ultisols. In Corrientes, the predominant soils are acid sandy soils and mollisols, with low contents of phosphorus and sodium, which limits the growth and quality of the forage as well as the productive and reproductive performance of the cattle in this region (Escobar et al. 1996).

Both provinces present areas with very different backgrounds in the use of their natural resources (Fig. 2.1).

The ecological subregions located in the departments in the center and North of Misiones (2 and 3), on the border with Paraguay and Brazil, have participated in national economic activities mainly through the exploitation of native forests and the plantations of yerba mate (*Ilex paraguayensis*), since the arrival of the immigrants in the early twentieth century (Günther et al. 2008).

The ecological subregion that covers the departments in the South of Misiones and the NE of Corrientes (1), predominantly grasslands and wooded formations in islets associated to water sources and others that mostly follow the course of the rivers, focused on animal husbandry since the arrival of the Jesuits in the seventeenth century; this became the typical activity to this day. Cattle breeding based on synthetic and indefinite breeds with a proportion of Indian blood prevails (Lacorte and Esquivel 2009).

The plantations of fast-growing tree species began in the early 1950s, first with the settlement of the cellulose industry and then promoted by tax incentive systems. There were some attempts to develop SPSs in the North of Misiones in the 1970s. Then several farmers started these systems with native forests (locally called "parquizados") or with trees planted on the coast of the Paraná river and in the North of Corrientes, that for several reasons did not succeed (Fassola

**Fig. 2.1** Province of Misiones and NE of Corrientes (1-2-3 Ecological Subregions, Papadakis 1974). Areas promoted for forestry

et al. 2004). These authors indicate that the technical bases for the management of these systems are laid in the mid 1980s with the first forest and livestock studies, such as the one by Navajas et al. (1992a), Goldfarb et al. (2007b) which contributed a large part of the knowledge available today. Following this partial integration of forest and livestock systems, **SPSs** were considered by livestock breeders as an alternative to diversify production and improve the profitability of the traditional system.

### 2.2.2 Productive Activities

In Corrientes, the main productive activity is cattle breeding with around 4.8 million head, followed by sheep; both occupy 71 % of the territory of the province. This province has 6 million hectares (15 million acres) of grasslands with pasture-fed cattle raising; the traditional system of cattle breeding prevails, and is currently evolving to integrated breeding, post-weaning and beef cattle systems. Between 2002 and 2013, the area planted with Pines and Eucalyptus increased from 283,028 to 500,000 ha and, as a result, the area dedicated to cattle was reduced. Following this partial integration of forest and livestock systems, **SPSs** were considered by livestock breeders as an alternative to diversify production and improve the profitability of the traditional system. Currently, almost 60,000 ha are under these systems with various degrees of applied technology (Esquivel com.pers. 2013).

The province of Misiones presents a variety of agricultural and livestock activities in its territory; it is an outstanding producer of quality timber (planted and native forests), yerba mate, tea, tobacco, citrus and beef. Around one fourth of its area, mainly in the North, is still covered by native rainforests, with various degrees of intervention and submitted to extractive management systems without replacement.

Compared to the rest of the country, both provinces make a high use of **SPSs**; in the province of Misiones there are 40,000 ha under this system (SIFIP 2010), which, added to the 60,000 ha in Corrientes, amount to around 100,000 ha (com. pers. 2013).

### 2.2.3 Evolution in the Use of Natural Resources

Fassola et al. (2009) identify several stages in the use of natural resources; the first stage started with the aboriginal agriculture before the Spanish colonization. The two Guaraní groups that lived and still live in the region (Keller 2001; Frey et al. 2012) used to practice (and still do) the slash-and-burn technique, which allowed them to obtain plenty of vegetable food for a certain period of the year (Carbonell de Masy 1992). Although the crops developed in wooded environments cannot be considered agroforestry, ethno-botanists have determined that the crops benefited from the forests due to the lower environmental demand they faced during the spring and summer and the protection against the late frosts that are common in the region. Just like in other parts of the world, it is possible to say that the Guaraní agroforestry systems took care of food safety, while the forests had the main function of recovering fertility once the land was abandoned.

During the colonial period, the main influence was observed following the arrival of the Jesuits from Peru to Santiago del Estero around 1585 and from Paraguay to the Guaraní territory in 1587 where they founded the missions (Fassola et al. 2009). They organized the guaraní villages with political, social, military and economic structures that were unprecedented at the time. The contact between these villages and other cities in the Viceroyalty of the Río de la Plata generated commercial exchanges. This period culminated with the domestication of the *Ilex paraguariensis* – a native tree species from the Paranaense Rainforest used to obtain the yerba mate, and the introduction of extensive cattle farming in the grassland region known as *Campos Misioneros/Correntinos*. This process was interrupted with the expulsion of the Jesuits by the Spanish Crown in 1767, although their influence remained for centuries.

Years later, the wars for Independence and then the civil wars and the War of the Triple Alliance -Argentina, Brazil and Uruguay against Paraguay, between 1865 and 1870- turned this region of low relevance for the country in economic and political terms into a region with a bigger influence in the national context. The end of the War of the Triple Alliance (1870) led to the development of navigation along the Paraná river, used to transport the products from the harvest of the natural yerba fields and the timber from the timber yards in Misiones to the port of Buenos Aires (Alcaráz 2006). At the same time, the Argentine North Eastern Railway, which operated a standard gauge railway network (1,435 mm), was expanded to reach Paso de los Libres (1894), Santo Tomé, in the NE of Corrientes (1901), and Posadas, Capital of the province of Misiones (1911).

These communication lines contributed to the integration of this region with the national economy, particularly the current province of Misiones, which was a National Territory in those days. This facilitated the arrival of the immigrants in the late nineteenth and early twentieth centuries, who were mostly dedicated to the exploitation of native tree species and growing yerba mate (*Ilex paraguariensis*). In the grasslands in the North of Corrientes, where immigration was very low, extensive cattle farming continued being the most relevant. This activity was similarly developed in the South of Misiones, where there were no colonies. Cattle breeding, initially Creole breeds and then predominantly with a high proportion of Indian blood, was the most significant (Lacorte and Esquivel 2009; Lacorte 2014).

On the other hand, World War I and II promoted the country's industrialization and the need to substitute imports such as paper and cellulose. Fast-growing tree species were first grown in this region in the late 1940s, with the settlement of cellulose industries and then the implementation of fiscal incentives. As a result, cattle breeding was concentrated in the fields in both provinces and single crops such as yerba mate, tea, conifers and annual species like tobacco, occupied the center/North of Misiones and North of Corrientes, and would then become the symbol of the region.

## 2.3 Current Status of Knowledge on Silvopastoral Systems

### 2.3.1 Initial Experiences in Misiones and Corrientes

The growth of the Argentine forestry industry sector in Misiones and Corrientes is one of the most promising events for the country's productive activity. This is due to the forest potential resulting from the high growth rates of various tree species, the climates and the large extensions of optimum soils, and the incentives for the development of the activity through the law that regulates the activity with a non-refundable financial support for planted forests.

The rapid acceptance and dissemination of these systems generated demands for knowledge on the interactions between their components, particularly the effect of spacing arrangements, densities and combinations of tree species on the productivity of the forage and the livestock. Several technologies under development focus on the operation of real silvopastoral production systems that integrate the soil-forage-tree-cattle components (Goldfarb 2009).

Simultaneously with the appearance of **SPSs** in the North of Misiones and NE of Corrientes in the 1970s, the initial experiences to evaluate the effect of grazing on the development of tree species were conducted by forestry companies that were initially involved in the activity. These led to results on the combination of forage species with *Melia azedarach* ("chinaberry") as the forest component; other experiences studied the use of *Pinus elliottii* and *Axonopus compressus* ("carpet grass") (Kozarik and Ruiz 1978).

Other species of summer or temperate grasses and forage legumes were then evaluated (Morales 1984); no negative effects were observed on the height growth of the *Pinus* ssp. in these combinations (Di Blasi 1989).

In Misiones, Kozarik and Varela (1989) used *Axonopus compressus* to determine that the differences observed in diameter at breast height (DBH) and total height compared to the sample without pastures 3 years after planting, were minimum or higher in the combination (Morales 1984). These experiences focused on obtaining results on the effects on the forest growth, rather than on the grasses and the interactions between them. In fact, the main purpose was to reduce costs in the cultural care of forest plantations, similarly to the South East of Asia where these practices started in the mid-nineteenth century at the Teak plantations (*Taungya* system, developed in Burma, now Myanmar) (described by Escalante and Guerra in this volume).

Furthermore, a forest cluster was consolidated, generated from a relevant cellulose production and the reconversion of the veneer and sawn timber industry, which mostly used raw materials from planted forests due to the decline in native forests. This process was accompanied by the expansion in the livestock activity in the center of the province of Misiones. In this context, **SPSs** were managed in the traditional manner, similar to the plantations in clusters, which required thinning; they were intended for cellulose production, when the prices of this product were and still are very low, delaying the production of logs for sawn timber or veneer timber. Subsequently, other areas in hilly regions were deforested to plant pastures, and in the rainforest the understorey was eliminated while keeping the canopy and planting pastures underneath. In many of these areas the forest and the soil were degraded; the isolated trees, without the support of the surrounding canopies, were uprooted by strong wind storms.

### 2.3.2 Tree and Forage Species Used in SPSs in Misiones and Corrientes

The preliminary results obtained from the **SPSs** demanded, particularly from the livestock sector, a deeper knowledge about the settlement and management of these systems. The argument that generated this demand was that these systems flexibilize the economy of the farms because the livestock provides the cash flow and forestry provides the capital increase.

These demands encouraged the search for new knowledge from a systemic perspective, which was favored by the availability of silvicultural tests carried out by the INTA based on the concept of "Direct Silvicultural Treatments" (Fassola 1991). This approach proposed a combination of pruning and thinning to concentrate growth in the best trees, discard the thinning for cellulose and evaluate the behavior of the forage, with grasses and forage species to form pastures under the canopy (Benvenutti et al. 2000; Goldfarb et al. 2012). Another proposal was to evaluate the livestock component in productive and reproductive terms, by comparing performance under the tree canopy and in open areas (Navajas et al. 1992a, b).

In the 1990s, the evaluations focused on the forest component and the others – forage, livestock and the environment- were only considered to the extent that they affected the forest. Several groups were formed since the year 2000 with professionals from various disciplines, both from the public sector, the INTA and private consultants from the CREA (Regional Consortia of Agricultural Experimentation) that directed research to the study of **SPSs** with a systemic approach. Some tests were started at the INTA's Experimental Stations and at private lands in order to obtain results on the interactions among the components, the development of management guidelines and the quality of the system's products. The evaluation of the **SPSs** in real production systems in both provinces revealed a diversity of models combining forest species, types of grasslands, pasture plantations, livestock systems and management. A survey was conducted in both provinces in order to characterize these models, and 82 different **SPS** models were detected, which confirmed their wide and rapid dissemination (Goldfarb et al. 2010).

The forest species in this region are used for timber, mainly high quality timber for decorating purposes. The most evaluated, disseminated and used in the SPSs in both provinces are: *Pinus taeda*, *Pinus elliottii*, *Pinus caribaea*, *Pinus elliottii x Pinus caribaea var. Hondurensis*, *Eucalyptus grandis*, *Grevillea robusta*, *Mellia azederach*, *Paulonia* sp. (Fassola et al. 2002a, 2007; Benvenutti et al. 1997; Alegranza et al. 1997; Colcombet et al. 2002; Pérego 2002; Pachas et al. 2008; Lacorte 2001; Lacorte et al. 2003).

The forage grasses combined with these forest species, besides the grasslands in both provinces and the agri-eco-regions in each, are: *Brachiaria brizantha* cvs Marandú, Toledo, Mulato and MG5; *Setaria sphacellata* var. *Sericea* cv Narok; *Brachiaria humidícola*; *Chloris gayana* cv Callide and *Panicum maximun* among those produced and propagated by seeds (Goldfarb et al. 2011). Those propagated by vegetative propagation are: *Axonopus catarinensis*, *Axonopus compressus*, *Acroceras macrum*, *Pennisetum purpureum* and the legume *Arachis pintoi* when there is no availability of seeds. This one and *Chamaescrista rotundifolia* are two of the few forage species incorporated to the experiences with **SPSs** given their promising performance in this and other regions, although the scarcity of seeds still limits their massive use in these systems (Lacorte et al. 2006; Pachas et al. 2010; Rossner et al. 2008, 2010; Goldfarb et al. 2008, 2009a, b; Skromeda 2013).

### 2.3.3 Interactions between Forest and Pastures; Forage Productivity

In **SPSs**, trees and forage compete for water, soil nutrients and light. The tree canopy captures the solar energy and, depending on the age, tree species, density, planting system and silvicultural management, imposes favorable or adverse conditions for the growth of the forage species beneath it (Andrade et al. 2002). The shade sensitivity varies with the species; in megathermic grasses, forage production decreases when the incident light beneath the canopy is below 40 % (Goldfarb et al. 2014c).

One of the variables under study is the quantity and distribution of the shade that reaches the forage and how it affects its productive persistence. The forage species listed above were incorporated to the **SPSs** because they have a higher shade tolerance as shown in the results from the tests conducted by the INTA and in real production systems (Vallset al. 2000; Rossner et al. 2011; Benvenutti et al. 2000; Lacorte and Esquivel 2009; Fassola et al. 2005, 2006; Lacorte et al. 2004, 2006; Goldfarb et al. 2007a, b, 2013a, 2014a, b, c).

The effect of *Pinus elliottii* var. *elliottii* x *P. Caribaea* var. *hondurensis* (hybrid pine) planted at two densities –400 and 250 trees/ha- and two spacing arrangements – single rows and double rows- was evaluated through the production of forage dry matter (kgDM.ha$^{-1}$) of *Brachiaria brizantha* cv Marandú. Both the density and the spacing arrangement affected forage production, which was more stable with double rows than with single rows at the same density (Pachas et al. 2009; Goldfarb et al. 2009c, 2014a). The same experiment measured the incidence of light and it was determined that at the same density, the incidence is higher in double rows than in single rows (Goldfarb et al. 2014d).

Another species that stood out for its shade and cold tolerance as well as for the quality of the forage is *Axonopus catarinensis* Valls (sp.nov. unpub.), a natural hybrid between *A. jesuiticus* (Araujo) Valls *x A. scoparius* (Flüggé) Kuhlm, known in the region as giant Jesuit grass (Pavetti and Benvenutti 2003; Pavetti et al. 2004; Pachas et al. 2004, 2008; Rossner et al. 2009).

In the first papers that evaluated grass growth under artificial shade, Pachas et al. (2004) worked with *Axonopus catarinensis* and determined a higher accumulation of dry matter with 50–65 % of photosynthetically active radiation (PAR) compared to the open areas. Similarly, Lacorte et al. (2004) worked with *A. compressus* and found out that with 50 % artificial shade, the dry matter accumulation was 837,033 kgDM.ha$^{-1}$, significantly different from the 2059 kgDM/ha achieved with 0 % shade.

Pachas (2010) determined a significant increase in the aboveground net primary production (ANPP) of *Axonopus catarinensis* under 38 % shade, of 41 % compared to the plants under the direct sunlight, whereas with a higher shade (53 % and 71 %) the ANPP was similar to the plants under direct sunlight. In *Arachis pintoi*, the increase was not significant: the ANPP was 12 % higher under 38 % and 53 % shade than in the direct sunlight, whereas with 71 % shade the ANPP fell by 13 % compared to the treatment with direct sunlight. This higher annual ANPP under shade was due to the increase in the growth rate in the seasons with a higher water deficit (end of summer and fall) and to a lesser extent to the winter growth.

In dark red soils *Eucalyptus* sp. is used for its fast growth and shorter rotation compared to *Pinus* sp. In these **SPSs**, forest management has an impact on the persistence and productivity of the forage. In *Brachiaria brizantha*, one of the forage species used with Eucalyptus, botanic composition and forage production (CB kg DM.ha$^{-1}$) were evaluated under the canopy, planted in double rows with 4 m spacing between rows ×2 m between plants ×19 m of alley. This test determined that up to the fifth year of the system, both variables were not affected by the tree canopy (Goldfarb et al. 2014d).

In areas with grasses as forage component of **SPSs**, the evaluation considered the effect of shade levels, the sensitivity of the forage species, the aboveground and root forage availability, weed invasion and livestock performance under different arrangements, densities and combinations with forest species (Goldfarb et al. 2007a, b, 2013b).

Aboveground forage availability (DM kg. ha$^{-1}$) and the botanic composition of the grasses in a soil with deficient drainage, locally known as "*malezales*", were modified in a **SPS** with *Pinus elliottii* compared to the same grassland in an open area. The predominant species in this grassland, outside the canopy, *Hypoginium virgatum* and *Sorghastrum setosum* were replaced under the Pine canopy with *Axonopus compressus* in 50 % of the total forage available. It should be noted that *Axonopus compressus* has a high forage value (Goldfarb et al. 2007a, b).

In a grassland with predominance of *Andropogon lateralis* and *Sorghastrum setosum* integrating a **SPS** with *Pinus elliottii* var *elliottii* x *P. caribaea* var *hondurensis* (hybrid Pine) and different densities, 646 trees/ha, 400 trees/ha and 250 trees/ha, the sensitivity of the forage species was evaluated in terms of total forage production (DM. kg.ha$^{-1}$) compared to the same grassland outside the canopy. The forage production of the grassland decreased significantly with the higher tree density (5000 kg.ha$^{-1}$ outside the canopy vs. 1600–2000 kg.ha$^{-1}$ under the canopy). This reduction was due to the fact that some species that are sensitive to low radiation, like *Andropogon lateralis* and *Sorghastrumsetosum*, disappeared while other species with a higher tolerance, like *Axonopus compresuss y Axonopus argentinuss*, were developed instead. Although the tree canopy provides protection during extreme climate events, it also competes for the incident light that is not captured by the canopy. The lower production of biomass is offset by the contribution of species with a higher forage value (Goldfarb et al. 2013b).

Fassola et al. (2005), in a test with different pruning and thinning treatments in *Pinus taeda* and *Axonopus jesuiticus*, established a high association between the biomass production of the grass and the section in the base of the live crown ($r = 0.71$, ns: $<0.0001$) and a low relation with environmental factors such as rainfall and temperature. The variables related to stand leaf biomass, tree size and stand age showed a higher potential to be used as predictors of annual production of the grasses.

### 2.3.4 Interactions between Trees and Pastures: Quality of the Forage

Lacorte et al. (2004) evaluated the forage quality under artificial shade and 1 year later, they observed an increase in the phosphorus content in the green forage expressed in dry matter. The increase in the gross protein was erratic with respect to shade levels (Table 2.1).

Pachas et al. (2004) conducted tests with *Axonopus catarinensis* and observed that under

**Table 2.1** Chemical analysis of the Green Forage of *Axonopus compresus* (Swartz) Beauv., 1 year after starting the experience

| Treatments | Shade (%) | Phosphorus (g.100 $gDM^{-1}$) | Nitrogen (g.100 $gDM^{-}$) |
|---|---|---|---|
| With fertilization (260 kgSPT $Ca.ha^{-1}$) | 0 | 0.231 | 10.3 |
| | 30 | 0.254 | 9.6 |
| | 50 | 0.254 | 9.3 |
| | 65 | 0.291 | 12.5 |
| Without fertilization | 0 | 0.170 | 9.4 |
| | 30 | 0.174 | 10.2 |
| | 50 | 0.201 | 9.2 |
| | 65 | 0.209 | 12.1 |

artificial shade conditions, gross protein content increased whereas soluble carbohydrate levels were reduced. This author (Pachas 2010), citing several authors, indicates that the increases in specific leaf area, plant height and, to a lesser extent, leaf area index, all contributed to ensuring a higher radiation capture when this was limited. The increase in specific leaf area under shade conditions is the most important factor in the maximization of the carbon gain per unit of leaf mass. The results of his experience indicate that although the leaf biomass increased by 11 %, the stem biomass was higher (44 %), because the plants in the shade generated more supportive tissue and therefore increased their height. Furthermore, the root biomass decreased significantly. The changes associated to a higher assimilate partition to the leaves and the architecture of the plants are strategies to improve the ability to intercept radiation in the plants growing under the shade. However, this response might lead to higher vulnerability in these plants when they are exposed to overgrazing and/or droughts.

In the same test, both in *A. catarinensis* and in *Arachis pintoi* the shade increased the concentration of some minerals (P, Cu and Fe); the concentration of Mg and Mn in the legume and of K and Zn in the grass. However, shade reduced the concentration of Ca and Mn in the grass (Pachas 2010). This author, as many others, indicates that the higher concentration of nutrients in forage species would be more related to the higher availability of nutrients in the soil and/or higher absorption due to the higher water availability. Citing Cruz (1997), he indicates that the absorption of P and K in the forage species is increased with the shade, mainly in the seasons with limited water availability. He also indicates that another reason for a higher P availability in the shade is the higher association of the plant with soil microorganisms that increase the solubility and absorption of that element.

Uguet and Lacorte (2014) observed that the phosphorus content in grass leaves was higher in the **SPS** treatment than in open areas (Fig. 2.2). They indicate that this difference in phosphorus content in leaves was determined and explained by several authors that also attribute this phenomenon to a better mineralization of the organic matter due to a higher microbial activity and/or an improvement in water availability. They also state that, for other authors, the shade would increase the phosphorus availability due to a higher association between the plants and the microorganisms that increase the phosphorus solubility and absorption.

Gross protein was higher in the **SPS** treatment than in open areas (Fig. 2.3). This 31.7 % difference reveals one of the strengths of **SPSs**. According to Uguet et al. (2014), these results are consistent with the responses documented in several studies with other grass species. Several hypotheses are presented to explain the positive effect of the shade on the nitrogen concentration in the DM. One of them attributes this to a reduction in cell size caused by the shade. Another hypothesis is based on the reduction in the photosynthesis and the consequent increase in nitrogen concentration.

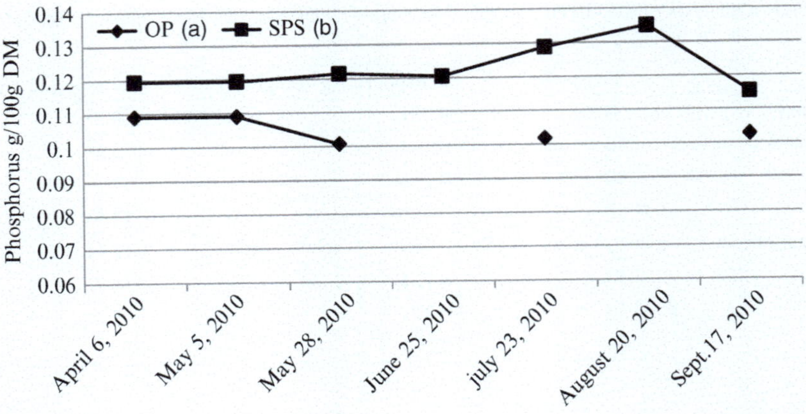


**Fig. 2.2** Phosphorus content evolution in sample leaves from the forage accumulated during the winter season in a typical grassland (Province of Corrientes) in an open area and under **SPS** conditions. * Dates with no cutting in the treatment in open areas due to no growth recorded (Uguet and Lacorte 2014)

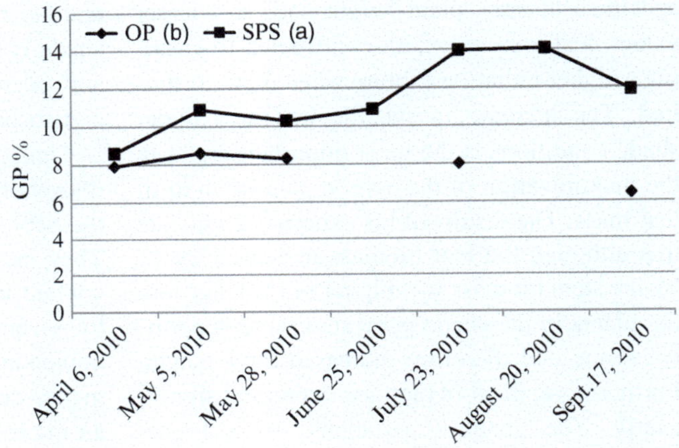


**Fig. 2.3** Evolution of the gross protein (GP) content in samples from the forage accumulated during the winter season in a typical grassland (Province of Corrientes) in an open area and under **SPS** conditions.* Dates with no cutting in the treatment in open areas due to no growth recorded (Uguet and Lacorte 2014)

Uguet and Lacorte (2014), based on several authors, explain that the increase in N and other minerals is due to the higher availability, given the higher mineralization of the organic matter in the soil under the canopy and the higher absorption of N. This results from a better efficiency in the use of water, because the shade generates physiological changes that increase that efficiency. On the other hand, the shade reduces evaporation and the temperatures are favorable to microbial development in the soil, which favors mineralization and the higher availability of nutrients.

Uguet and Lacorte (2014) observed differences in ADF (acid detergent fiber), with an average of 34.16 % in **SPSs**, significantly lower than the average of 36.14 % in open areas. Similar results were recorded in the NDF (neutral detergent fiber), which was of 63.10 % in **SPSs**, significantly lower than the 68.38 % in open areas. No significant differences were recorded in the Lignin content, which was of 5.14 % in open areas and the 4,36 % in **SPSs** (Figs. 2.4 and 2.5).

The growth of the forage between the establishment of an **SPS** and the first grazing is a situation that has to be solved, for instance, with mechanical weeding or rotations with high stocking rates. In traditional grazing systems, the accumulated DM is usually eliminated by burning, very rarely prescribed, which obviously cannot be done in the early stages of **SPSs**.

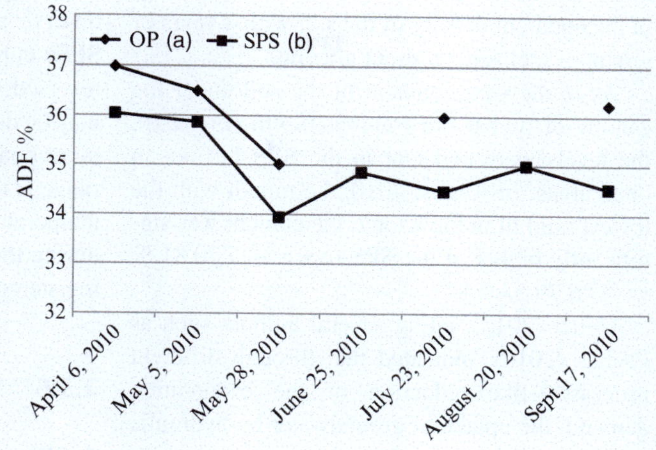


**Fig. 2.4** Evolution of acid detergent fiber in composite samples from a the forage accumulated during the winter season in a typical grassland (Province of Corrientes) in an open area and under **SPS** conditions.* Dates with no cutting in the treatment in open areas due to no growth recorded (Uguet and Lacorte 2014)

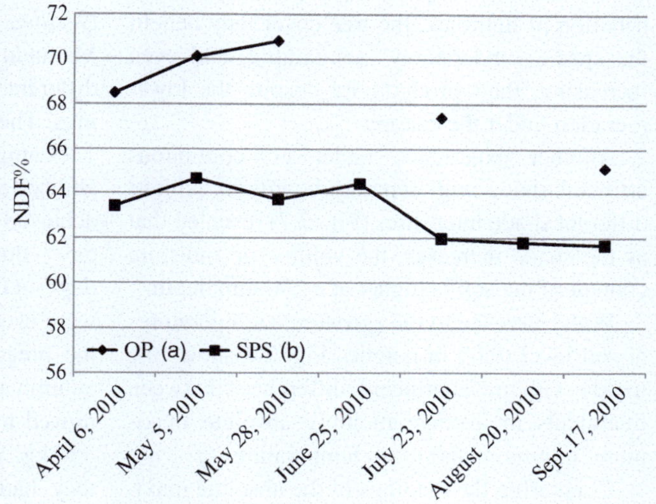


**Fig. 2.5** Evolution of neutral detergent fiber in composite samples from typical grassland, in open areas and under **SPS**. The data is significantly different with the lowest mean for the treatment in open areas in all the cuttings.* Dates with no cutting in the treatment in open areas due to no growth recorded (Uguet and Lacorte 2014)

Samarakoon et al. (1990) and Uguet (2013) reported increases in NDF and ADF under the shade, and this was associated to an increase in the proportion of the cell wall due to a reduction in the percentage of non-structural carbohydrates.

Some soil analysis in tests conducted in the South of Misiones detected a higher concentration of assimilable phosphorus under the canopy of *Grevillea robusta* than under hybrid pine (*P. elliottii* x *P. caribaea*) and *P. taeda*; although the total phosphorus concentration was similar in the three species. On the other hand, the leaves of *Toona ciliata* growing under the canopy of *Grevillea robusta* present more phosphorus than the leaves of the *Grevillea robusta* itself, which would indicate that this element is made available for the accompanying crops and the forage (Cordel et al. 2008).

### 2.3.5 Effect of Tree Canopy and Grazing on the Soil

A study on the interactions among the components of an **SPS** determined that the growth of the tree component (Grevillea) did not present significant differences under closure or continuous grazing, being slightly higher in the latter treatment (Lacorte et al. 2003). The same study established that there were no significant differences in the apparent density of the soil with 4 years of grazing vs. closure without grazing.

As to the water content of the soil under the canopy of 6-year old *Pinus helliottii* (Fig. 2.6), the levels were constant in the **SPS** but not in open areas. In August 2010, the month with the lowest rainfall in the winter, the content was significantly higher in the **SPS** (open area 70.81 % vs. **SPS** 79.53 %).

Uguet (2013), citing several authors such as Pachas (2010), indicated that through different processes like reduction of the evaporation demand, the uptake of groundwater by hydraulic lift, the attenuation of high temperatures, the improvement of edaphic conditions and the contribution of nutrients, the tree cover may benefit the species understorey maintaining, and even increasing, their productivity despite the lower radiation under the canopy.

Another experience under non-continuous artificial shade with separate boards to achieve different shade intensities (Fig. 2.7) revealed that as the shade increases, the volumetric moisture content of the soil is higher at a 200 mm depth.

With reference to the agronomic temperatures at soil level (0.05 m height), Fig. 2.8 shows the trends with no significant differences between treatments in average monthly absolute maximum, minimum and mean temperatures.

Comparing the readings of the absolute maximum and minimum temperatures, it is possible to observe a trend for higher temperatures under **SPSs** compared to open areas, but this trend is reversed for absolute mean temperatures. This may be due to the fact that mean temperatures are more constant under **SPSs**, and the temperature range is more reduced than in open areas, where temperatures are higher during the day and lower during the night. Temperatures below 0 °C were registered in open areas (Uguet 2013).

### 2.3.6 Livestock Productivity

A **SPS** was evaluated at a farm in the South of Misiones for 9 years; its components were *Gevillea robusta*, *Brachiaria brizantha* cv Marandú as grass, young categories of cattle with different degrees of Brangus and Braford crossings. The stocking rate ranged between 0.63 and 1.95 animal/ha, depending on the forage availability and supply. The daily weight gains achieved were above 0.492 kgLiveWeight.an$^{-1}$.day$^{-1}$ during six cycles and 0.250 kgLW.an$^{-1}$.day$^{-1}$ in the three remaining cycles. When heifers were used in this same experience, they reached the breeding weight (300 kgLW) in their first autumn after weaning; as to the calves, they were moved to the steer category in 27 months with 450 kgLW.an$^{-1}$. Although frosts were registered, they had no effects on the forage during the experiment, except in the forest edges and in

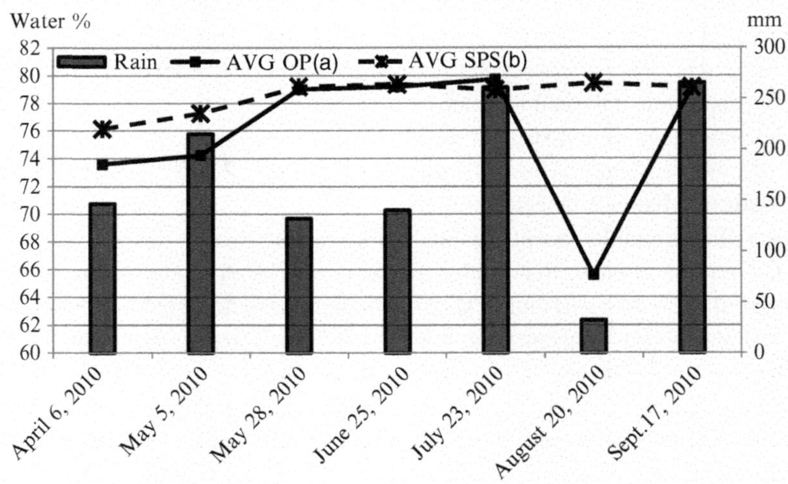


**Fig. 2.6** Volumetric soil water content in soil and rainfalls events (average of four weather stations) during the 2010 winter season in soils in the NE of Corrientes, in open areas and under a tree cover (**SPS**). The letters between parentheses indicate significant differences (Uguet 2013)

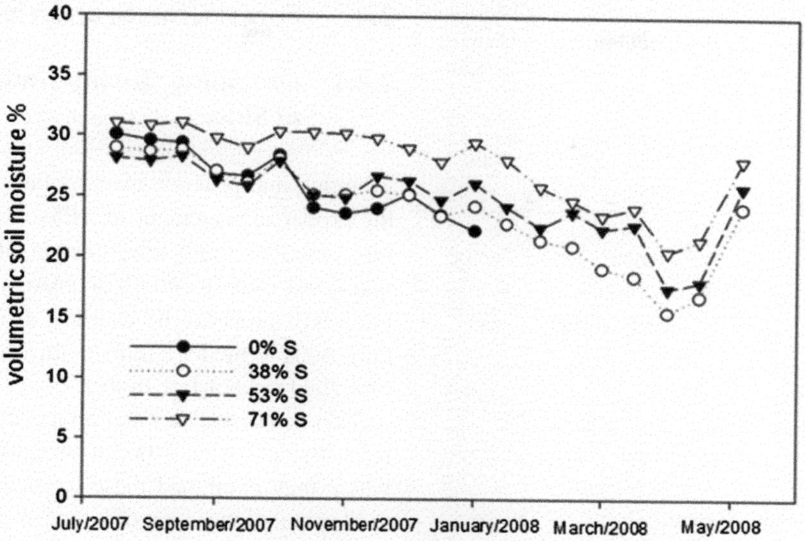


**Fig. 2.7** Average volumetric soil moisture content from the first 200 mm of depth in a pure crop of *Axonopus catarinensis* for four shade treatments (interrupted curves due to data losses) (Pachas 2010)

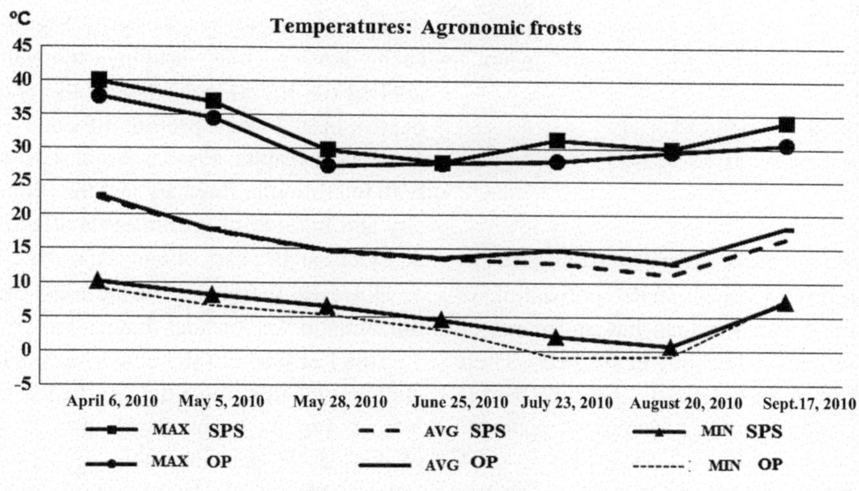


**Fig. 2.8** Maximum, minimum and mean temperatures registered at a 0.05 m height in open areas and **SPSs** during the winter season in the NE of Corrientes (Uguet 2013)

open areas within the forest with little o no canopy protection (Lacorte et al. 2009).

The productivity of the heifers was compared in open areas and under a canopy of *Pinus elliottii* and *Axonopus compresuss* grasses, in terms of weight gains, body condition and genital development. Performance was higher in heifers in **SPSs** compared to those in open areas. This would allow to breed them in the first autumn, when they are 18 months old, thus advancing the useful life of the wombs. In the case of fattening steers on grasslands similar to the above, the same trend was observed in daily weight gains (Table 2.2). It should be noted that in the SPSs the grassland was closed until the start of the experience whereas in the open areas there were strategic burns (Goldfarb et al. 2007a, b).

**Table 2.2** Post-weaning grazing of heifers and fattening of steers in **SPSs** with *Pinus elliottii* and predominant grass *Axonopus compressus*, in the North of Corrientes, Argentina

| Treatments/stocking rates | Open areas TO/low | SPS T1/low | SPS T2/high |
|---|---|---|---|
| **Steer fattening system** | | | |
| Initial average weight March/2005 (kg) | 350 | 225 | 309 |
| Final average weight Dec/2006 (kg) | 475 | 398 | 457 |
| FW – IW difference (kg.an$^{-1}$) 630 days | 125 | 173 | 148 |
| Daily live weight gain (kg. an$^{-1}$.day$^{-1}$) | 0.198 | 0.275 | 0.235 |
| **Heifer post-weaning system** | | | |
| Initial average weight May/05(kg) | 191 | 186 | 197 |
| Final average weight March/06(kg) | 271 | 287 | 283 |
| FW – IW difference (kg. an$^{-1}$) 315 days | 80 | 101 | 86 |
| Daily live weight gain (kg. an$^{-1}$.day$^{-1}$) | 0.255 | 0.320 | 0.273 |
| Body condition (1–9 scale) | 3,5 | 5 | 4 |
| Genital development (1–4) | 2 | 3 | 3 |

Grazing under a tree cover contributes to the thermal comfort of the animals, preventing or reducing caloric stress, which has an impact on production efficiency (Lacorte et al. 2009). There is a lower loss of energy to dissipate the heat in the summer and a lower energy consumption to increase it in winter (Pérego 2002; Lacorte et al. 2003). On the other hand, since the forage is less damaged by the frosts, the animals' winter supplement is eliminated or reduced, which is a financial benefit of **SPSs** compared to open areas (Lacorte and Esquivel 2009).

The modification of the environment under the cover also generates a higher receptivity in the pasture, and therefore it is possible to increase the stocking rate (Lacorte et al. 2003). In fact, in the **SPS** described above (Grevillea+Brachiaria), the stocking rate was tripled compared to the animal performance outside the tree cover (Lacorte et al. 2009).

## 2.4 Forest Growth and Yield

### 2.4.1 Obtaining Quality Timber at SPSs

Thinning and pruning are essential practices for the proper management of **SPSs**. Both silviculture practices imply the removal of foliar biomass, and consequently both growth and harvest yield will probably be affected. However, their purpose is to model a forest with a higher industrial yield in knot-free timber.

In order to evaluate the effects of these practices, Fassola et al. (2002b) set up several trials where they combined these treatments in *Pinus taeda* plantations in the province of Misiones and NE of Corrientes. In Corrientes they established a trial at a 3 year plantation with 1,666 plants/ha, in which four different densities were generated through selective thinning: 0 % thinning, 50 % of the original density, 75 % and 87.5 % thinning. In each density, four pruning intensities were applied (0; 30; 50 and 70 % of the green crown depth), in 2, 3 and 4 pruning lifts at 1-year intervals. As a results, Figs. 2.9 and 2.10 show that both the thinning intensity and the pruning intensity and number of pruning lifts affected growth and yield at 19 years of age (Pezzutti 2011).

However, the quality of the logs differed significantly, if we consider diameter as an indicator.

This test leads to the conclusion that the precommercial thinning intensity had a significant impact on the growth of the *Pinus taeda* plantation. Higher intensities generated higher DBH and lower basal area and total volume. The highest value reached in DBH was 45.7 cm (208 pl/ha), and the highest values for basal area and volume were 55.4 (m$^2$/ha) and 765.7 m$^3$/ha (1666 pl/ha), respectively. When increasing the percentage of crown removal, the diameters of the trees were reduced and consequently the basal area and total volume/ha were also lower. When increasing the number of pruning treatments, the growth in DBH, basal area and total volume decreased gradually as a result.

The pruning and thinning treatments also had an effect on the internal properties of the

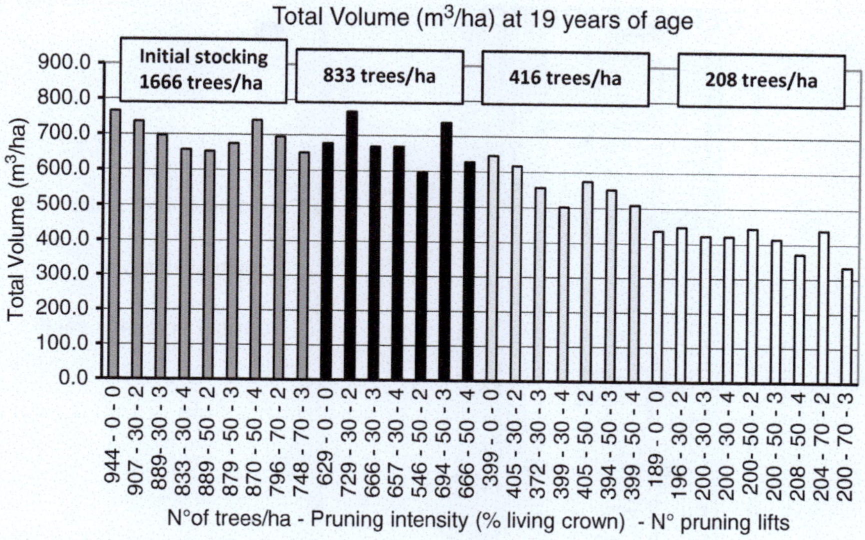


**Fig. 2.9** Volume performance (m³/ha) of the different thinning and pruning treatments, at 19 years of age, applied to *Pinus taeda*

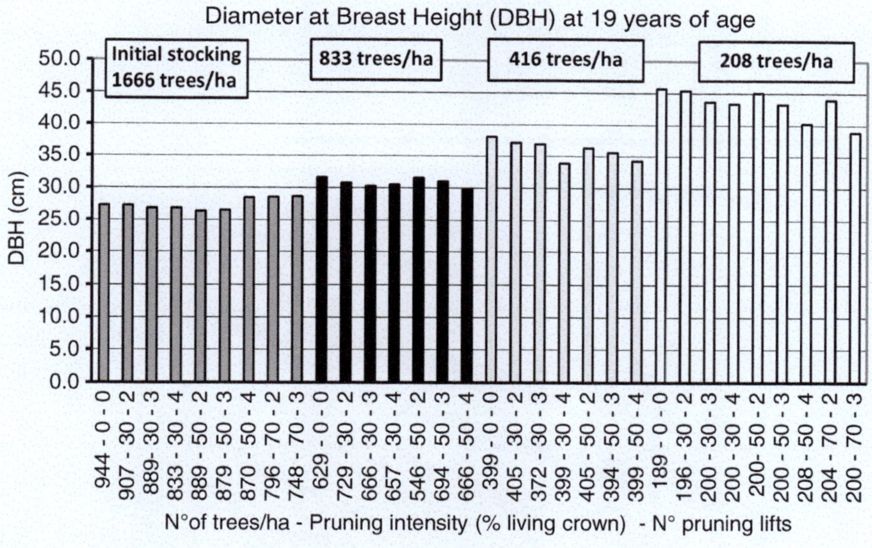


**Fig. 2.10** Growth in diameter (cm) in the different thinning and pruning treatments, at 19 years of age, applied to *Pinus taeda*

pruned logs. The dimensions of the log, combined with the diameter of the cylinder that contains the defects (defect core) (Park 1982), are the main factors that determine the yield in quality timber. The cylinder that contains the defects may be defined as the cylinder that contains the sinuous pith, branch stubs and the pruning occlusion (Fig. 2.11). When evaluating the variable maximum diameter over stubs (MDOS) in each pruning lift, a whorl always has a larger diameter; the results obtained are presented in Fig. 2.12 (Fassola et al. 2002c; Pezzutti 2011).

The results obtained show that with pruning treatments with intensities of 50 % green crown removal it was possible to maintain a better con-

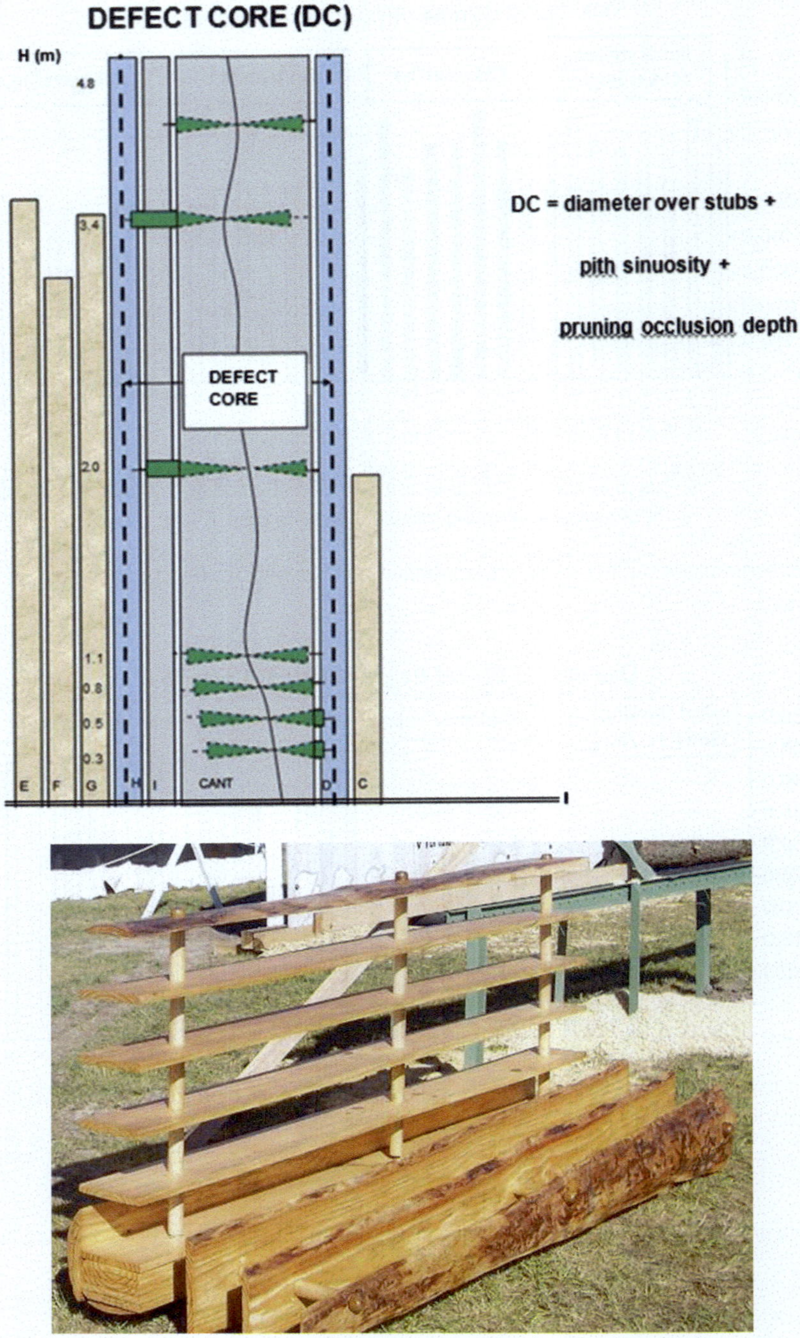


**Fig. 2.11** Diagram and photograph showing the reconstruction of the cylinder that contains the defects (defect core) in the sawn log of a 17-year-old *Pinus taeda* (Fassola et al. 2002b)

trol of the maximum diameter over stubs, the ideal being to maintain the same diameter in each lift. However, this objective in *Pinus taeda* demands intervals between lifts of less than 1 year.

The production of quality timber in **SPSs** demands constant monitoring and the permanent control of the **MDOS** variable to ensure the quality of the logs to be harvested. Other factors, like pruning tools and workers' training, are also important.

Furthermore, the removal of foliar biomass also modifies the photosynthetically active radiation (PAR) that reaches the grass layer. In the test

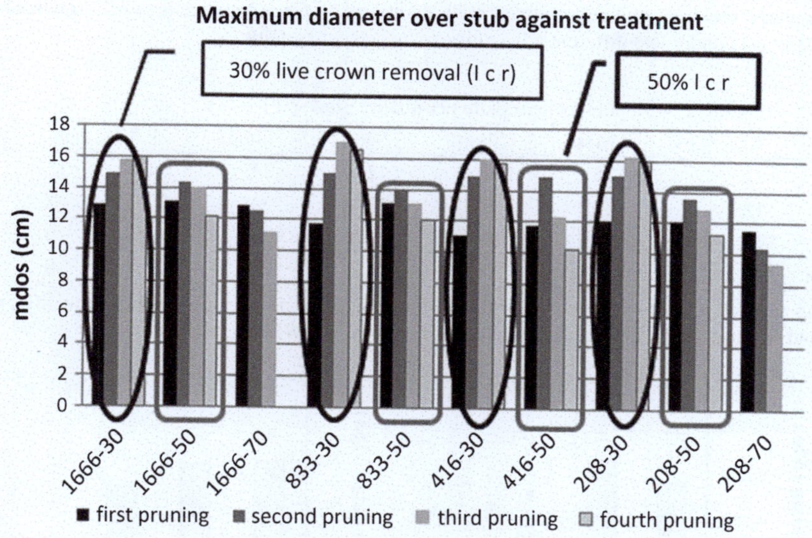


**Fig. 2.12** Average maximum diameter over stubs obtained in the different thinning and pruning treatments, at 3, 4, 5 and 6 years of age, first, second, third and fourth obtained in *Pinus taeda*

conducted at 10 years of age with 208 plants/ha and 70 % crown removal in 3 pruning lifts, the most extreme situation as to the crown coverage, there was no more forage production of *Axonopuss compressus*. In a test of thinning intensities and opportunities in *Pinus taeda*, (Crechi et al. 2014), in the treatments with 100 or fewer plants/ha at that age, they obtained PAR conditions that were compatible with forage production.

These concepts are applicable to *Pinus elliottii* x *Pinus caribaea*, although their canopy is more permeable than *Pinus taeda* and do not thicken their branches as the latter. As a result, it is possible to maintain more trees, and this makes them very suitable for **SPSs**; the same may be said of **SPSs** with *Eucalyptus grandis* and *Grevillea robusta*, presenting similar trends.

The most studied **SPSs** in the region as to the quality of the timber are those that use *Pinus taeda* as a forest component. This species was the most widely used until the introduction of hybrids of *Pinus elliotti* x *Pinus caribaea*. These are gaining relevance for their better shape and small branches and crown that allow the passage of radiation.

A comparative study of the yield of sawn logs between two stands of *Pinus taeda* at the same age, one managed under **SPS** with 200 plants/ha and the other one with an intensive forest management for sawn logs, with 400 plants/ha at 11 years and 300 plants/ha at 15 years, revealed a better performance in quality among those trees from the **SPS**. When classifying production under appearance grading, specifically those used for remanufacture (Factory grade), the silvicultural treatment applied in the **SPS** yielded a higher percentage of high quality grades in the sample under analysis (Fassola et al. 2007, 2012) (Table 2.3).

Table 2.3 shows that both at 11 and 15 years of age, the proportion of M&B – the highest-priced quality grade – was higher in trees from stands with pruning and thinning treatments, although these were not as intense as those applied to the **SPS**.

Winck (2013) and Winck et al. (2013), however, determined that a dramatic early reduction in density in *Pinus taeda* stands, from 1960 to 245 plants/ha at 3 years, resulted in a reduction of the physical and mechanical properties such as the static modulus of elasticity (MOE) and modulus of rupture (MOR) (Fig. 2.13), which may be attributed to an increase in the microfibril angle (MFA) (Fig. 2.14). This might affect the possibility for some uses of the timber from stands of the species planted at very low densities or subject to an intensive thinning at an early age.

However, the silvopastoral systems in the region imply pruning treatments. In an evaluation of logs

**Table 2.3** Percentage yield in volume per appearance quality grade for remanufactures of a sawn sample of *Pinus taeda* of 11 and 15 years under two different silviculture management systems

| Silvicultural management | Age/years | Percentage share per quality grade (%) M&B | Shop1 | Shop2 | Shop3 | P99 | nc | Total sample % |
|---|---|---|---|---|---|---|---|---|
| **SPS** | 11 | 25 | 12 | 12 | 23 | 14 | 15 | 100 |
| **SPS** | 15 | 31 | 8 | 17 | 34 | 8 | 2 | 100 |
| Intensive forest | 11 | 18 | 10 | 20 | 38 | 15 | 0 | 100 |
| Intensive forest | 15 | 13 | 25 | 28 | 29 | 5 | 0 | 100 |

M&B: Moulding and better, quality grade used for moulding and better products; Shop 1-2-3, quality grades used for parts of doors and windows; P99: quality grades used for finger joints; *NC* not classified

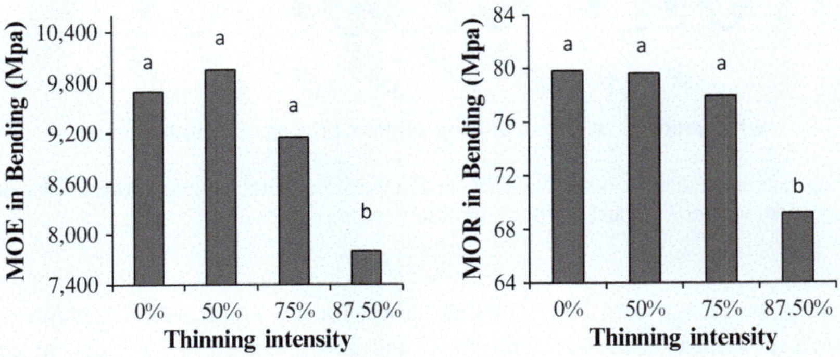


**Fig. 2.13** Variation of the average static modulus of elasticity (MOE) and modulus of rupture (MOR) of saw logs of *Pinus taeda* with four thinning treatments (0 %: 1960 plants/ha; 50 %: 980 plants/ha; 75 %: 490 plants/ha and 87 %: 245 plants/ha)

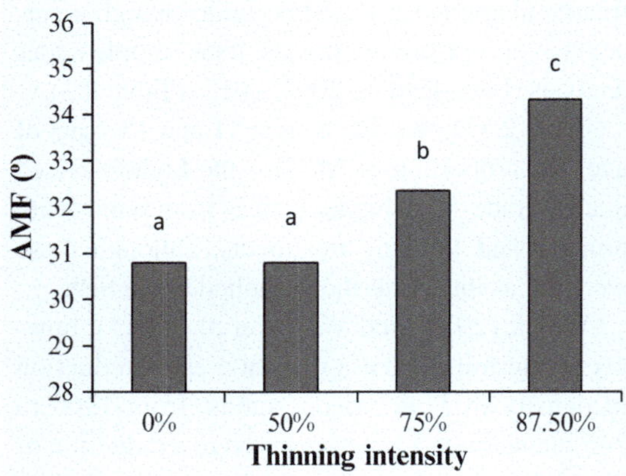


**Fig. 2.14** Variation of the average microfibril angle (MFA) of saw logs of *Pinus taeda* with four thinning treatments (0 %: 1960 plants/ha; 50 %: 980 plants/ha; 75 %: 490 plants/ha and 87 %: 245 plants/ha)

from a thinning treatment and different pruning treatments through acoustic methods, Fassola et al. (2014a) found out that the pruning treatments tend to improve the MOE of the basal logs. Figure 2.15 shows the dynamic MOEs of logs from trees with and without pruning treatments.

Figure 2.19 shows that the MOE of the logs up to 6 m in 20-year-old *Pinus taeda* without pruning does not exceed a dynamic MOE of 8 GPa, whereas 4 of the basal logs from fields with the same number of trees but pruned up to a height of 6–6.50 m in three lifts, eliminating 50 % of the green crown in each log, presented MOE values above 8 GPa. This behavior was repeated with other pruned and unpruned stands (Fassola 2013, unpublished), which partially offset the negative effect of very drastic early thinning treatments.

*Grevillea robusta*, (Barth, pers. comm. 2014) showed a different behavior as to density in a test of initial planting densities, which revealed that at 19 years of age the physical and mechanical properties were significantly higher with 125 plants per hectare than with higher densities (350, 700 and 1600 plants per hectare respectively).

As to the other popular species, *Eucalyptus grandis*, Caniza (2010) recommends maintaining a regular spacing between the trees in a stand in order to reduce the stress and consequently the cracks in the timber.

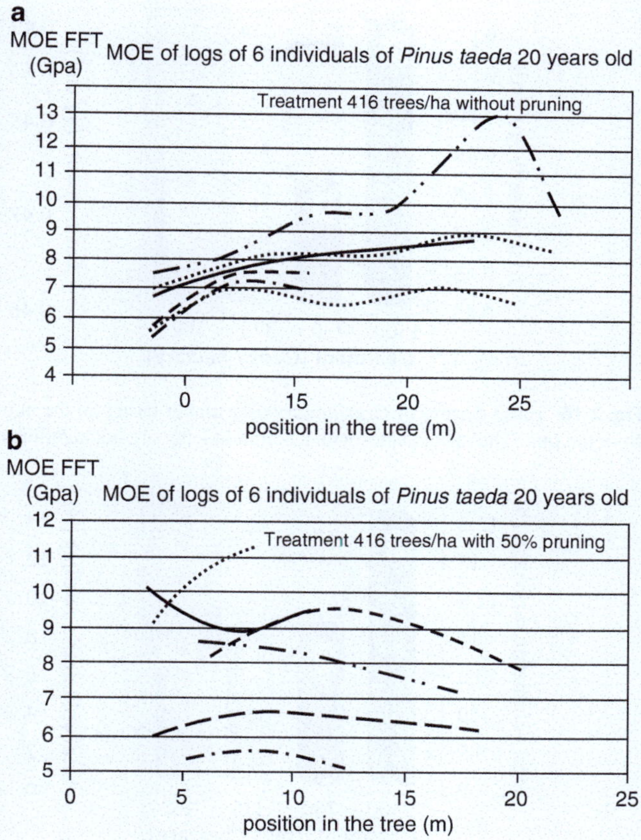


**Fig. 2.15** MOE of logs from 20-year-old trees of *Pinus taeda* under different pruning treatments determined through resonance with the Fast Fourier Transform (FFT). **Note**: each curve belongs to a different tree. (**a**) *Pinus taeda* without pruning and (**b**) *Pinus taeda* with 50 % pruning in three lifts

### 2.4.2 Agroforestry System with *Grevillea robusta* as an Alternative in the Production of Quality Timber

Several tests conducted in the South of the Province of Misiones revealed that *Grevillea robusta* responded differently in growth terms at different planting densities (Fassola et al. 2004). However, at lower densities (that led to a higher diameter growth), the timber obtained did not present significant changes in its physical properties.

The tests conducted in sawn timber of *Grevillea robusta* revealed a basic density of 0.48–0.52 g/cm$^3$, without significant differences between 750 and 162 plants per hectare, but big differences between these densities and 1500 plants/ha. With reference to the existence of radial variations (from the core to the periphery of the log) no differences were found in basic density of the timber, except for the pith section where there were also cracks in sawn sections. Basic density ranges from a mean value of 0.48 g/cm$^3$ in the pith to 0.49 g/cm$^3$ in the rest of the section (Fig. 2.16).

As to total volumetric shrinkage (radial, tangential and axial, from saturated to dry state), no differences were found between the treatments of 750 and 162 trees/ha, but differences were detected between those and the treatment of 1500 trees/ha with values of 11–13.4 % respectively (Fig. 2.17).

These characteristics make *Grevillea robusta* an interesting species for Agroforestry Systems or **SPSs**. But in Misiones the agricultural activity was traditionally developed in open areas, and given the climate, topography and soil characteristics, this led to problems of exhaustion of the soil nutrients and hydric erosion (Cabrera 1976; Ligier et al. 1990; Píccolo et al. 2012; Toledo et al. 2014). The incorporation of a canopy cover or tree windbreak to the agricultural activity provides a more stable environment, protecting the soil and

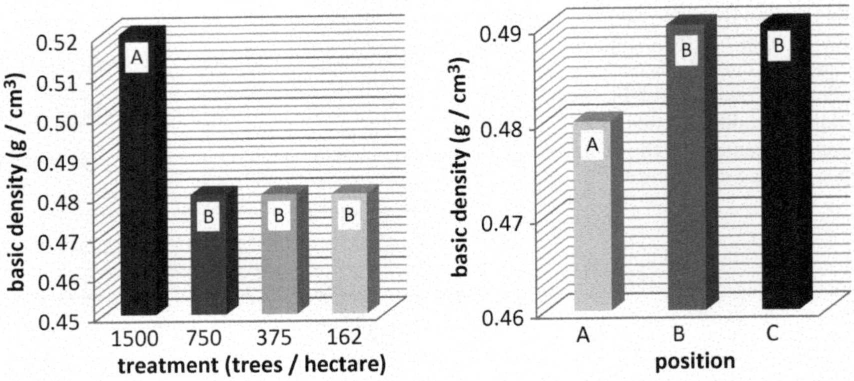


**Fig. 2.16** Basic density of *Grevillea robusta* timber based on the number of trees per hectare and the radial distance from the pith. The mean values with a common letter are not significantly different (p>0.05)

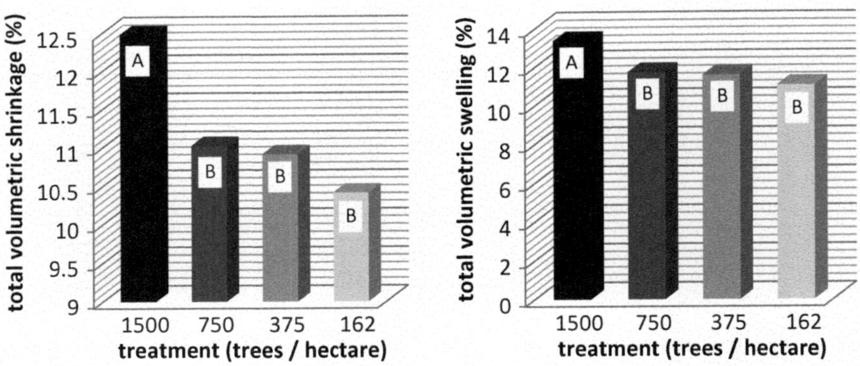


**Fig. 2.17** Total volumetric shrinkage and swelling in *Grevillea robusta* timber at different planting densities. The mean values with a common letter are not significantly different (p>0.05)

promoting a higher recycling of the nutrients. The planting of *Grevillea robusta* associated to traditional agricultural activities like livestock, yerba mate, tea and others, offers the farmers the ability to increase their income with quality timber with similar physical properties to that obtained in pure forest plantations with more trees per hectare.

### 2.4.3 Evaluation of the Logs and Their Relation to SPSs

Among the forest species used in the **SPSs** in this region, *Pinus taeda* is used for appearance timber, for structural use, pulp milling and boards. The quality of the timber for these various uses will be defined by a combination of external and internal properties. A proper silviculture management – pruning and thinning – contributes to improving the appearance characteristics, as well as the properties of the log.

To know the effect of thinning and pruning on the appearance and structural properties, tests were conducted with trees from silvicultural tests in the North of Corrientes and North of Misiones (Fassola et al. 1997, 2002a; Pezzutti 2011). The analyzed thinning intensities ranged between 0 %, 50 %, 75 % and 87.5 % of the original planting density (1960 pla.ha$^{-1}$), applied on year 3 in both cases. Also on year 3, both tests applied different pruning intensities and opportunities, which varied between 0 %, 30 %, 50 % and 70 % of live crown removal applied in 1, 2, 3 and 4 pruning lifts at 1-year intervals. Table 2.4 presents the results corresponding to the treatment with 87 % thinning and different pruning levels from the test located in the Province of Corrientes. The same treatment was applied in the basal logs with different thinning intensities, although without pruning, in trees felled at the test located in the province of Misiones (Table 2.5).

The Factory Grade classification was used. This showed that pruning treatments provided a higher percentage of high quality grades in the samples under analysis (Tables 2.4 and 2.5).

In the treatment with 208 plants per hectare in the Province of Corrientes the yield was higher (21 %) in moulding timber (Moulding and Better) and in superior products in the treatments with pruning, whereas the moulding quality was almost non-existing (1 %) in the treatments without pruning (Table 2.1, Fig. 2.18). The Shop3 quality also obtained higher percentages (57 %, 39 % and 49 %). However, when applying 30 % and 50 % pruning intensities, the percentages of Shop3 and Shop2 quality timber decreased.

In the treatments without pruning in the province of Misiones (Table 2.5), regardless of the thinning intensity, no moulding quality (M&B) timber was found, whereas the timber in lower

**Table 2.4** Yield in the saw logs of 20-year-old *Pinus taeda*, in percentages according to Factory Grade per thinning and pruning treatment, in the province of Corrientes

| Treatment | | Volume % under Factory Grade-Province of Corrientes | | | | | | |
|---|---|---|---|---|---|---|---|---|
| Thinning | Pruning | M&B | Shop1 | Shop2 | Shop3 | P99 | NC | Total vol. (%) |
| 87.5 % | 0 % | 1.08 | 5.08 | 22.02 | 56.73 | 4.79 | 10.30 | 100.00 |
| 87.5 % | 30 % | 20.81 | 8.76 | 18.17 | 39.44 | 3.75 | 9.07 | 100.00 |
| 87.5 % | 50 % | 20.96 | 3.68 | 13.81 | 48.82 | 4.18 | 8.55 | 100.00 |

**Table 2.5** Yield in the sawn basal logs of 20-year-old *Pinus taeda*, in percentages according to Factory Grade per pruning treatment, in the province of Misiones

| Treatment | | Volume % under Factory Grade-Province of Misiones | | | | | | |
|---|---|---|---|---|---|---|---|---|
| Thinning | Pruning | M&B | Shop1 | Shop2 | Shop3 | P99 | NC | Total vol. (%) |
| 0 % | 0 % | 0.00 | 0.00 | 5.05 | 39.90 | 46.97 | 8.08 | 100.00 |
| 50 % | 0 % | 0.00 | 0.00 | 2.28 | 40.30 | 53.61 | 3.80 | 100.00 |
| 75 % | 0 % | 0.00 | 0.00 | 9.73 | 43.78 | 46.22 | 0.27 | 100.00 |
| 87.5 % | 0 % | 0.00 | 4.93 | 8.70 | 59.11 | 25.94 | 1.31 | 100.00 |

M&B: Moulding and better, quality grade used for mouldings and superior products; Shop1-2-3, quality grades used for parts of doors and windows; P99: quality grades used for finger-joint; *NC* not classified

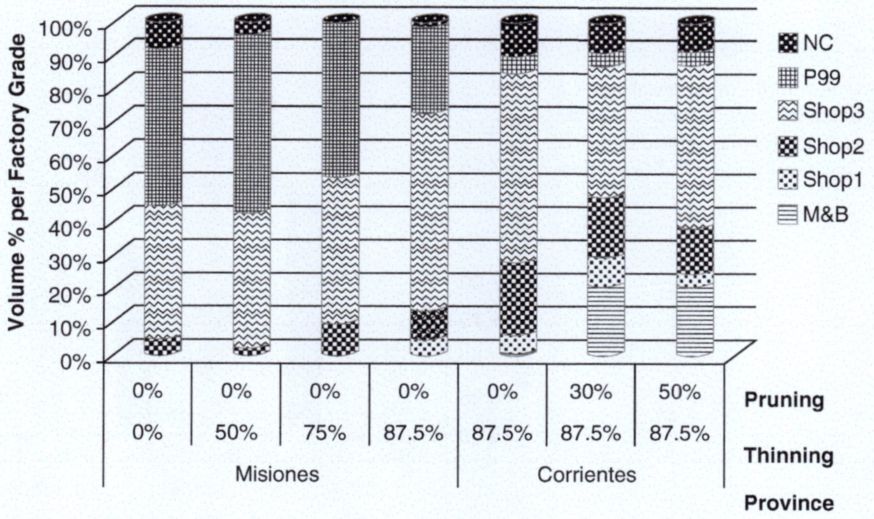


**Fig. 2.18** Percentage yield of the boards per quality grade (%) under Factory Standards according to the treatment received

quality grades, like finger-joint (P99), increased, ranging between 26 % and 54 % (Table 2.5).

It may be concluded that the percentage yield in moulding timber classified under Factory Grade standards was higher in the treatments with pruning.

### 2.4.4 Strength Properties with Acoustic Methods

The timber from **SPSs** is mainly intended to be used as appearance timber. The use as structural timber must take into account the mechanical properties. These were determined through acoustic methods that identified the dynamic modulus of elasticity ($MOE_d$) of the logs and boards from these tests.

Figure 2.19 shows the average $MOE_d$ of the logs from trees felled in the 87.5 % thinning treatment and 3 pruning levels in the test with 20-year-old *Pinus taeda* in the province of Corrientes. It shows a higher $MOE_d$ in the treatments with pruning, because this improved the quality of the basal logs. Figure 2.20, which depicts the basal logs up to a height of 5.5 m and above, clearly shows this effect (Fassola et al. 2014b).

The logs from the test were sawn and in the boards obtained the MOE was determined through vibration; the results obtained were similar to those with the logs. The boards from the sawn logs contributed to improving the MOE values in the treatments with pruning compared to those without pruning (Fig. 2.21).

Consequently, the pruning improves the mechanical properties of the boards. As to the thinning, static methods were used to determine that very intense thinning treatments at early ages generate lower mechanical strength (Winck et al. 2013a) (Fig. 2.22). Winck (2013) attributes this behavior to the larger microfibril angle that is generated with very intense thinning treatments at early ages in this species.

Since the application of drastic thinning at once reduces the strength properties of the timber, the recommendation is not to reduce density below 400 pla.ha$^{-1}$ at early ages.

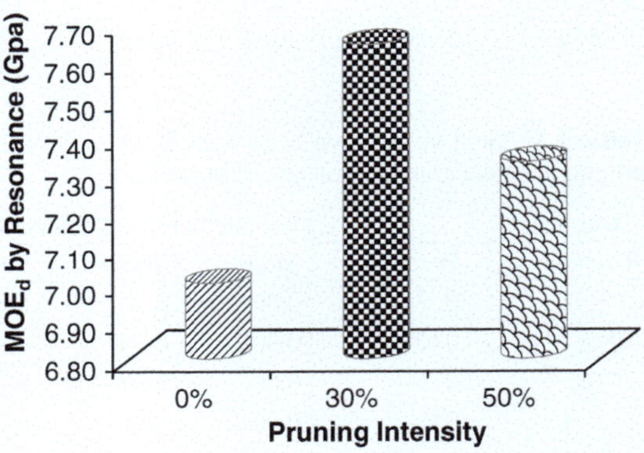


**Fig. 2.19** Average $MOE_d$ of the logs from felled trees of *Pinus taeda* at 20 years of age with 208 plants/ha and three levels of pruning in the Province of Corrientes

**Fig. 2.20** Average $MOE_d$ of the basal and upper logs from felled trees of *Pinus taeda* at 20 years of age with 208 pla.ha$^{-1}$ and three levels of pruning in the Province of Corrientes

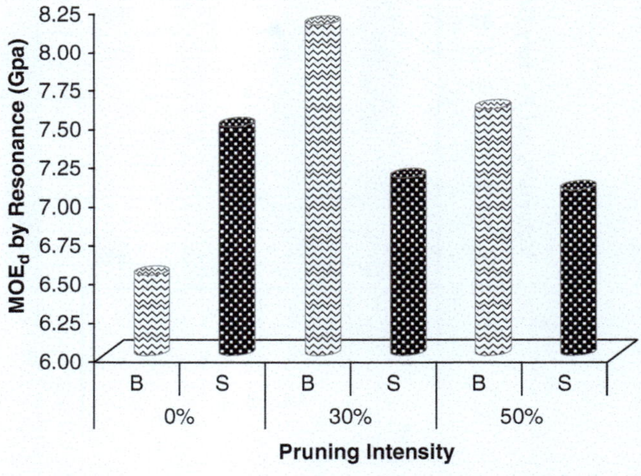

**Fig. 2.21** Average MOE of boards from basal logs (*B*) and upper logs (*S*) from felled trees of *Pinus taeda* at 20 years with 208 pla. ha$^{-1}$ and three pruning levels in the Province of Corrientes

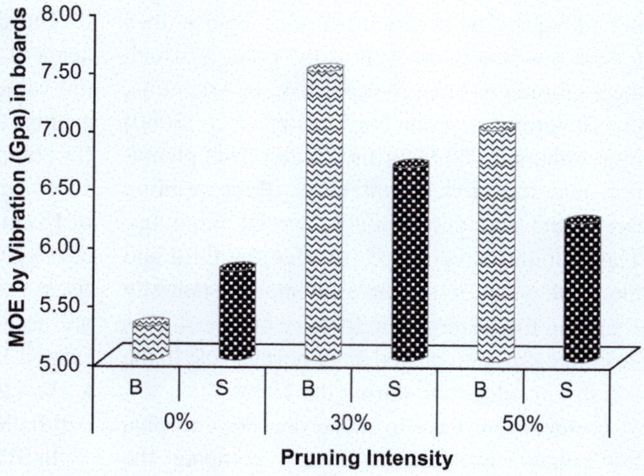


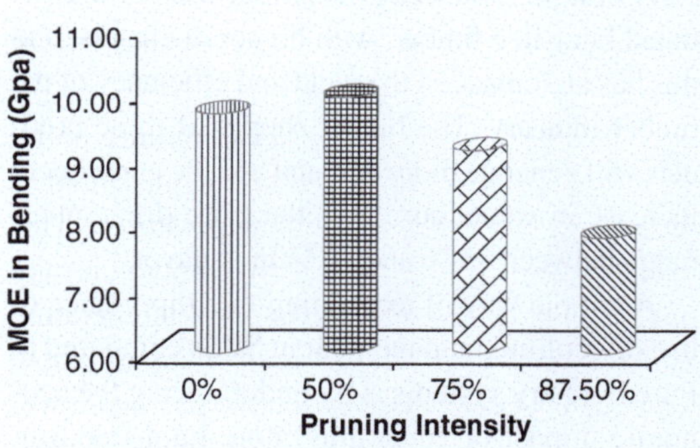


**Fig. 2.22** Average static MOE in felled trees of *Pinus taeda* at 20 years with different thinning intensities in the Province of Misiones

## 2.5 Economic and Financial Analysis in Silvopastoral Systems in Misiones and Corrientes

In the NE of Argentina, the incorporation of trees to the grazing lands generated situations ranging from the total displacement of the livestock to the highest complementation, as is the case with **SPSs**. Cattle ranchers are not used to analyzing the forest business in economic and financial terms, and the time factor discourages them from being involved in the activity. However, they see it as an alternative to diversify production, considering forestry as a "savings account" while they continue with cattle raising as their main activity that provides them with the "petty cash". As their income from forestry increases with time, they are encouraged to making longer-term investments in this activity.

There are several methods to conduct a financial analysis of **SPSs** based on the different alternatives for the use of the circulating capital. The methods normally applied to forest models use interest rates to discount an uneven cash flow until an initial moment (Net Present Value or NPV) or they calculate the interest rate of that cash flow when the NPV equals zero (Internal Rate of Return or IRR). In the case of livestock or agricultural systems, the unit of economic analysis is the result expressed in $/ha/year. In order to compare an activity with annual results with another activity that has multi-annual production times, it is necessary to express their economic result in the same unit. The Equivalent Annual Income (EAI), that is, the NPV multiplied by an annuity factor, transforms the multi-annual cash flow into a single constant amount that represents the business with uneven incomes and expenses. Thus, it is possible to compare corn production with calf production or Eucalyptus timber production.

However, it is also necessary to conduct the analysis from the point of view of the immobilization of the circulating capital. Investments in forestry, that imply a transfer of circulating or livestock capital to the trees, have a recovery period of at least 6 years and a maximum of 18 years. The correct financial planning prevents the farmer's cattle decapitalization or the early sale of the trees. Consequently, it is necessary to know the amount to be invested and the opportunities for the investment, as well as the point at which the accumulated balance is positive. In Argentina, the Government promotes forestry (Act 25080) by reimbursing 80 % of the value of the plantation and the silvicultural tasks (three pruning treatments and one non-commercial thinning). This amount is recovered between the third and the sixth year of the investment, substantially changing the result of the forestry business. This incentive program played an important role in the forestry development during the 1990s.

Farmers that have to make decisions to plan tree-related investments, always compare the cash flow of a forestry model and an **SPS**. The comparison between both types of management, pure forestry and **SPS**, shows that the NPV calculated with a reference interest rate of 8 % is 1,596 U$S/ha for the Forestry model and 1,342 U$S/ha for the **SPS**. The Equivalent Annual Income of the forest investment is 170 U$S/ha/year; whereas for the **SPS** it is 162 U$S/ha/year (considering livestock income at 30 U$S/ha/year). Finally, the IRR is higher in the **SPS** (34.5 %) than in the forestry cash flow (25.9 %).

In agroforestry systems and **SPSs**, production diversification provides more economic stability or, at least, reduces instability. In this region, the forest component -Pine, Eucalyptus, Grevillea, Chinaberry, Kiri- is integrated with other components -Yerba Mate (*Ilex paraguariensis*), grasses/grasslands, cassava (*Manihot esculenta*), tobacco (*Nicotiana tabaco*) and other annual crops at a lower scale. Small farmers have more productive diversification, whereas large companies present more diversification but within the same activity (different kinds of timber, sale of different categories of cattle). When working with a portfolio of products it is possible to redirect sales and purchases depending on the price relations between these products. Within this system integration, decisions related to the forest component may be delayed when facing adverse market conditions. When integrated with other activities that share the same soil, these decisions are less flexible, since the lack of light may affect the development of the lower layers.

The change relationships among the components of a **SPS** may vary with the fluctuations in the value of the forest and cattle products. The average in the past 11 fiscal years (2002/2012) for companies in the Province of Corrientes was 13.46 kg of calf bought with 1 t of standing Pine of 18 cm top diameter. This value ranged between 5.53 kg/t in fiscal year 2011/1012 and 22.7 kg/t in fiscal year 2008/2009. The kilograms of calf that are necessary to plant 1 ha with Pines in a **SPS** ranged between 80 kg/ha (2010/2011) and 230 kg/ha (2008/2009), the average being 174 kg of calf/ha/**SPS** planted.

In **SPSs**, the underuse of the soil due to a lower density of the forest component that allows the entry of light is offset by the production of differentiated knot-free timber, with bigger diameters, and this has an impact on the yield and efficiency of the timber industry. The higher supply of these products will generate more demand and the consequent increase in value; currently, the price differentials range between 35 % and 70 % in its favor.

A simple way of explaining the importance of the silviculture management in **SPSs** compared to Pure Forestry systems is by underscoring the economic impact of generating more knot-free timber (100 % more), even when the total volume of timber is 30 % lower. If we take the value of the knot-free and gross timber and we add the livestock income, the equation is very interesting.

The incorporation of forestry into cattle raising estates generates three situations: *first*, there is a period when the forested land must be closed – 2–3 years in pine plantations and 1–2 years in eucalyptus plantations – and cattle must be excluded to prevent damage to the very young plants. *Second*, grazing is normally conducted under the plantations, and the management of light is important (planting arrangements, thinning, pruning). Finally, the *third* situation is determined by the shading of the forage component. The cutting rotation and the years under closure will determine the proportion of the soil used for cattle.

In a Pine forest under an intensive forest management it is possible to graze under the forest for 5 years and therefore with a cutting rotation of 16 years, grazing takes up 31 % of the time. These values go up with the application of thinning and pruning treatments that extend the grazing period. In a transitory **SPS** (during the final years of tree growth, shading prevents grazing), cattle use may reach 44 % (*7 years of grazing in a 16 year cutting rotation*). On the other hand, in a **SPS** with an initial closure of 3 years and final rotations of 16 years, cattle use may be as high as 81 % (*13 years of grazing in a 16 year cutting rotation*). If it were possible to use the soil during the first waiting years, with haymaking, pasture seed harvesting, farm leasing or grazing with electric fencing, soil use would amount to 100 %.

The livestock production systems in the Provinces of Corrientes and Misiones have different characteristics. In Corrientes they are developed over different types of soils, from deep soils to permanently flooded soils, slightly hilly areas or completely plain lands. They have a long cattle raising tradition, mainly for breeding; the region is mostly covered by grasslands and natural ponds, with small areas with pasture and mostly infrastructure for livestock. The limitation for the development of **SPSs** is not the area but the circulating capital; these systems have the highest acceptance where the capital investment is ensured. Since the livestock infrastructure already exists, the grassland shading does not affect the system, since it is possible to move the animals to another land or temporarily reduce the stocking rate without significantly modifying the final result (Temporary **SPSs**).

On the other hand, the production system in Misiones is based on smaller estates, marked slopes, more fertile soils from clear felled areas, little infrastructure for livestock and the impossibility for the cattle to drink water from the streams due to erosion-prone shores and pollution. Having no livestock tradition, both farmers and the staff have an elementary knowledge of cattle management. The growth of pastures (all agamic and planted) and trees is much higher than in Corrientes. Since the limitation is the available area, the selected forest species must yield higher quality timber. Shade management is paramount, since it should not affect the growth of the pastures. Livestock must be a production activity in itself and, in some cases, provide material for composting to be used as organic fertilizer for industrial or intensive crops.

In Corrientes there is a trend to developing **SPSs** starting with a lower planting density and an arrangement in paired rows with 12 m alleys, particularly with hybrid pine, with three pruning levels up to 5.5 m high and two thinning treatments (one non-commercial or ¨lost¨ and the other one commercial).

The forestry management described above would simply consist of planting 500 plants/ha, making a non-commercial thinning on the third year (DBH 10 cm) (usually known as ¨lost¨) leaving a density of 350 plants/ha, making the first pruning of all these trees (until a height of approximately 2.2 m), and the second pruning between the fourth and fifth year (depending on the soil) up to 3.5 m in the existing 350 trees/ha. Between the fifth and sixth year, the pruning height is increased to 5.5 m for the best 210 trees/ha. The 140 trees/ha that are not pruned are subject to a commercial thinning between the year 10 and 11, and the 210 trees/ha that had three pruning treatments will remain for the cutting rotation on year 16.

As opposed to the above, **SPSs** in Misiones start with a higher density of trees, three prunings (up to 6 m), one lost thinning and four commercial thinnings, cutting the trees based on the shading and not the forest competition. In this case the hybrid pine is also suggested for its crown architecture; *Pinus taeda* is not recommended since the thickness and length of its branches complicates shading management. Densities decrease every 2 or 3 years until the cutting rotation with densities of 80–100 trees/ha. Since they grow in better forest sites, pruning is more frequent and reaches a bigger height.

Continuing with the economic analysis, and considering the restrictions to be taken into account when complementing two or more productive activities, Linear Programming is an adequate tool for that analysis, and more expeditious than traditional financial equations (Berger 2006; de Mello et al. 2005; Céspedes Trujillo 2005).

The choice of the forest species to be planted and the management to be applied would differ a lot if cattle restrictions were not taken into account. The same applies to circulating capital, since the return of **SPSs** is higher than in forestry systems per unit of money invested. If we also consider the receptivity of grazing lands, animal requirements, the possibility of intercropping, the limitations in the number of hectares to be forested, etc., through Linear Programming it is possible to know the optimum solution, the opportunity cost of the limiting resources and the replacement cost of the activities not selected by the solution.

## 2.6 Adoption by the Productive Sector

One approach to the advantages of **SPSs** in Misiones, considering particularly the components of an **SPS** and the System as such, is presented by Pachas (2010):

*Animal component*: prevention and reduction of caloric stress by tree shading; possibility to manage animals with a proportion of British breeds under the tree canopy, thus obtaining a higher growth rate and very tender cuts.

*Forest component*: timber of large dimensions and high quality in shorter rotations; grazing reduces the risk of fire by reducing the forage biomass under the forest or in fireroads.

*Forage component*: the forage species adapted to the shade increase their aboveground net primary productivity under the tree canopy; as the shade increases, there is a change in the floristic composition of the grasses, increasing the proportion of species with a higher forage value (*Axonopus compressus*, *Paspalum sp*); the tree canopy reduces the effect of long frosts and droughts on the pasture or grass; forage species improve their nutritious value (higher phosphorous and protein content) under the tree cover.

The System in itself: it flexibilizes the economy of small and medium-sized farms; the introduction of livestock slightly increases the net present value and reduces the internal rate of return compared to the forestry activity.

On the other hand, in a detailed study Frey et al. (2011) analyze the advantages and disadvantages of these systems from the farmer's perspective and vision. They conducted a survey with 35 farmers with estates ranging between 20 and 14,000 ha. They were classified into three groups: 13 small farms (up to 50 ha), 10 medium-sized farms (90–1,000 ha) and 12 large farms (over 1,000 ha).

A large proportion of all the groups stated that the benefits of **SPSs** are the possibility of obtaining two products per unit of area and weed control. Small farmers tended to give more emphasis to the first answer, whereas medium and large farmers favored weed control and fire control in the forests.

Between the adoption of the system and the moment of the survey (around 10 years) the advantages detected changed, and the microclimate gained more importance (shade protection for cattle, higher proportion of British breeds, lower impact of frosts over the grasses/pastures) followed by a higher cash flow, particularly for small farmers.

The disadvantages detected initially about the complexity of the management and negative interactions between forage and trees decreased with time, indicating the higher experience in the management of these systems, despite the intensive planning required.

The highest disadvantage perceived today is the need for capital investment to start the activity. Initially this was not a limitation since there was more economic stability in the country and it was easy to have access to sources of funding or to the supply of materials like seedlings and wire fences (Gregory et al. 2012).

With reference to the expectations to continue with **SPSs**, most farmers answered that if the incentives continued, they would increase the areas destined to these systems. However, small farmers, despite recognizing the advantages of these systems, might increase this activity up to a point, since the highest value soils are used for more profitable crops (Yerba Mate, tea, tobacco).

Finally, most of the farmers interviewed stated that they would continue implementing these systems even without the government's support, which is an indication of the benefits they see in **SPSs**.

## 2.7 Conclusions and Future Prospects

The region formed by the provinces of Corrientes and Misiones has comparative and competitive advantages compared to other regions for the harmonious expansion of the forest border and higher livestock efficiency through **SPSs** without competition between both activities.

These systems were first developed in the 1990s, particularly by cattle farmers that adopted them as an alternative to diversity production, increase the profitability of their farms and increase the productivity of traditional systems.

Both provinces have a climate and soil characteristics that result in high growth rates for various tree species. Besides, the forest activity enjoys incentives for development through the law that regulates it with non-reimbursable financial support for planted forests.

As a result of some technological demands, several multidisciplinary groups were formed with the support of institutions and joint actions from farmers.

The technology and background available on **SPSs** and the interactions between their components, specifically the effect of planting arrangements, densities and combinations of trees on the productivity of the forage and animal components, favor an expansion of these systems at different production scales, from small farms to corporate levels.

The impact is highly promising on: (a) the social environment, particularly as a result of higher incomes and the generation of genuine jobs; (b) the environment, given the sustainability of resources; and (c) the economy of the territories, given the generation of products with a present or future differential value and the efficient use of resources.

Future research should focus on identifying other factors and more accurately interpreting the interactions among the forest, forage, cattle, soil, microclimate and fauna components. The information generated will provide technologies and practices to the production, commercial and industrial sectors for an environmentally friendly development of the territory.

## Bibliography


Alcaráz A (2006) La navegación y las actividades económicas en el Alto Paraná (1880–1920). CEML, Montecarlo, p 120

Alegranza DA, Torres E, Reboratti H, Fassola HE (1997) Efecto de la densidad del *Pinus caribaea* var. *caribaea* sobre la oferta forrajera. Inf.Tec. N° 18. INTA EEA Montecarlo, 10 p

Andrade C, Carneiro J, Valentín J (2002) Reuniao Anual Soc. Brasileira de Zootecnia 39. Recife. Anais.2002.1CD-ROM

Benvenutti M, Pavetti D, Pérego J, Correa M (1997) Evaluación de especies forrajeras en distintos niveles de iluminación bajo monte forestal de pino. Curso de capacitación Suplementación en Ganadería, organizado por INTA. Eldorado, Misiones

Benvenutti M, Pavetti D, Correa M, Pérego J (2000) Evaluación de especies forrajeras gramíneas tropicales, en distintos niveles de iluminación bajo monte forestal de pino, para uso en sistemas forestoganaderos. INTA EEA Cerro Azul. Informe técnico N°70, pp 1–18

Berger A (2006) Efecto de variaciones en la relación de precios del ternero y del novillo sobre la integración y el resultado de diferentes planteos ganaderos. Asoc. Argentina Economía Agraria, Mineo, 21p

Cabrera A (1976) Regiones Fitogeográficas Argentinas. En: Enciclopedia Argentina de Agricultura y Jardinería, 2ª Edición, tomo II, Fascículo 1, Buenos Aires, 85 pp

Cameron C, Drance S, Edwards D, Jones D (1994) Árboles y pasturas: un estudio sobre los efectos del espaciamiento. Agroforestería de las Américas: 8–9

Caniza F (2010) Efecto de los estados de competencia post-raleo en los caracteres de importancia de la madera del *Eucalyptus grandis* Hill ex Maiden para uso sólido. Tesis Mag. Sc. Facultad de Cs. Forestales-UNaM. Eldorado, Misiones – Argentina, p 151

Carbonell de Masy R (1992) Cap. Técnicas y tecnologías apropiadas. En: Estrategias de desarrollo rural en los pueblos Guaraníes. Antonio Bosch Ed.-Barcelona, pp 135–160

Carranza C, Ledesma M (2009) Bases para el manejo de sistemas silvopastoriles. XIII Congreso Forestal Mundial [Buenos Aires, Argentina], p 18–23

Céspedes TN (2005) En: Programación Lineal: su aplicación a los negocios agropecuarios. Ctro. Univ. Las Tunas. Dpto. Matemática. Ed. Universitaria, La Habana, Cuba, 55p

Colcombet L, Lacorte S, Fassola H, Pachas N, Ferrere P, Alegranza D (2002) Resultados iniciales de un sistema silvopastoril en el Norte de Misiones, Argentina, entre *Pinus elliottii* x *Pinus caribaea*var. *Hondurensis* (F2) y *Brachiaria brizantha* (Hochst) Stapf. X Jornadas Técnicas Forestales y Ambientales. Facultad Cs. Forestales-UNaM – EEA Montecarlo. Eldorado, Misiones, Argentina, 8p

Cordel J, Simard S, Bauhus J, Seely B, Prescott C, Hampel H (2008) The influence of nurse-trees on biomass production and allocation in understorey plantings of *Toona ciliata* Roem. in Argentina. IPEF-IUFRO-Process controlling productivity in tropical plantations. Oral Presentation. Porto Seguro, BA Brazil. IPEF. IUFRO, pp 55–56

Crechi E, Keller A, Fassola H, Fahler J, Maletti R (2014) Efectos de la intensidad y oportunidad de raleo en Pinus taeda L. sobre la producción de madera en el noroeste de Misiones, Argentina- Actas XVI Jornadas Técnicas Forestales y Ambientales – Ecología y Biodiversidad Forestal, pp 140–148

Cruz P (1997) Effect of shade on the growth and mineral nutrition of a C4 perennial grass under field conditions. Plant Soil 188:227–237

de Mello A, Carnieri C, Arce J, Sanquetta C (2005) Planejamento florestal visando à maximização dos lucros e a manutenção do estoque de carbono. Cerne, Lavras 11(3):205–217

Di Blasi A (1989) Experiencia sobre la evolución de cuatro variedades de cubiertas verdes y su influencia sobre el crecimiento del pino recién implantado. Actas V Jornadas Técnicas: Uso múltiple del bosque y sistemas agroforestales. Tomo II. Facultad de Ciencias Forestales. UNaM. Eldorado, Misiones, pp 136–137

Escobar E, Ligier D, Melgar R, Matteio H, y Vallejos O (1996) Mapa de Suelos de la Provincia de Corrientes.1:500.000. INTA: EEA Corrientes, 432 p

Fassola HE (1991) Respuesta del *Pinus caribaea* var. Caribaea a los tratamientos silvícolas directos en Misiones, Arg. Informe de Avance 2. Actas Jornadas Sobre Pinus Caribaea. CIEF, pp. 151–163

Fassola HE, Gelid de Ruibal M (1997) Diámetro máximo sobre muñones al momento de la primer poda en arboles dominantes de *Pinus taeda* L. del depto. Iguazú, Misiones. Actas del II Congreso Forestal Argentino y Latinoamericano, Posadas, pp 6

Fassola H, Ferrere P, Lacorte S, Rodríguez A (2002a) Predicción de la producción de un pastizal bajo distintas estructuras de canopia de *Pinus taeda* en el Noreste de Corrientes, Argentina. Rev de Investig Agrop, INTA, Argentina 31(2):73–96

Fassola HE, Moscovich FA, Ferrere P, Rodríguez AF (2002b) Evolução das principais variáveis de árvores de Pinus taeda L. origem Marion submetidos a diferentes tratamentos silvícolas no nordeste da Provincia de Corrientes, Argentina. 2002. Rev Cienc Florestal 12(2):51–60. U. Fed. de Santa María. Brasil

Fassola H, Fahler J, Ferrere P, Alegranza D, Bernio J (2002c) Determinación del cilindro con defectos en rollizos podados de Pinus taeda L. y su relación con el rendimiento en madera libre de nudos. 2002. Rev de Investig Agrop 31(1):121–137

Fassola HE, Lacorte SM, Pachas N, Keller A (2004) 3er Simposio Latino Americano sobre manejo Florestal. Univ. Fed. de Santa María. RS. Brasil 23–24 Set. 2004 – Disertación. Panel sobre Gestão de sistemas silvipastoris

Fassola H, Lacorte S, Pachas N, Pezzuti R (2005) Factores que influencian la producción de biomasa forrajera de *Axonopus jesuiticus* Valls, bajo dosel de *Pinus taeda* L. en el nordeste de Corrientes. Argentina. Rev de Investig Agrop 34(3):21–38

Fassola H, Lacorte S, Pachas N, Pezzutti R (2006) Efecto de distintos niveles de sombra del dosel de *Pinus taeda* L. sobre la acumulación de biomasa forrajera de Axonopus compressus (Swartz) Beauv*. Rev Argent de Prod Anim 26:101–111

Fassola H, Crechi E, Videla D, Keller A (2007) Estudio preliminar del rendimiento en el aserrado de rodales con distintos regímenes silvícolas de *Pinus taeda* L. Disponible en CD: Actas 13as Jornadas Técnicas Ftales. y Amb. Facultad de Cs. Forestales-UnaM – INTA. Eldorado, Misiones – Argentina, p 7

Fassola H, Lacorte S, Pachas A, Goldfarb C, Esquivel J, Colcombet L, Crechi E, Keller A, Barth S (2009) Los sistemas silvopastoriles en la región subtropical del NE argentino. Actas del XIII Congreso Forestal Mundial, pp 1–6. Disponible en http://inta.gob.ar/documentos/los-sistemas-silvopastoriles-en-la--region-subtropical-del-ne-argentino/at_multi_download/file/INTA-sistemas-silvo-NE-argentino.pdf prensa. Acceso 6-8-2014

Fassola HE, Videla D, Keller AE, Crechi EH, Winck RA, Barth SR, De Coulon E (2012) Rendimiento y valor bruto en el aserradero de árboles de *Pinus taeda*. Bajo manejo silvopastoril: Estudio de caso. 15as Jornadas Técnicas Forestales y Ambientales. Facultad de Ciencias Forestales, UNaM – EEA Montecarlo, INTA. 7, 8 y 9 de Junio de 2012 -Eldorado, Misiones, Argentina, p 15

Fassola H, Videla GD, Winck R, Área M, Crechi E, Keller A, Maestri G, Mastropaolo J (2014a) Relación entre métodos acústicos con propiedades anatómicas y físico mecánicas de *Pinus taeda* L. sometido a raleos. Actas XVI Jornadas Técnicas Forestales y Ambientales -Ecología y Biodiversidad Forestal-.Facultad de Cs. Forestales-UnaM – INTA. Eldorado, Misiones – Argentina, pp 611–618

Fassola HE, Videla D, Pezzutti R, Winck A, Barth S, Crechi E, Keller A (2014b) 11° Informe de avance – 3er trimestre 2014: PIA 10004. Efecto de la silvicultura sobre la acumulación de biomasa, la calidad de rollizos, la calidad y propiedades de la producción aserrable en rodales de *Pinus taeda* del NE de Corrientes. INTA EEA Montecarlo, 5 pp

Fernández RA, Lupi AM, Pahr NM, Domecq C (1996) Respuesta del Kiri (Pawlonia ssp) a la fertilización y el encalado. Inf. Técn. N°7. INTA – EEA Montecarlo

Frangi J (2008) ¿Qué corno son los ciclos de nutrientes y qué pito tocan el N y el P? LISEA, UNLP. Conferencia. FCF-UNaM. Eldorado, Misiones. 11 Nov 2008

Frey GE, Pachas AN, Noellemeyer E, Balmelli G, Fassola HE, Colcombet L, Stevenson HD, Hamilton J, Hubbard W, Cubbage FW (2011) Resumen y comparación de los sistemas silvopastoriles enseis regiones del mundo.1er Congreso Nacional de Sistemas

Silvopastoriles, Posadas Misiones, Estudio de Casos, pp 435–446

Frey GE, Fassola HE, Pachas AN, Colcombet L, Lacorte SM, Pérez O, Renkow M, Warren ST, Cubbage FW (2012) Perceived advantages and disadvantages of silvopasture systems among adopters of varying scales in Northeast Argentina. Agric Syst 105(1):21–32

Goldfarb MC (2009) Las oportunidades del sistema silvopastoril a la ganadería. En: Pag.V Congreso Internacional de Braford. [Abril 2009, Punta del Este, Uruguay]

Goldfarb M, Lacorte S, Gimenez L, Núñez F, Fassola H, Pachas A, Pinazo M (2007a) Producción animal en sistemas silvopastoriles del Norte de Corrientes. 5° Reunión de Producción Vegetal y 3° de Producción Animal del NOA. [San Miguel de Tucumán 26 y 27 de Abril de 2007]

Goldfarb M, Lacorte S, Gimenez L, Núñez F, Fassola H, Pachas AY, Pinazo M (2007b) Efecto del sombreamiento sobre la disponibilidad forrajera y composición botánica del pastizal en sistemas silvopastoriles. IV Congreso Nacional para el manejo de pastizales naturales y I del MERCOSUR. [Villa Mercedes, San Luis], p 66

Goldfarb M, Lacorte S, Giménez L, Núñez F, Quiroz O (2008) Manejo inicial del pastizal en un Sistema Silvopastoril para optimizar el ciclo ganadero. XXII Reunião do Grupo Técnico em Forrageiras do Cone Sul – Grupo Campos. 2008. Lavallejas, Uruguay, p 202

Goldfarb M, Lacorte S, Esquivel J, Aparicio J, Giménez L, Núñez F, Quiroz OG (2009a) Producción forrajera en un sistema silvo pastoril con diferentes esquemas de plantación II. Pastizal de *Andopogon lateralis*. I Congreso Nacional de Sistemas Silvopastoriles [Posadas Misiones, 14–16/Mayo/2009], p 350

Goldfarb M, Lacorte S, Esquivel J, Aparicio JL, Giménez LI, Núñez F, Quirós OG (2009b) Producción forrajera en un sistema silvopastoril con diferentes esquemas de plantación. I.*Brachiaria brizantha* cv Marandú. II. I Congreso Nacional de Sistemas Silvopastoriles. [Posadas, Misiones, 14–16/Mayo/2009], p 346

Goldfarb MC, Aparicio JL, Pachas ANA, Esquivel JI, Núñez F, Quirós OG (2009c) Caracterización dasométrica y lumínica con diferentes diseños de plantación de *Pinus elliottii* var *elliottii* x *P. caribaea* var *hondurensis* enSistemas Silvopastoriles. Caso Estudio I Congreso Nacional de Sistemas Silvopastoriles [Posadas, Misiones, Argentina, 14–16/Mayo/2009], p 456

Goldfarb M, Esquivel J, Giménez L (2010) Caracterización de los componentes forrajeros, arbóreos y ganaderos en Modelos Silvopastoriles difundidos en la Mesopotamia Argentina.VI Congreso Latinoamericano de Agroforestería para la producción pecuaria sostenible. [Panamá, 28–30/09/2010] Formato Online

Goldfarb M, Giménez L, Núñez F, Quiroz O, Ponce de León L (2011) Implantación de gramíneas forrajeras megatérmicas bajo dosel arbóreo. Comunicación.34 Congreso AAPA. [Mar del Plata, Argentina, 04/07/Oct/2011], p 124

Goldfarb MC, Lacorte SM, Núñez F, Quirós OG (2012) Manejo de un pajonal para mejorar la oferta forrajera en un sistema silvopastoril con *Pinus elliottii*. Actas 2° Congreso nacional de sistemas silvopastoriles [Sgo. del Estero, Argentina, 9–11/Mayo/2012], pp 28–33

Goldfarb M, Esquivel J, Núñez F, Quiroz O (2013a) Variación en la humedad gravimétrica en suelos arenosos forestados con pino hibrido. 4to Congreso Forestal Argentino y Latinoamericano [Iguazú, Misiones, Argentina, 28–30/09/2010]. En CD

Goldfarb M, Esquivel J, Núñez F, Quiroz O (2013b) Cambios en la composición botánica y producción forrajera del tapiz vegetal en lotes forestados con *Pinus elliottii*. 4to Congreso Forestal Argentino y Latinoamericano [Iguazú, Misiones, Argentina, 23–27/09/2013]. En CD

Goldfarb M, Núñez F, Quiroz O, Aranda R (2014a) Producción forrajera de *Brachiaria brizantha* cv Marandú en un sistema agroforestoganadero de Pino Híbrido F2. Aceptado en: 37 Congreso Argentino de Producción Animal – 2nd Joint Meeting ASAS-AAPA y XXXIX Congreso de la Sociedad Chilena de Producción Animal [a realizarse en ciudad de Buenos Aires, 20–22/Octubre/2014]

Goldfarb MC, Núñez F, Quirós O, *y* Aranda R (2014b) Producción radical de *Brachiaria brizantha* cv Marandú en un sistema agroforestoganadero de Pino Híbrido F2. Aceptado en: 37Congreso Argentino de Producción Animal – 2nd Joint Meeting ASAS-AAPA y XXXIX Congreso de la Sociedad Chilena de Producción Animal [a realizarse en ciudad de Buenos Aires, 20–22/Octubre/2014]

Goldfarb MC, Núñez F, Quirós O, *y* Aranda R (2014c) Luz incidente bajo el dosel arbóreo en un sistema agroforestoganadero de Pino Híbrido F2. Aceptado en: 37Congreso Argentino de Producción Animal – 2nd Joint Meeting ASAS-AAPA y XXXIX Congreso de la Sociedad Chilena de Producción Animal [a realizarse en ciudad de Buenos Aires, 20–22/Octubre/2014]

Goldfarb M, Esquivel J, Núñez F, Quiroz O, Aranda R (2014d) Descripción del componente forrajero en un sistema agroforestoganadero de *Eucalyptus grandis*. Aceptado en: 37Congreso Argentino de Producción Animal – 2nd Joint Meeting ASAS-AAPA y XXXIX Congreso de la Sociedad Chilena de Producción Animal [a realizarse en ciudad de Buenos Aires, 20–22/Octubre/2014]

Günther D, Correa M, Lysiak E (2008) Zonas Agroeconómicas Homogéneas – Misiones. Estudios socioeconómicos de la sustentabilidad de los sistemas económicos y recursos naturales. N°5. INTA EEA-Cerro Azul, p 117

Keller H (2001) Etnobotánica de los guaraníes que habitan la selva misionera. Disponible: http://www1.unne.edu.ar/cyt/2001/6-Biologicas/B-008.pdf. Acceso 20/5/09

King KFS (1987) The history of agroforestry. In: Steppler HA, Nair PKR (eds) Agroforestry a decade of development. ICRAF, Nairobi, pp 1–10

Kozarik J, Ruiz J (1978) Efecto de la cobertura verde en el crecimiento volumétrico de una plantación de *Pinus elliottii* en Misiones. Informe interno Celulosa Argentina. S. A. Puerto Piray, Misiones, 4 p

Kozarik J, Varela H (1989) Resultados obtenidos al cuarto año de implantada una cobertura verde con *Pinus elliottii* en Misiones. Actas V Jornadas Técnicas. Facultad de Ciencias Forestales. UNaM. Eldorado. Página 52

Lacorte S (2001) Árboles fuera del bosque en la República Argentina. Proyecto información y análisis para el manejo forestal sostenible: integrando esfuerzos nacionales e internacionales en 13 países tropicales en América Latina, pp 4–14

Lacorte SM (2014) Perspectivas de la ganadería bovina misionera. Revista Braford, Año 30, N° 74, pp 32–34

Lacorte S, Esquivel J (2009) Sistemas silvopastoriles en la Mesopotamia Argentina. Reseña del conocimiento, desarrollo y grado de adopción. 1° Congrceso nacional de sistemas silvopastoriles. [Posadas, Misiones, Argentina, 14–16/Mayo/2009], pp 70–82

Lacorte S, Fassola H, Domecq C, San José M, Hennig A, Correa M, Ferrere P, Moscovich F (2003) Efecto del pastoreo en el crecimiento de *Grevillea robusta* Cunn. y la dinámica del pastizal en Misiones, Argentina. RIA 32(2):79–97

Lacorte S, Fassola L, Pachas N, Colcombet L (2004) Efecto de diferentes grados de sombreado con y sin fertilización fosfórica, sobre la producción de un pastizal modificado con predominio de *Axonopus compressus* (Swartz) Beauv. en el Sur de Misiones, Argentina. Actas XI Jornadas Técnicas Forestales y Ambientales. Eldorado, Argentina, pp 1–9

Lacorte S, Hennig A, Domecq C, Pachas A, Fassola H, Keller A (2006) Intersiembra de leguminosas forrajeras cultivadas en pastizales bajo dosel arbóreo de *Pinus elliottii* var. *elliottii* x*Pinus caribaea* var. *hondurensis* en el Sur de Misiones, Argentina. XII Jornadas Técnicas Forestales y Ambientales. FCF UNaM – INTA, Eldorado – Misiones, p 17

Lacorte S, Domecq C, San José M, Hennig A, Fassola H, Pachas A, Colcombet L, Hampel H, Espíndola F (2009) 1° Congreso nacional de sistemas silvopastoriles. [Posadas, Misiones, Argentina, 14–16/Mayo/2009], pp 400–411

Ligier D (2002) Análisis ambiental-productivo de sistemas alternativos a la ganadería en la provincia de Corrientes. Memorias de la Reunión de Grupo Técnico en Forrajeras del Cono Sur. Zona Campos. XIX- FAO, Mercedes, Corrientes, Argentina, 22–24/Oct./2002, pp 60–66

Ligier HD, Matteio HR, Polo HL, Rosso J. (1990) Provincia de Misiones. In: Atlas de Suelo de La República Argentina, Vol II. Secretaría de Agricultura, Ganadería y Pesca, Proyecto PNUD ARG. 85/019 INTA

Morales A (1984) Coberturas verdes en *Pinus elliottii*. Informe interno. Celulosa Argentina. S. A. Puerto Piray, Misiones. 2 páginas

Müller y Bopp, serie 1927–2007 Información inédita de registros pluviométricos. INTA Montecarlo, Misiones

Navajas S, Lacorte SM, Fahler J, Casanova D (1992a) Pastoreo de pasto elefante CV "Panamá" (*Pennisetum purpureum* Schum.) bajo cubierta de un monte de *Pinus elliottii* Engelm. Var. *elliottii* en el NE de Corrientes. Yvyraretá (UnaM-Fac. Cs. Ftales.), Julio/1992, pp 72:78

Navajas S, Fahler J, Casanova D, Lacorte SM (1992b) Pastoreo de pasto elefante CV Panamá (*Pennisetum purpureum* Schum.) bajo cubierta de un monte de *Pinus elliottii* Engelm. var. *elliottii* en el NE de Corrientes. YVYRARETA, año 3, N° 3, Julio 1992, 72:78

Pachas A (2010) *Axonopus catarinensis* y *Arachis pintoi*- Alternativas forrajeras en sistemas silvopastoriles de la Provincia de Misiones, Argentina. Tesis Mag. Sc., Escuela para Graduados Ing. Agr. Alberto Soriano, Facultad de Agronomía-Universidad de Buenos Aires, p 99

Pachas N, Keller A, Fassola H, Lacorte S, Pinazo M (2004) Producción, morfología y calidad nutritiva de *Axonopus catarinensis Valls* bajo diferentes condiciones lumínicas e hídricas. XI Jornadas Técnicas Forestales y Ambientales, FCF- UNaM, INTA-EEA Montecarlo. Eldorado, Misiones, Argentina, 7–9 Oct 2004

Pachas A, Colcombet L, Correa M, Henning H (2008) Producción forrajera de *Axonopus catarinensis Valls* bajo diferentes densidades de *Eucalyptus Grandis* en sistemas silvopastoriles. XIII Jornadas Técnicas Forestales y Ambientales -Facultad de Ciencias Forestales-UNAM, INTA-EEA Montecarlo. Eldorado, Misiones, Argentina, 5–7 Jun 2008

Pachas N, Goldfarb M, Pinazo MA (2009) Caracterización de la radiación fotosintéticamente activa disponible en el sotobosque de plantaciones de Pino Híbrido con diseño en callejones. 1° Congreso nacional de sistemas silvopastoriles. [Posadas, Misiones, Argentina, 14–16/ Mayo/2009], p 205

Pachas AN, Colcombet L, Fassola HE (2010) Los sistemas silvopastoriles en Argentina. Oportunidades para pequeños productores de producción de leche en sistemas silvopastoriles en la provincia de misiones Argentina. Resúmenes III Congreso sobre Sistemas Silvopastoriles Intensivos para la ganadería sostenible del Siglo XXI. 2 y 3 de Marzo de 2011, Morelia y Tepalcatepec, Michoacán, México

Papadakis J (1974) Ecología, posibilidades Agropecuarias de las Provincias Argentinas. Fascículo 3. Enciclopedia Arg. de Agric. y Jardinería. Ed. ACME. Buenos Aires, Argentina, p 82

Park J (1982) Occlusion and the defect core in pruned radiate pine. New Zealand. Forest Service. Fri Bulletin N° 2. 15 p

Pavetti D, Benvenutti M (2003) Jornada de presentación Jesuita Gigante por INTA-EEA Cerro Azul. Jardín América, Misiones, Argentina

Pavetti D, Benvenutti M, Moreno F (2004) Memoria Anual Año 2004. INTA-Centro Regional Misiones. Argentina, pp 31–35

Pérego J (2002) Sistemas Silvopastoriles en el Centro Sur de la Provincia de Corrientes. Reunión del grupo técnico en forrajeras del cono sur zona campos, XIX. Mercedes, Corrientes, Argentina. Memorias. INTA-EEA Mercedes

Pezzutti RV (2011) Efeitos biológicos e econômicos de tratamentos silviculturais em plantios de *Pinus taeda*L. no Nordeste Argentino. Tese apresentada a o Curso de Doutorado do Programa de Pós-Graduação em Engenharia Florestal, Área de Concentração em Manejo Florestal, da Universidade Federal de Santa Maria (UFSM, RS). Santa Maria, RS, Brasil, p 179

Píccolo GA, Goya JF, Grangi JL, Arturi M, y Vaccaro S (2012) Revista Compartiendo Tecnología. Año II n°1. INTA Cerro Azul. Misiones. Argentina. Evaluación de indicadores de calidad de suelo en ecosistemas de Misiones

Rossner M, Houriet JL, Pavetti D (2008) Descripción de pasturas evaluadas en sistemas silvopastoriles del Centro Sur de la Provincia de Misiones. EEA Cerro Azul. Miscelánea N°60. ISSN 0327–2052. 32 p

Rossner M, Guerrero D, Domínguez M (2009) Evaluación de Técnicas de Multiplicación de *Axonopus catarinensis Valls* para Implantación en Sistemas Silvopastoriles. Actas 1er. Congreso Nacional de Sistemas Silvopastoriles. Aspectos relacionados a pastizales y especies forrajeras. [Posadas, Misiones, Argentina, 14–16/Mayo/2009], pp 1–3

Rossner M, Goldfarb M, Lacorte S (2010) *Chamaecrista rotundifolia* – an outstanding legume in silvopastoril systems. XXVI Buiatrics Congress. [Santiago de Chile, 15/Nov/2010]

Rossner M, Goldfarb M, Lacorte S (2011) Utilization of shade cloth strips to evaluate forages species in fluctuating light regimes. IX International Rangeland Congress. [Rosario Argentina], p 439

Samarakoon SP, Wilson JR, Shelton HM (1990) Growth, morphology and nutritive quality of shaded *Stenotaphrum secundatum*, *Axonopus compressus* and *Pennisetum clandestinum*. J Agric Sci 114:161–169

SIFIP (2010) *Pennisetum clandestinum*. J Agric Sci 114: 161–169. Sistema de información Foresto Industrial de la Provincia de Misiones. Disponible en: http://extension.facfor.unam.edu.ar/sifip/index.html. Acceso 9-10-2010

Skromeda M (2013) Efecto de diferentes niveles de sombreo sobre la implantación y persistencia productiva de forrajeras en sistemas silvopastoriles. Tesis de Grado. Facultad de Ciencias Agrarias –UNNE, Corrientes, Argentina. 45 pp

Sotomayor G, Moya I, Teuber O (2009) Manual de establecimiento y manejo de sistemas silvopastoriles en zonas patagónicas de Chile. INFOR 46 p

Toledo DM, Galantini JA, Ferreccio E, Arzuaga S, Gimenez L, y Vázquez S (2014) Indicadores e índices de calidad en suelos rojos bajo sistemas naturales y cultivados. Ciencias del suelo. Ciudad Autónoma de Buenos Aires. V. 31, n 2. Disponible en: http://www.scielo.org.ar/scielo.php?pid=S1850-20672013000200006&script=sci_arttext. Acceso: en 05 sept. 2014

Uguet Vaquer Piloni JP (2013) Producción y calidad de un pastizal y evolución del peso de vaquillonas en el período invernal a cielo abierto y bajo un sistema silvopastoril. Tesis Mag. Sc., Facultad de Ciencias Agrarias, UNNE, Corrientes, Argentina, pp 95

Uguet Vaquer Piloni JP, Lacorte SM (2014) Producción y calidad de un pastizal en el período invernal a cielo abierto y bajo un sistema silvopastoril. Livestock Research for Rural Development 26 (7), pp

Valls J., Santos S., Tcacenco F., Galdeano, F. 2000. Grama missioneira gigante: híbrido entre duas forrageiras cultivadas do gênero Axonopus (*Gramineae*). In: Congresso Nacional de Genética, 46, Águas de Lindóia. Genetics and Molecular Biology, vol. 23 p 388.

Winck R (2013) Influencia del raleo sobre las características anatómicas de la madera y las propiedades físico-mecánicas del *Pinus taeda*L. de la región NE de la Argentina- Tesis Mag. Sc. Maestría en Ciencias de Madera, Celulosa y Papel, FCF-UNaM, Eldorado, Misiones, Argentina

Winck R, Fassola H, Tomazello M, Area M (2013a) Case study: microfibril angle and its relationship with basic density in *Pinus taeda* L. wood from silvopastoral systems. O PAPEL 74(5):55–61

Winck RA, Fassola H, Tomazello M, Area M (2013b) Empleo de ultrasonido para determinar propiedades físico mecánicas de *Pinus taeda* podados. 4to. Congreso Forestal Argentino y Latinoamericano, Iguazú. 24 al 27 de Octubre del 2013

# 3 Silvopastoral Systems in the Delta Region of Argentina

Edgardo A. Casaubon, P.S. Cornaglia, P.L. Peri, M.L. Gatti, M.P. Clavijo, E.D. Borodowski, and G.R. Cueto

**Abstract**

The increase in cultivated area worldwide has brought about a change in land-use pattern. In the temperate-wet regions, the silvopastoral systems (SPS) that integrate afforestation using deciduous species with cattle raising create diversified systems, which demand management practices to capitalise on the beneficial effects and to minimise negative impact on the environment. The increase in silvopastoral activity on Salicaceae plantations in the region of the Delta of the Paraná River in Argentina raised the challenge of developing technologies for a forest-forage system that leads to sustainable cattle raising in this kind of environment. The acquired knowledge on SPS establishment aimed at producing quality wood of *Populus* sp. and *Salix* sp. for sawn wood, veneers, crushed wood, wood pulp and energy, and foliage as cattle feed. The use of bare (unrooted) pole cuttings of 1, 2, or 3 years of age as planting material, instead of cuttings, necessitates wide spacing in nursery planting (1 × 1 m and 1,2 × 1,2 m). This should reduce the time between field planting of trees and letting cattle into the system without causing damage to the trees during the first and second years of plantation establishment. The wider spacing of trees in the plantation (5 × 5 m and 6 × 6 m) allows the establishment of a dense, productive understory consisting of native species of recognized nutritive value as well as shade-tolerant exotic species such as *Dactylis glomerata*

E.A. Casaubon (✉)
EEA INTA Delta del Paraná,
CC 14, (2804) Campana, Buenos Aires, Argentina
e-mail: casaubon.edgardo@inta.gob.ar

P.S. Cornaglia • M.L. Gatti • M.P. Clavijo
Facultad de Agronomía, Universidad de Buenos Aires (UBA), Buenos Aires, Argentina

P.L. Peri
Universidad Nacional de la Patagonia Austral (UNPA), Instituto Nacional de Tecnología Agropecuaria (INTA), CONICET, CC 332, 9400 Río Gallegos, Argentina

E.D. Borodowski
Facultad de Agronomía, Universidad de Buenos Aires (UBA), Buenos Aires, Argentina

Dirección de Producción Forestal-MINAGRI, Buenos Aires, Argentina

G.R. Cueto
Facultad de Ciencias Exactas y Naturales, UBA, Instituto IEGEBA (CONICET-UBA), Buenos Aires, Argentina


P.L. Peri et al. (eds.), *Silvopastoral Systems in Southern South America*, Advances in Agroforestry 11, DOI 10.1007/978-3-319-24109-8_3

that increase the understory carrying capacity. These management techniques, which optimise system productivity, should ensure proper use and balance between SPS components.

**Keywords**
Salicaceae installation • Productive understory • Native and exotic species

## 3.1 Introduction

The development of silvopastoral systems (SPSs) in several regions of Argentina has increased in the last 15 years, because of the expansion of the agricultural frontier. The integration of forestry and cattle raising systems poses the challenge of attaining a sustainable use of the soil. Knowledge of the interaction among components of both systems is necessary in order to develop optimal management plans. From the point of view of the environment and production, one of the main advantages of SPSs is the use of land with multiple goals, aiming at increasing efficiency in the use of resources in terms of space and time, reducing risks, and improving system stability (Mosquera-Losada et al. 2006).

Temperate wetlands occupy the southern portion of the American continent (Neiff 1999; Fig. 3.1). The wetlands associated to the Paraná River conform one of the most important fluvial wetland corridors in the world (Zoffoli et al. 2008). Compared with similar types of temperate wetlands in the northern hemisphere, some of those in South America temperate have not been subjected to extreme and massive hydrologic alterations, and thus have retained the capacity to support much of the original biodiversity (Brinson and Malvárez 2002).

The Lower Delta of the Paraná River is one of the main regions afforested with Salicaceae since mid-nineteenth century in Argentina as well as the one with the greatest potential for silvopastoral use. It has been carried out continuously for over a century and it handles natural resources in a sustainable management: It preserves a wide diversity of flora and fauna: 632 autochthonous and naturalized plant species, 50,9 % of them possesses some type of use (Kalesnik 2010). More than 280 sp of birds (Haene and Pereyra 2003; Fracassi 2014); 200 sp of fish (Minotti 2010); 50 sp of mammals, 260 sp of birds not *Paseriformes*, 37 sp of reptiles, 27 sp of amphibians (Quintana and Bó 2010), and it exhibits unique features due to its adaptation to floods and its continuous production (Borodowski 2006). Area systematisation by means of polders is estimated at 48,000 ha (Gaute et al. 2007). These are the most apt areas for the establishment of SPSs, for which specific management aimed at sustainable production of wood for multiple uses, grass and beef is essential. SPSs are an important alternative for large, medium and small farms in the region, since they offer diversification and efficiency in the use of available natural resources with no need for profound transformations in the production systems. The economic advantages of establishing SPSs include an increase in higher investment owing of greater product diversification, which leads to products with different timing and operation scales. This also lowers risks inherent to the market. In the area of the Delta of the Paraná River, for a 10 % rate, annual income estimated to be 17.47 % higher for silvopastoral systems than for plain forestry (Luccerini et al. 2013).

Even though cattle entry into Salicaceae plantations on the islands of the Delta is an ancient technique, the systemic view on silvopasture management is relatively recent. Forest plantations originally were grazed to reduce spontaneous herbaceous vegetation, thus minimising the risk of grassland and forest fires. The new agricultural scenario, mainly due to the expansion of soybean (*Glicine max*) cultivation (more rentable) in the Pampean region, has displaced cattle to others regions. This land use change has promoted

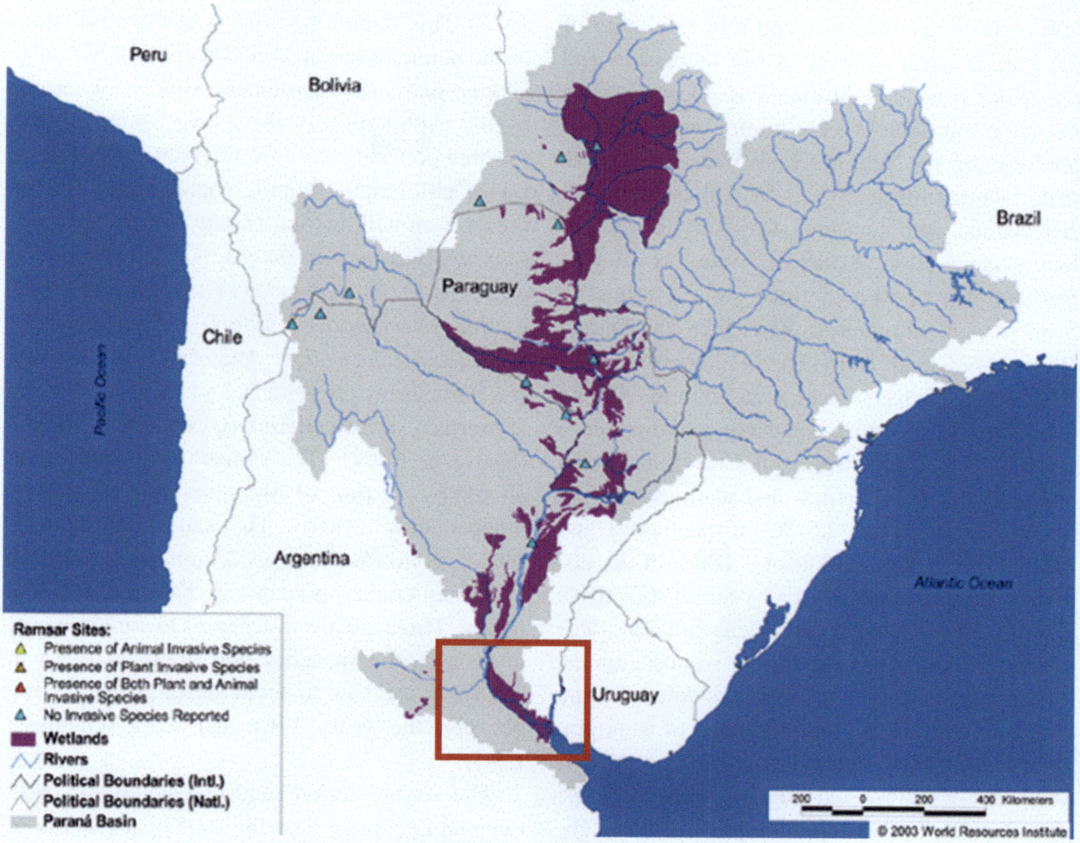


**Fig. 3.1** Location of the wetlands associated to the Paraná River afforested with Salicaceae

the entry of beef cattle into environments traditionally used for forestry in the Lower Delta of the Paraná River over the past decade. Therefore, it has been necessary to use a new kind of silviculture, which should consider such changes in the region. Closeness of concentration, marketing and consumption centres, together with knowledge of the origin of livestock and wood produced in the region, provides competitive advantages for the final product as well as certifiable differential quality (Casaubon and González 2008).

## 3.2 Description of the Delta Zone

### 3.2.1 Geographical Location

The Delta of the Paraná River is located in the final portion of the Paraná River basin, spreading along 300 km (186 mi), between 32°5′S, 58°30′W, and 34°29′S, 60°48′W (Bonfils 1962). Due to its expanse, its flow, and the size of its basin, the Paraná River is the second most important river in South America after the Amazon River. It converges with the Uruguay River at the Río de la Plata estuary. The Delta of the Paraná region is a floodplain defined by Malvárez (1997) as a vast macromosaic of wetlands, its heterogeneity being due mainly to its climatic regime, past and present geomorphological processes, and hydrological regime. Its area has been estimated at 1,750,000 ha, 83.7 % of it belonging to the province of Entre Ríos, and the remaining 16.3 % to the province of Buenos Aires (Bonfils 1962). The area and the permanence of these wetlands depend mainly on superficial water input (by rainfall and water discharge of the rivers), as well as on the timing of the flooding/non-flooding cycle (Neiff and Malvárez

2004). The characteristics and fluvial activity of the Paraná River as well as sea ingression and regression processes during the Holocene may be two relevant factors determining the geomorphology of the region (Malvárez 1997). The materials that formed the Delta soil are very heterogeneous, with intermediate, loamy-sandy, and loamy-sandy-silty textures. These are imperfectly or very poorly drained soils, originating from suspended materials carried by the Paraná River from Brazil, Bolivia and Paraguay.

The region has a temperate humid climate (Malvárez 1997) with a mean annual temperature of 16.5 °C and narrow seasonal variation (mean temperature of the coldest and warmest months being 11.5 °C and 22.5 °C, respectively) and mean annual precipitation of 1,100 mm distributed evenly throughout the year (1960–2010; INTA Delta Agrometeorological Station). The warmest months are generally the wettest months. However, there are cases of hydric deficit during the summer due to longer hours of sunlight, higher heliophany, depending as well on the different water inflow and outflow pathways. The annual average relative humidity is high throughout the year (76 %) owing to closeness of the Río de la Plata and to the large number of rivers and streams forming the Delta. Wind direction modifies water inflow and outflow dynamics, leading to considerable river risings (Kandus 1997).

### 3.2.2 Historical References

The Guaraní people originally settled the Lower Delta region. The oldest evidence of human presence in the Low Delta dated to 1,700 years ago. Archaeological sites on the mainland, on the riverbank lowlands and on the southeast of the Entre Ríos province are located on the "*cordones medanosos*" (sanddune chains), whereas those on the islands are located on the "*albardones*" (higher parts) (Loponte 2013). The earliest Creole and European settlement of this region took place between the eighteenth and the nineteenth centuries with the Jesuits (stated by Liborio Justo's introduction to Sarmiento's "El Carapachay", 1974). This started the process of transformation of the natural environment. This period, virtually lacking permanent settlement, was characterised by the exploitation of the native Monte Blanco for firewood and coal. It is a Sylvan ecosystem of Low Delta Paraná Region, characterized for the presence of willow (*Salix humboldtiana*), Curupí (*Sapium haematospermum*), Ceibo (*Erythrina crista-galli*), Aliso del río (*Tessaria integrifolia*), Laurel (*Nectandra falcifolia*), Palo amarillo (*Terminalia australis*), Anacahuita ó Arrayán (*Blepharocalyx tweediei*), y Palmera pindó (*Syagrus romanzoffianum*) between others (Malvárez 1999). The native forest had been altered by the time of Argentina's declaration of independence (1816). The sauce colorado or sauce criollo (*Salix humboldtiana*) was the only South American spontaneous Salicaceae in the region. There are also references to fruit tree production exotic (peaches and oranges) (Darwin 1978), as well as to areas destined to grazing between the years 1761 and 1762 (Galafassi 1995).

The second period (mid-nineteenth and early twentieth century) sees the start of a process of transformation of the natural environment, with permanent settlements by immigrants from various communities, the intensive cultivation of species for *Salix* sp. ó willow for wickerwork, fruit trees and vegetables, and aviculture and apiculture in small family units. The weeping willow (*Salix babylonica* L.) was introduced into the country in the mid-nineteenth century. Its wood was used as firewood, as well as for coal and for ox yokes (Galafassi 1995). French immigrants in the late nineteenth century are credited with the first plantations of *carolino* poplar trees (*Populus deltoides* ssp. angulata cv. carolinensis), used mainly for firewood (Borodowski and Suárez 2004).

The third and final period of colonization of the region starts in the mid-twentieth century, when wood became the main product. The growth in population correlates with pome and stone fruit production, peaking in the 1940s and subsequently dropping drastically due to competition from markets in other regions of the country. Fruit growing first gave way to forestry, complemented in small farms by willow for wickerwork. Extraordinary floods (1940, 1958, 1959 and 1982–1984) sped up the transformation of the landscape which pushed the Lower Delta to forestry with *Salix* sp. and *Populus* sp., a change which had started to occur in the 1950s (Galafassi 1995).

### 3.2.3 Environmental and Socioeconomic Characteristics of the Region

Bonfils (1962) differentiated the Delta of the Paraná River in four geomorphological regions with a focus on soil analysis: the Ancient Delta (700,000 ha), apt for cattle raising and forestry; the Predelta (600,000 ha) and the Riverbank Lowlands (80,000 ha), fit for the most part for cattle raising; and the Lower Delta (350,000 ha), fit for forestry and fruit cultivation. These broad regions have developed two relevant production and management systems and their own respective agro-industrial chains: the cattle raising wetland and the diversified forestry wetland (Geomorphological aspects, risk level and flood profile have shaped production type, socioeconomic organization of the rural population, farm size, and land distribution and occupation).

**Cattle Raising Wetland** (including the Ancient Delta, the Predelta and the Riverbank Lowlands) farmers undertake "island-style" cattle raising, carried out on large farms (over 5,000 ha) focusing on fattening cattle: long *invernada* (winter period) and/or *veranada* (summer period), feeding the steers for meat production for export. On smaller farms, cow/calf operations are the main economic activity. As a complement, it includes wetland apiculture, mostly with transhumant beekeepers, and, on a much lower scale, plantations of trees of the Salicaceae family (poplars, willows) under special conditions of infrastructure to fight floods with a small area covered by native and/or planted forest, which does not usually receive typical forest management. There are two types of exploitation of these lands: leasing to small and medium-sized "*pastajeros*" [pastoralist] farmers; or through capitalisation. It is a contract of finance is that a person to acquire the place in a farm for a certain time, the owner which has the obligation to handle and feed, delivering at the end of the contract gains, calves or kg of fat, with the owner land. Grazing management done exclusively on natural grassland (of high forage quality such as limpo grass [*Hemarthria altissima*]) hinder the inclusion of improvements (fencing, watering points). In addition, inadequate infrastructure and lack of systematisation (such as polders) create a big technological and organisational gap (PTR 2009).

**Diversified Forestry Wetland** (including the Lower Delta of the Buenos Aires and Entre Ríos provinces) comprises most of the Salicaceae plantation activity in the country. This region is characterised by a high percentage (98 %) of island farmers who incorporate diversification alternatives ranging from cattle raising (mainly heifers) to products strongly identified with regional production and economy, such as wicker, pecan nuts, fruits, apiculture and agro-ecotourism. There is more communications infrastructure and a lower risk of floods, except for the effect of "*sudestadas*" (southeastern winds). Agroforestry farms manage water efficiently, resulting in differentiated, higher-quality products (PTR 2009).

### 3.2.4 General Characteristics of the Lower Delta or Low Delta

The Lower Delta is the zone with the largest number of rivers and streams. It is a recent area, constantly growing towards the Río de La Plata (60 m per year, approximately) because of the great quantity of sediments that drags the river Parana from tropical and subtropical zones of Brazil, Paraguay, Bolivia and north of Argentina. The relief in this zone is flat-concave, with a higher edge known as "*albardón*", which periodically receives contributions of new sediment, and a central, lower, flooded partially area known as "*pajonal*", which is more receptive, since the water remains retained in it refilling with the sediments in suspension the sub overwhelmed area. These two sectors account for 20 % and 80 % of the total area of the island, respectively. The intermediate land between the two former situations is called "*semialbardón*" or "*caída de albardón*" (Bonfils 1962). In its natural state, owing to exposure to flooding and high soil acidity, this land is not economically profitable for growing Salicaceae. In order to make it possible, farmers modify land structure by managing the amount of water flowing in with systematisation projects such as polders, inner canals, and drainage ditches, as well as with the installation of sluice gates and water pumps. Poplar trees are generally planted on higher ground or "*albardones*", on intermediate ground or "*semialbardones*", and on poldered lowland (Casaubon et al. 2004). Instead, willows grow better on lower, previously systematised "*bañado*", "*estero*" or "*pajonal*" ground (Casaubon et al. 2013). *Albardón* soil is predominantly loamy-silty or loamy-clayey, and *bañado* soil is mostly loamy-silty, due to the action of frequent, long-lasting floods. Organic matter content is low (between 4 % and 8 %), with a C/N ratio generally under 14, and an acidic pH (between 5 and 6). The soil in "*pajonales*" or "*bañados*" is semi-swampy, with high contents (on occasions up to 40 %) of organic matter, much of which is not yet decomposed, contributing to very high C/N ratios (over 16). Soil pH is low – between 4 and 5 (Bonfils 1962). Ceballos et al. (2012) reported a neutral C balance following wetland drainage. Afforestation resulted from the opposing effects of aeration, favouring decomposition in the organic layer, root colonization and organic matter stabilization, favouring its accumulation in the mineral soil. In *albardón*, as well as in high and mid-high sectors, there exists more species richness, trees (22 %), herbaceous terrestrial (53 %), and herbaceous wetlands (11 %), than in low sectors, trees (only 9 %), herbaceous terrestrial (48 %), and more herbaceous wetlands (30 %); wealth percentage of shrubs and creepers are similar (3 % and 4 %, and 11 % and 9 % respectively) (Malvárez 1999; Casaubon et al. 2004; Bó et al. 2010).

When summers are very dry, good maintenance of drainage networks as well as sluice gate and pumping systems facilitate water inflow into stands, and during wet period's water excess, is eliminated easily. Proper water management in the stand and a lower tree density optimise tree and grass development, thus providing livestock with a larger forage supply and with higher animal comfort. In such a system, the small volume of combustible material under plantations minimises in turn the risk of surface fires typical of the Delta region. Vegetation communities in the understory of Salicaceae plantations in the region exhibit high biodiversity provide data with numbers of species of grasses, forbs, etc. (Kandus and Malvárez 2004). The growth of numerous species of forage (annual and perennial grasses as *Bromus catharticus*, *Phalaris angusta*, *Lolium multiflorum*, *Leersia hexandra*, *Panicum sp. Poa sp. and forbs as Trifolium repens*, among others) is frequent (Casaubon et al. 2005a, b; Rossi et al. 2006). Although appearance with low coverage (20 %), especially under adult or very dense plantations, because of the canopy shade (Cornaglia et al. 2009; Clavijo et al. 2010). With respect to fauna, Salicaceae plantations with understories and native forest patches (secondary forest or unmanaged forest) such as grasslands, integrated into the forest matrix might favour environment conditions for the presence of some wildlife birds and mammals such as dusky-legged guan (*Penelope obscura*) and marsh deer (*Blastocerus dichotomus*) (Kalesnik 2010; Fracassi 2012).

## 3.3 Salicaceae Production

The area cultivated with Salicaceae in the Lower Delta is estimated at 80,000 ha (75 % of it under management), 64,000 ha of which are planted with willows (*Salix* spp.), with average yields of 15–20 $m^3$/ha/year, the remaining hectares being planted with poplar trees (*Populus* spp.), with average yields between 20 and 25 $m^3$/ha/year. Poplar trees are cut down at 12–16 years, and willows at 10–14 years, depending on production goals and on the market exigencies. Practically 95 % of willows are destined to the paper cellulose industry. Expected yield at felling time is between

200 and 400 $m^3$/ha for poplar trees, and from 120 to 250 $m^3$/ha for willows (MAGyP 2011). The poplar trees most widely planted for commercial purposes in the region are *Populus deltoides* 'Australiano 129/60', 'Australiano 106/60' and 'Mississippi Slim', known as "Stoneville 67". In the case of willows, *Salix babylonica* var. Sacramenta, *Salix babylonica x Salix alba* 'Ragonese 131-25 INTA', *Salix babylonica x Salix alba* 'Ragonese 131-27 INTA', *Salix matsudana x S. alba* 'Barret 13-44' and *Salix nigra* 'Alonzo nigra 4' (Cerrillo 2010; Borodowski et al. 2014).

Silvicultural management traditionally used for *Populus deltoides* is mostly focused on wood production for solid uses (sawn wood and/or veneers) and, to a lesser extent, for milling (crushed wood and wood pulp) and for energy uses. In general, one single clone is used, increasing the risk of attack by pests and diseases. Currently, willow (*Salix* spp.) wood is mainly destined to the newspaper industry and crushing (95 %), and, less known, to solid uses, such as fruit and vegetable crates. Wood could be potentially used for furniture and for house-building in the region (Cerrillo 2010). Poplar trees are planted on high *albardón* sites, *caídas de albardón* and on low systematised poldered ground, with a loamy coarse texture, moderately well drained, and deep, with a slightly acidic to neuter pH. Willows are planted on *bañados*, often flood-prone, with a phreatic zone permanently on the surface, with aerated, moving water, and loamy-clay-silty, well-drained soils, with acidic to neuter pH (Casaubon et al. 2004; Casaubon et al. 2013b).

## 3.4 Silvopastoral Systems with Poplar Trees

### 3.4.1 Forestry

Silvopasture livestock production is done mainly under poplar plantations. Such activity requires a new kind of silviculture with a more diversified use of trees and of the environment than traditional silviculture. The objective is to produce a larger volume of wood per plant for multiple uses. This demands land use management, identifying first the most suitable site for each clone to be cultivated. This type of silvicultural management including SPS provides: (i) a larger volume of quality wood for various uses (sawn wood, veneers, crushed wood, wood pulp, forage and other industrial uses); (ii) greater individual growth per plant; (iii) better tree homogeneity; (iv) higher percentages of cylindrical shafts; (v) a longer planting period (May-August); (vi) greater plant rooting; (vii) greater access to phreatic zone water. It is also expected to obtain (i) a decrease in diseases and pest attacks due to higher plantation aeration, and (ii) less weed competition and a lower risk of forest fires since forage stays green and available for livestock (Casaubon 2014c).

Establishing a silvopastoral system demands management strategies different from those used for traditional tree monocultures. The combination of tree size and foliage palatability may determine success in establishing a SPS, where palatable tree species such as Salicaceae demand higher protection than non-palatable species (Eason et al. 1996; McAdams 2003). In this sense, the fact that poplar tree leaves and tender branches are highly appetizing to livestock (Lefroy et al. 1992; Taranaki 2001) poses a problem in establishing a SPS, namely, higher establishment costs (Carvalho et al. 2003). In addition, potential damage by livestock to the shafts of trees planted in the SPS decrease when understory forage availability and quality is high (Simón et al. 1998; Casaubon 2013a).

Commencing grazing at the early stages of plantation may cause damage to young plants through browsing, trampling, bark stripping, breakage or tipping (Casaubon et al. 2009b). These are direct effects on trees from livestock. Damage results in higher tree mortality, shaft quality loss, or smaller growth due to defoliation. The decision on tree age at livestock entry may vary according to initial tree growth (site quality), to propagation materials used, and to forage availability in terms of volume and quality. Farmers should weigh potential damage that is more serious in the early period of plantation against the benefits derived from lower weed control costs and a higher potential for livestock production during the first years, when forage is abundant. In this respect, selection of animal species or of livestock production type can help regulate direct damage levels. For example, the feeding habits of cows result in lower levels of tree damage than those of goats or horses. On the other hand, the greater size and weight of bulls compared to developing calves may result in differences in damage due to plant tipping or breakage, or soil compaction, affecting tree growth (Somarriba 1997). To minimize or to avoid damages in the young plants, it is proposed changes in silvicultural management.

In the new type of silvicultural management with SPSe (more apt for SPS establishment and operation), the poplar plantation is started using 1-, 2- (and up to 3-) year-old unrooted pole cuttings between 3 and 8 m high as planting material (FAO 1957; Sanhueza 1998; Casaubon and González 2008; Casaubon et al. 2014d). The nurseries providing these materials demand a spacing plantation greater than traditional ones, as well as good site quality, and intensive site management in order to produce tree propagation materials with greater stem straightness and good conicity index, providing the harvestable poles with greater resistance to external mechanical factors (Casaubon 2013a). In sites most suitable for poplar trees, such pole cuttings exhibit good rooting percentage, surpassing cuttings in numerous *Populus deltoides* commercial clones (Casaubon et al. 2001; Casaubon 2014a). In the traditional system, with plantation densities equal or greater than 625 plants per hectare, many farmers bring livestock into the plantation at the fourth, fifth or sixth year (Fig. 3.2). Beef cattle remain in the system as long as there is forage availability, possibly up to the ninth or tenth year of plantation (depending on site quality). With larger spacing between plants and between rows, approximately 5×5 and 6×6 m, and with higher-quality plantation materials, the entry of beef cattle into the system may occur at the first or second year, and cattle permanence may be longer than with more narrow spacing (Casaubon 2013a). In addition, poplar tree leaf palatability and forage value for different types of livestock is well known (Carou et al. 2010a, b; Thomas 2011; Casaubón et al. 2012b). This makes losses due to herbivory and/or mechanical damage frequent in SPSs (Casaubon et al. 2009a). The use of propagation materials of larger diameter at breast height (≥6 cm) brings forward the entry of beef cattle into the SPS, minimising plant loss due to knockdown, breakage and/or bark-stripping (Casaubon 2013a).

One silvicultural management practice is pruning, undertaken preferably in spring-summer, in order to minimise the growth of epicormic shoots on tree shafts. Pruning height is variable, reaching up to 50 % of total tree height in excellent sites, and up to 30 % of total height in good sites, in order not to reduce annual tree volume increase drastically. When pruning up to half of total tree height, two interventions are enough in good plantation sites. When only the first third is pruned, three interventions at most are needed to remove all branches on the first 7 m of the shaft (Casaubon et al. 2005a). Another silvicultural practice is thinning, consisting of tree density reduction, allowing for the passage of sunlight (Signorelli et al. 2011), and promoting the establishment and growth of herbaceous understory with forage value and greater forage productivity (Cornaglia et al. 2011a; Clavijo et al. 2012). In a thinning pattern, depending on the objectives of production, it is possible to achieve a balance between stand growth and individual tree growth, and to obtain the development of an herbaceous understory, which should optimise the forage component in the silvopastoral system (Fernández Tschieder et al. 2011b; Cornaglia et al. 2011a; Clavijo et al. 2012). Fernández Tschieder et al. (2011a) adjusted equations to estimate the total volume for the most common *P. deltoides* clones in the region.

Salicaceae foliage may be used as forage or as a regular component of livestock feed (Rossi et al. 2005; Benavides 2006). Poplar tree (*Populus* sp.) foliage has good nutritional value and it is a valuable diet supplement (Ball et al. 2005; Mead 2009). Poplar young branches and leaves have high forage potential (Table 3.1). Because of their palatability and nutritional qualities, they can also make good feeding supplements in a SPS, improve grazing cattle diet, and contribute forage volume in times of shortage (Mead 2009; Carou et al. 2010b; Thomas 2011). However, such preference is not consistent throughout tree

**Fig. 3.2** Silvopastoral system with 6-year-old *Populus deltoides* 'Australiano 129/60' planted from 2-year-old unrooted pole cuttings, and Aberdeen Angus heifer, on a natural fodder grassland in the Lower Delta of the Paraná River

**Table 3.1** Average values (± standard deviation) for dry matter (DM), digestibility (D), gross protein (GP), phosphorus (P) and potassium (K) in *Populus deltoides* 'Australiano 106/60' leaves during vegetative period

| Month | DM% | Digestibility % | GP % | P (mg/kg) | K (g/100g) |
|---|---|---|---|---|---|
| October | $20.17 \pm 2.55^a$ | $63.73 \pm 3.94^b$ | $30.26 \pm 2.07^a$ | $500 \pm 123.34^a$ | $1.95 \pm 0.24^a$ |
| December | $36.59 \pm 1.04^b$ | $67.36 \pm 2.64^a$ | $15.71 \pm 2.44^b$ | $183.64 \pm 21.49^b$ | $0.96 \pm 0.31^b$ |
| March | $42.3 \pm 1.05^c$ | $62.02 \pm 3.03^b$ | $16.52 \pm 2.01^b$ | $176.94 \pm 23.93^b$ | $1.07 \pm 0.33^b$ |

Different letters indicate significant differences ($p <= 0,05$) between months

vegetative period, but decreases considerably halfway through the summer and in autumn. Gross protein (GP) values of spring tender leaves double those of summer and early autumn leaves, while estimated digestibility (ED) increases in spring and decreases in late summer (Casaubon 2012b).

Nutritional values of poplar leaves and tender shoots in October, December and March are linked to heifer higher preference for these materials observed on the field in spring (% DM 20.17 %) and early summer (% DM 36.59 %). The decrease or lack of livestock preference for such leaves as from February is related to poplar tender leaf higher nutritional value and forage quality in spring and early summer, and with quality loss of grasses, at that time of year, at reproductive stage (Casaubon et al. 2012b). It is probable that the high values of gross protein (GP), phosphorus (P) and potassium (K) recorded in poplar leaves and tender shoots in spring, as well as greater estimated digestibility (ED), relate directly to beef cattle being more avid for spring and early-summer poplar foliage (Casaubon et al. 2012b). Analysing mineral composition in poplar and willow leaves with nutritional interest for livestock, Carou et al. (2010) determined that such values are higher than those known for feed grasses are similar to those reported for feed legumes. In both genera, these nutritional values meet beef cattle requirements.

### 3.4.2 Understory Forage Production

The Delta of the Paraná River grasslands possesses great species richness. This is a natural source of forage for wild fauna such as marsh deer (*Blastocerus dichotomus*), capybaras (*Hydrochaeris hydrochaeris*), and the coypu (*Myocastor coypus*). It is also the main forage source for the various meat production cattle-raising systems (breeding, rebreeding and *invernada*) in the region (Rossi 2010). A variety of Delta grassland species are apt for good livestock development, since they surpass forage good quality values in terms of gross protein and digestibility (Rossi et al. 2006, 2009; Casaubon et al. 2010; Carou et al. 2010a). *Lolium multiflorum* is an exotic cool season grass naturalised to the Argentinean Pampean region grassland. It has nutritional values similar to those of the Delta of the Paraná River native plants: 60.3–76.6 % leaf digestibility and 7.5–14.6 % leaf gross protein (at reproductive and vegetative stages, respectively) (González et al. 2008). Forage species with the greatest abundance-dominance in the natural grassland of the Lower Delta of the Paraná River SPS have good GP content. Of the 15 species studied, 14 surpass the 7 % GP critical value, considered the limit of N content in beef cattle diet (Rossi et al. 2012).

When these grasslands are under afforestation, understory vegetation tends to be scant, depending on forest density, but usually not over 50 % of the land (Clavijo et al. 2005; Rossi et al. 2006). The poplar tree (*Populus deltoides*) is a deciduous species which remains leafless during autumn and winter, allowing for the growth of an understory consisting of forage-value native and cultivated herbaceous species (Casaubon et al. 2005b; Rossi et al. 2006; Pincemin et al. 2007). Although this combination of deciduous-leaf trees and winter forage species allows for complementariness of production cycles (Monlezun et al. 2008; Nordenstahl et al. 2011), tree shade directly affects underlying species productivity (Jose et al. 2004; Peri et al. 2007). In addition, the falling of poplar senesced leaves in autumn constitutes a physical barrier for annual herbaceous species reestablishment (Cornaglia et al. 2009; Clavijo et al. 2010). Control of tree canopy opening and the inclusion of planted temperate feed grasses (Dallisgrass) synergically improved forage stability, composition and production in the understory (Cornaglia et al. 2011a; Clavijo et al. 2012).

### 3.4.3 Livestock Production

Dominant production systems in the region are *invernada* in the Upper and Middle Delta (1,400,000 ha), and breeding and full cycle in the Low Delta (350,000 ha). The proportion of each system depends of wetland environmental features and to farmers' socioeconomic features (PTR 2009). The region's topography and rivers block the entry of infectious and parasitic livestock diseases, with an ideal microclimate for the development of high-quality natural forage resources. This favours breeding cow longevity as well as adult cow marketing status. Temperature is lower in the shade than in the open, creating a very favourable microclimate for livestock, reflected in animal behaviour. During sunny days under tree canopies surface soil temperature can increased 8–12 °C by midafternoon, and remain nearly constant (1.5–2 °C) at unshaded sites depending on soil water levels (Feldhake 2001). The presence of cattle simultaneously with cultivated forests allows for product diversification, thus reducing biological and marketing risk factors. Knowledge of cattle origin provides the final product with certifiable premium quality, positioning it favourably in both regional and worldwide beef consumer markets (Casaubon et al. 2012a). Production is mainly aimed at the local beef market. The Delta's calves and heifers have high meat-quality, since they possess certain qualities that allow them to enter almost directly into product certification systems or quality protocols such as Certificate of Origin (González et al. 2006).

The features of the region's spontaneous forage prevent the development of an economically profitable *invernada*. Even if breeding and *invernada* would have very low production costs in the "Zona Núcleo Forestal", since the basic feeding resource is grasslands, beef-cattle raising seems to be the most appropriate activity to be integrated into Salicaceae afforestation under SPSs in the open. The abundance of tender grass, with a high nutritional value, a low percentage of dry matter, and high degradability to balance nutrient requirement (PTR 2009), restricts consumption. As a result, animals evolve much more slowly, mainly during *invernada* and full cycle. This leads to inefficient products, since animals are finished over a very long period.

In general, agricultural farms in the region combine a variable proportion of non-afforested areas with afforested areas of different ages. As a result, there is a tree coverage gradient (grassland-afforestation) (Borodowski 2006). Thus, cattle do not graze exclusively under plantations but rather alternate with grassland grazing (Laclau 2012). British cattle breeds predominate in the area, especially Aberdeen Angus and Hereford. Animal density ranges from 0.2 to 0.5 livestock units/ha, it is very heterogeneous and closely linked to natural forage availability on the site. Livestock production is at between 70 and 100 kg $ha^{-1}$ $year^{-1}$ (Casaubon, in Peri 2012).

Species heterogeneity and consequently seasonal variation of forage production in the various sites of livestock use enable complementary site exploitation. Estimated minimum and maximum carrying capacity levels (in animals/ha) for grasslands (G) and under poplar plantations (F), either understory natural (US) or enriched by *Dactylis glomerata* plantations (S) were: G: 0.25 (autumn) to 3.42 (spring), FS: 0.3 (winter) to 2.07 (spring), and FUS: 0 (autumn) to 1 (spring). The presence or absence of annual winter species, especially annual ryegrass (*Lolium multiflorum* Lam.) might explain the variability in cattle receptivity levels found in non-afforested natural grasslands (G). Under afforestation, the differences might be explained in terms of the presence of orchard grass (*Dactylis glomerata*) on forests sown (FS) and of the absence of perennial forage species on forests unsown (FUS). These differences call for strategic, differential grazing management planning in order to guarantee efficient livestock production in each silvopasture in the region (Clavijo et al. 2014).

## 3.5 Advances in the Knowledge of System Components

### 3.5.1 Effect of Nursery Planting Spacing on Poplar Pole Cutting Production and Quality

The establishment of silvopastoral systems (SPSs) in poplar (*Populus deltoides* 'Australiano 106/60') plantations requires proper management strategies. In this context, the combination of the type of propagation material and its palatability may determine the success of system establishment. The general objective of the trial was to create a new technology for the establishment of a SPS with poplars aiming at the production of wood for multiple uses. The first specific goal was to assess changes occurring at different planting spacings in the size and morphological characteristics of poplar pole cuttings, multiplication materials that are potentially apt for establishing a SPS. Results show that the largest spacings tested (1 × 1 m and 1.2 × 1.2 m) resulted in better 1-, 2- and 3-year old pole cuttings, with higher DBH, total height, aerial biomass, straightness, conicity, and shaft stability (Casaubon 2013a).

In addition, it was determined that it is possible to bring forward the entry of cattle into the SPS by using pole cuttings with a DBH equal or greater than 6 cm, with a positive response during the first years after plantation (Table 3.2). Poplar leaves and tender branches obtained after spring pruning provided a good supplement to animal diet due to their higher GP, P and K values, and their digestibility. It was concluded that poplar pole cuttings as multiplication material may lead

**Table 3.2** Average diameter at breast height (DBH) and total height (Ht) of *Populus deltoides* 'Australiano 106/60' pole cuttings, growing at different densities in nurseries during 3 years

| | | 2007 | | 2008 | | 2009 | |
|---|---|---|---|---|---|---|---|
| Spacing (m) | N | DBH (cm) | Ht (m) | DBH (cm) | Ht (m) | DBH (cm) | Ht (m) |
| 0.6×0.6 | 36 | 2.15(0.62)h | 4.60(0.82)h | 3.42(1.02)f | 6.48(1.54)f | 4.46(0.35)e | 7.97(2.17)e |
| 0.8×0.8 | 36 | 2.73(0.50)g | 5.09(0.55)gh | 4.38(0.79)e | 7.79(1.02)e | 5.47(0.44)cd | 9.60(1.88)c |
| 1.0×1.0 | 36 | 3.14(0.69)f | 5.22(0.64)g | 5.22(0.93)d | 8.41(0.93)de | 6.94(0.69)b | 10.88(1.50)b |
| 1.2×1.2 | 36 | 3.49(0.38)f | 5.28(0.43)g | 5.72(0.68)c | 8.94(0.61)cd | 7.78(0.27)a | 11.95(1.03)a |

Source: Casaubon (2013)
Different letters indicate significant differences P<=0.05 for a single variable

to early production of poplar wood for multiple uses, spontaneous natural pastures, and beef in the SPSs on the Lower Delta of the Paraná River.

### 3.5.2 Effects of Thinning and Sowing the Forage on Forest and Forage Productivity

The effects of plantation density and enrichment with perennial gramineae on forest growth and forage productivity in the understory were assessed to evaluate the effects of tree canopy shade over composition and productivity in the understory (Table 3.3). In a commercial poplar (*Populus deltoides*) plantation, a randomized complete block experimental design was used (with four replications); and paired plots with (S) and without (NS) added $C_3$ grasses were compared. Thinning intensities were T 0 % (control): original density (400 trees/ha), 30 % (280 trees/ha) and 60 % (160 trees/ha). The added forage species were *Festuca arundinacea* and *Dactylis glomerata*. Hemispherical photography was used to estimate the fraction of radiation reaching the understory of a deciduous forest and to describe canopy structure. In autumn, when tree canopy is determined by branches and not by foliage, thinning treatments were more similar to each other with respect to light transmission to the ground, and exhibited significant differences with the control (unthinned). Instead, in spring, when tree canopy has foliage, the control and the 30 % thinning treatment of the BA showed similar LAI and light transmission percentage to the ground, whereas the 60 % thinning treatment of the BA exhibited lower LAI and a higher light transmission percentage than the others (Signorelli et al. 2011). With respect to stand growth, after 3 years of treatment application, intense thinning allowed for a 60 % increase in diameter growth (in cm year$^{-1}$), but with a 70 % decrease in volume growth (in m$^3$ ha$^{-1}$ year$^{-1}$). Accumulated volume was 206 vs 247 m$^3$ ha$^{-1}$ in the control treatment (Fernández Tschieder et al. 2011b). Thinning and seed incorporation synergically improved forage stability, composition and production in the understory: The accumulated production of dry matter from introduced forage species varied significantly from one thinning value to the other. The treatment with 30 % thinning doubled control production ($p<0.001$), whereas the treatment with 60 % thinning tripled it ($p<0.001$). In the treatments with seed incorporation, species changes, which promoted forage production, were more abrupt, and the situation achieved initially remained relatively stable over time. As regards the unsown treatments, there was positive evolution of forage coverage in thinned treat-

**Table 3.3** Tree growth (PIA, Periodic Annual Increment)) and understory productivity (ANPP) as associated to thinning and forage seed addition

| Thinning treatment | Forestry canopy | | Understory | |
|---|---|---|---|---|
| | PIADBH (cm/year) | PIA VOL (m$^3$/ha) | ANPP (kg DM/ha/year) | |
| | | | NS | S |
| Control | 1.3 a | 28 a | 400 a | 1.800 a |
| T 30 % | 1.7 ab | 21 ab | 800 b | 3.800 b |
| T 60 % | 2.1 b | 16 b | 1.700 c | 4.400 c |

Different letters indicate significant differences in each column, p<0.05

ments (30 and 60 %). This was mainly due to the establishment of winter annual grasses (Cornaglia et al. 2011a; Clavijo et al. 2012).

### 3.5.3 Effects of Shading and Defoliation on the Morphogenetic, Structural and Functional Characteristics of Three Perennial Temperate Forages

Complementarity in the use of resources between herbaceous species and the wood component is the key to SPS success. SPSs can be favoured with the choice of deciduous trees and forages with complementary phenology. An alternative to the widespread implantation of annual crops -winter annual grasses- to meet demand of winter animal forage, is the introduction of perennial grasses. This would contribute to system sustainability by avoiding both annual stirring of these fragile soils and production instability of annual species during either dry or very wet years.

Pastures dominated by perennial grasses have better ecosystem function than those dominated by annual ones (Garden and Bolger 2001; Lazenby and Tow 2001). They also have a deeper root system. For this reason, they make use more efficient of water and capture nitrates from deep strata. This reduces soil salinization and acidification (Kemp et al. 2000; Kemp and King 2001) through better leaching control (Garden and Bolger 2001). They also have less production variability throughout the year and provide more complete soil coverage reducing weed invasion (Kemp et al. 2000; Lazenby and Tow 2001). However, slow or poor establishment of some perennial grasses (Robson et al. 1988) as well as their low initial growth rates entail lower potential productivity, mainly under shaded conditions.

Shading, associated with changes in the intensity and quality of light reaching the understory, varies in time and space, depending primarily on forest features and on the growing season. Under these light conditions, plants often express shade tolerance and evasion mechanisms: specific leaf area expansion and an increase in internode elongation, leaf/stem ratio reduction, and shoot/root ratio increase, respectively (Ballaré et al. 1990). These mechanisms are advantageous since they improve grass competition for light, but they might hinder species persistence in these systems by restricting root growth. In addition, these modifications demand changes in defoliation management. Therefore, there is some uncertainty over which perennial grasses sown and over how to manage defoliation in these shaded systems.

We aimed at establishing if it was possible to maximize simultaneously aerial productivity and persistence, root growth and tillering, of $C_3$ perennial grasses in silvopastoral systems (Table 3.4). To that end, we assessed morphogenetic, structural and functional changes associated with poplar shading and with defoliation frequency. We carried out an experiment with pots in field conditions, in a factorial arrangement with three repetitions and screening three $C_3$ perennial grasses with different shade tolerance: *Phalaris aquatica* – *Pha*; sensitive-, *Dactylis glomerata* – *Dg*; tolerant- and *Festuca arundinacea* – *Fa*; moderately tolerant to shade-. We probe two defoliation frequencies: at optimal – Optimal-(ODF) and High (HDF), and two sites: UNDER (90 % reduction of photosynthetically active radiation -PAR- and 50 % reduction of Red:Far-Red ratio -R:RL-) during poplar growth season, and OUTSIDE (in the open). ODF for each species was the necessary time to expand the maximum number of living leaves per tiller, and HDF was half that time. Optimal defoliation interval was in 4 leaves.tiller$^{-1}$ for *Dg* and *Pha*, and in 3 leaves.tiller$^{-1}$ for *Fa* (Rawnsley et al. 2002; Sinclair et al. 2006). Defoliation methodology was adapted from Berone et al. (2008). Defoliation intensity included total lamina removal. Experimental period was 263 days between sowing and final harvest. Response variables were: (a) morphogenetic: phyllocron and lamina longevity or leaf lifespan (LLS) (both expressed in GDD.leaf$^{-1}$), (b) structural: total blade length per tiller – cm.tiller$^{-1}$; indicating tiller size- and tiller density (number or tillers per plant, and (c) functional: aerial, root and total

**Table 3.4** Results of variance analysis comparing the effect of site (two levels: UNDER and OUTSIDE), defoliation frequency (two levels: ODF and HDF) and their interactions, on certain morphogenetic and structural variables of Dactylis glomerata (Dg), Festuca arundinacea (Fa) and Phalaris aquatica (Pha) (n=3)

| | Species | Site effect | | Defoliation frequency effect | | Site*frequency interaction | |
|---|---|---|---|---|---|---|---|
| Variable | | F | P | F | P | F | P |
| Phyllocron (GDD.leaf$^{-1}$) | *Dg* | 62.64 | *** | 7.54 | ** | 0.0047 | ns |
| | *Fa* | 0.09 | ns | 0.38 | ns | 0.30 | ns |
| | *Pha* | 28.71 | *** | 0.02 | ns | 1.86 | ns |
| Tiller size (cm.tiller$^{-1}$) | *Dg* | 193.84 | *** | 2.17 | ns | 0.81 | ns |
| | *Fa* | 197.72 | *** | 0.16 | ns | 0.13 | ns |
| | *Pha* | 97.53 | *** | 8.13 | ** | 5.88 | ** |
| Tiller density (tillers. plant$^{-1}$) | *Dg* | 0.63 | ns | 1.43 | ns | 0.36 | ns |
| | *Fa* | 6.35 | ** | 6.35 | ** | 0.46 | ns |
| | *Pha* | 7.48 | ** | 0.01 | ns | 0.01 | ns |

*ns* no significant effect, *ODF and HDF* Optimal and High Defoliation Frequency, *GDD* growing degrees days
The results presented are F and P values: *** $p<0.01$; ** $p<0.05$; * $p<0.10$

biomass of individual plants (mg.plant$^{-1}$) and biomass shoot/root ratio (mg/mg).

In general, shading increased species phyllocron, with the exception of *Fa* (Table 3.1). *Fa* and *Pha* leaf lifespan decreased, accelerating leaf tissue turnover under shaded conditions. In *Fa*, HDF increased LLS only OUTSIDE (site*frequency interaction). *Dg* was the only species also exhibiting a phyllocron increase as a response to defoliation frequency (Table 3.1), whereas *Dg* leaf lifespan showed no changes associated to any of the factors under evaluation (Cornaglia et al. 2010). UNDER poplar trees, total blade length per tiller was higher than OUTSIDE (Table 3.1). *Pha* showed bigger tillers especially when it was defoliated with ODF (site*frequency interaction; Table 3.1). UNDER shading *Fa* increased tiller number per plant, *Pha* reduced it and *Dg* exhibited no changes. AFD only affected tillering in *Fa* (Cornaglia et al. 2011b). Tested defoliation frequencies did not affect the biomass of the species ($p>0.05$); it is possible that current defoliation intensity was excessive. Total species biomass (aerial + radical) did not differ between sites ($p>0.05$). Aboveground biomass was higher in UNDER and root biomass, higher in OUTSIDE in *Pha* and *Fa*; being biomass shoot/root ratio higher UNDER poplar trees. *Dg* behaviour was indifferent between sites (Gatti and Cornaglia 2011).

According to these results, the species most tolerant to shade, *Dg*, preserved the morphogenic, structural and functional parameters associated with its growth and persistence in UNDER. High light transmission during winter UNDER would only favour tillering in *Fa*. UNDER, only *Fa* and *Pha* root biomass decreased. Thus, UNDER, *Pha* persistence would be totally affected, and *Fa* persistence would be moderately affected. *Pha* and *Fa* accelerated leaf tissue turnover under shaded conditions, for that reason, defoliation with respect to OUTSIDE was expected to perform earlier. Frequencies higher than species physiological optimum – HDF- affected the phyllocron in *Dg*, tillering in *Fa* and tiller size in *Ph* only UNDER. It would only be possible to maximize aerial productivity and persistence of *Dg* simultaneously in these systems.

## 3.6 Silvopastoral Systems with Willows

### 3.6.1 Traditional Willow Silviculture

Willow silviculture traditionally practiced in the Delta of the Paraná River is aimed at the production of a higher volume of wood per hectare, rather than individual plant volume (Casaubon

2012). Consequently, such wood is mainly destined to the milling industry (crushed wood and wood pulp) (Cerrillo 2010). Predominant plantation densities are very high (between 1,100 and 1,600 plants per hectare) when the intended use is crushed wood. Afforestation is undertaken in sites of very diverse quality and lacking in proper water management, which alters plantation growth. Water excess may result in wood quality loss due to colour changes (purplish stains) and even tree death. Water shortage, in turn, can cause growth loss and even diseases. In addition, there are vast areas with monoclonal plantations, increasing plantation vulnerability to insect pest attacks (Casaubón 2014b). Plantations are installed from 0.70- to 2-metre-long stakes as propagation material, and, in a lower proportion, from 3- to 5-metre-long pole cuttings when planting on land that has not been adequately prepared for afforestation. These are 1- or 2-year-old nursery materials. Depending on their size as well as on soil type, materials are planted between 0.40 and 0.60 m deep, usually in rectangular patterns (Casaubon 2014b).

In general, traditional willow plantations are thinned of pole cuttings. This removes all basal branches, leaving the best-positioned pole on the stake standing. Later, formation pruning is carried out on the shaft. Pruning can be repeated at the second or third year of age in order to facilitate the movement of machinery to control weeds and insect pests around the stand. In high-density plantations, if the goal is to produce wood for sawing, selective thinning is used, removing oppressed, defected or damaged trees, or trees with shaft diseases (Casaubon 2014b).

### 3.6.2 Proposed Management for Willow Silvopastoral Systems

The larger quantity of beef cattle in the region in the last years has led to the need to move from the traditional plantation system to a more intensive production system with the aim to achieve a comprehensive production of wood, grasses and beef (Casaubon and Gonzalez 2009). Willow SPSs currently focus on wood production for various uses in an environmentally friendly fashion. The new willow silviculture is designed to produce a higher percentage of wood for solid uses (sawn wood and veneers), in addition to crushed wood, wood pulp and/or wood for energy uses (Casaubon 2014b). In the case of plantations under Agrosilvopastoral Systems particularly, it is designed to produce high-quality beef, forage and/or honey according to the demands from both the domestic and the international markets (Fig. 3.3). Predominant plantation plots are square-shaped, such as 4×4 and 5×5 m. Their silvicultural management demands organisation of the plantation area, identifying the most suitable site(s) for the clones to be cultivated. High-quality wood production calls for the greatest possible clone diversity under the "single-clone stand mosaic" model, with heterogeneous distribution of ages and of areas occupied by each clone under cultivation, in order to try to maintain biological diversity and to minimise potential disease and pest risk. In well-managed willows plantations, with good water availability, it is common to see a greater diversity of native fauna species. In mammals: Marsh deer (*Blastocerus dichotomus*), capybara (*Hydrochoerus hydrochaeris*), neotropical otter (*Lontra longicaudis*), coypu-nutria (*Myocastor coypus*), common armadillo (*Dasypus novemcinctus*), lutrine opossum (*Lutreolina crassicaudata*), white-eared opossums (*Didelphis albiventris*), common fox (*Cerdocyon thous*), Islander colilargo (*Oligoryzomys delticola*). In birds: dusky-legged guan (*Penelope obscura*), epaulet oriole (*Icterus cayanensis*), solitary cacique (*Cacicus solitarius*), green kingfisher (*Chloroceryle americana*), variable antshrike (*Thamnophilus caerulescens*), and Buff-throated Warbling-finch (*Poospiza lateralis*) (Fracassi 2012). Increased bird and mammal diversity of predators contributes in turn to biological control of Willow Sawflies (*Nematus oligospilus*) in leaf, and of ambrosia beetle (*Megaplatypus mutates*) in stem. These are pests of willows that affect the growth and the wood quality (Casaubon 2014b).

It is essential to implement good water management, which requires proper land systemati-

**Fig. 3.3** A silvopastoral system with 5-year-old *Salix babylonica x Salix alba* 'Ragonese 131/25' planted from 1-year-old pole cuttings, and spontaneous *Lolium multiflorum* and *Brommus catharticus* natural fodder grassland in the Lower Delta of the Paraná River

sation facilitating river water movement using drainage channels into the plantation during both droughts and periods of heavy rain (Casaubon et al. 2013b). In order to do so, to control phreatic zone water depth in each stand by means of phreatic water level meters guarantee an oscillating water level close to the zone with the greatest root development. This ensures proper water availability for forage and tree growth, while improving water quality in drainage networks as livestock drinking water.

In order to yield quality materials, nurseries require planting spacings greater than traditional ones, such as 1 × 1 m, as well as excellent sites and more intensive nursery management. In the most apt plantation sites, there have been good experiences of planting 1- and 2-year-old, 3.5- to 7-m long, unrooted pole cuttings, especially in *Salix babylonica* x *Salix alba* 'Ragonese 131/25' and 'Ragonese 131/27' and *Salix matsudana* x *Salix alba* 'Barret 13/44' willow cultivars, obtained from nurseries specially designed to that end. Formation pruning, also known as branch thinning, should be carried out as from the first year, leaving the highest, straightest, most conical and best-positioned pole cutting from the mother strain standing. At the second or third year, the first systematic shaft pruning is carried out, ensuring the absence of larger knots. Such pruning should be carried out gradually, removing at least the first third of basal branches in good-quality sites, or to a height not exceeding 50 % of total tree height, especially in excellent-quality sites, in order to reduce growth loss (Casaubon et al. 2006).

This silviculture accelerates SPS establishment, with livestock entry at the second or third year of plantation, facilitating natural grass consumption and the simultaneous production of beef and wood for various uses. It also favours plant individual growth with a higher percentage of cylindrical shafts, as well as augmenting planting period (May–August), spontaneous forage and wood quality by reducing dark stains and standing tree death due to excess water. There is less competition among plants and a lower risk of surface and forest fires, since dry grass volume is smaller, and water is always present in drainage networks and available for livestock. It also increases the diversity of animals and plants associated with water availability (Casaubon 2014b. In addition, willow leaves are very palatable and nutritious for livestock, and thus a good complement to livestock feed (Rossi et al. 2005; Carou 2010b).

### 3.6.3 Experiment to Study Dasometric Response and Interaction between Willow Clones and Beef Cattle

The objective was to assess dasometric behaviour in a silvopastoral trial with six willow clones, and to characterise Aberdeen Angus heifer behaviour with respect to herbivory. The hypothesis was that willows under study exhibit different behaviour towards herbivory. Two-year-old pole cuttings with 1-year-old roots (PC2 R1) were planted in 2005. Experimental design was randomised complete block design with four 25-plant repetitions of *Salix babylonica x Salix alba* 'Ragonese 131-25 INTA', *Salix babylonica x Salix alba* 'Ragonese 131-27 INTA', *Salix matsudana x Salix alba* 'Barret 13-44 INTA', *Salix matsudana* x *Salix alba* '26992', *Salix matsudana* x *Salix alba* '26993' and '*Salix nigra Alonzo Nigra 4*'. Annually, in winter, all plants were pruned of epicormic shoots up to the first three metres of shaft height, and diameter at breast height (DBH) and total heights were measured. At year 4 after planting, four 3-year-old 300-kg calves entered each block. Even if there are no significant differences between clones for the diameter, height or volume variables, there are "marginal" differences for the height variable (*Salix nigra* 4 is higher than the '26992' clone, but these differences are not significant). The percentage of browsed leaves and tender shoots was assessed with respect to beef cattle entry into the system. Even if all clones with branches at animal height were browsed, no trees exhibited shaft damage due to browsing. There appeared differences as regards browsing intensity: whereas 'Barret 13/44 INTA' and '26992' clones showed 100 % of browsed shoots, the '26993' clone exhibited a lower proportion (median 88 %). However, not all the plants from this clone had epicormic shoots at animal height – some branches were higher. Willows under silvopastoral systems can produce quality wood for various uses, as well as forage and shade for cattle (Casaubon et al. 2011).

## 3.7 Environmental Impact Related to Beef Cattle in Silvopastoral Systems with Salicaceae. Some Suggestions to Mitigate Situations of Extreme Management

Livestock production constitutes a land management system, which can optimise food (beef) production as well as maintain ecosystem productivity. As a stimulant to ecosystem processes, grazing contributes to manure distribution and incorporation, to maintaining soil fertility and physical characteristics, and to forage persistence through seed distribution by cattle. Controlled grazing on the original vegetation ("*pajonal*") increases species diversity. It gives rise to spontaneous forage species with a higher nutritional value than original species such as *Phalaris angusta*, *Lolium multiflorum*, *Bromus catharticus*, *Leersia hexandra*, and *Trifolium repens*. However, when animal density surpasses carrying capacity, deterioration sets in and affects all system components. Excessive grazing (in terms of both intensity and frequency) may lead to vegetation degradation, resulting in a reduction of desirable forage species, an increase in forage species less appetising to livestock, and loss of vegetation cover. Soil denudation causes erosion, as well as soil-fertility decline and soil- structure deterioration. Trampling exerts mechanical pressure on the soil, especially in sites where it occurs most frequently, such as trails, water-drinking areas and resting areas. With soil surface compaction, soil's bulk density increases, and porosity and water infiltration rates decrease. The presence of compacted soil reduces forage production and negatively affects the coexistence of livestock and fauna, since it increases competition for plants or for water. Water quality declines when sediment increases. Erosion may lead to severe fertility loss when water removes or carries away soil surface layer. Surface runoff carries fine soil particles and organic matter particles, resulting in soil and nutrient loss, water pollution by suspended solids, and sediment deposition.

Burning is an ancient technique used by island farmers in order to manipulate the vegetation for the purposes of cattle grazing. Burning is used to control unwanted vegetation and high weeds, to remove less appetizing herbs and plants, and to favour the growth of the most palatable, digestible or nutritious plants. Nonetheless, burning might damage the forest, the vegetation and the soils, and cause higher erosion levels. Using chemical products to control pests and diseases, or herbicides to control weeds when preparing the site for plantation may have negative environmental impact, leading to water pollution and, therefore, negative effects for livestock, fauna, (both superficial and phreatic) water sources, and vegetation.

The recommended management techniques to increase productivity and to control erosion are mechanical and physical intervention with respect to the soil or the vegetation, and other soil and water conservation techniques. Increasing the number of water sources, placing them strategically and maintaining the soil covered with native herbaceous vegetation and/or planting shade-tolerant forages may reduce soil erosion. It is essential to estimate carrying capacity in each individual situation, as well as implementing grazing control to guarantee proper use of and balance between SPS components.

## Bibliography

Ball J, Carle J, Del Lungo A (2005) Contribución de álamos y sauces a la silvicultura sostenible y al desarrollo rural. Unasylva 56(221):3–9

Ballaré CL, Scopel AL, Sánchez RA (1990) Photocontrol of stem elongation in plant neighbourhoods: effects of photon fluence rate under natural conditions of radiation. Plant Cell Environ 14:57–65

Benavides J (2006) Árboles forrajeros en América Central. In Seminario Centroamericano y del Caribe sobre Agroforestería con Rumiantes menores. Memorias. Conferencias Magistrales. San José de Costa Rica

Berone GD, Lattanzi FA, Agnusdei MG, Bertolotti N (2008) Growth of individual tillers and tillering rate of *Lolium perenne* and *Bromus stamineus* subjected to two defoliation frequencies in winter in Argentina. Grass Forage Sci 63:504–512

Bonfils C (1962) Los suelos del Delta del Río Paraná. Factores generadores, clasificación y uso. Revista de Investigación Agrícola. INTA. T.XVI, N°3. Buenos Aires Argentina

Borodowski E (2006) Álamos y sauces en el Delta del Paraná: situación del sector y silvicultura. *En:* Jornadas de Salicáceas. Buenos Aires, Argentina

Borodowski E, Suárez R (2004) El cultivo de álamos y sauces: su historia en el Delta del Paraná. SAGPyA For 32:5–13

Borodowski E, Signorelli A, Battistella A (2014) Salicáceas en el Delta del Paraná: situación actual y perspectivas. Disertación. Actas Jornadas de Salicáceas. La Plata, Buenos Aires, Argentina

Brinson M, Malvárez A (2002) Temperate freshwater wetlands: types, status, and threats. Environ Conserv 29(2):115–133

Carou N, Casaubón E, De Loof E, Bajerski R, Rossi C, González A, González G, Dallorso M (2010a) Contenido estacional en macro y microminerales del pastizal natural de un sistema silvopastoril con álamos y sauces. 33° Congreso Argentino de Producción Animal. Comarca Viedma Patagones. Río Negro. Rev Argent Prod Anim 30(1):234

Carou N, De Loof E, Casaubón E, González A, Dallorso M (2010b) Mineral composition of willow and poplar leaves of nutritional interest for cattle in silvopastoral systems at the delta of the Paraná River, Argentina. Rev Livest Res Rural Dev 22, N°1. ISSN 0121-3784. www.lrrd.org/lrrd22/1/cont2201.htm

Carvalho M, Castro C, Yamaguchi L, Alvim M, Freitas V, Ferreira X (2003) Two methods for the establishment of a silvopastoral system in degraded pasture land. Livest Res Rural Dev 15(12). http://www.cipav.org.co/lrrd/lrrd15/12/carv1512.htm

Casaubon E (2012) Silvopastoral systems with *Salix* in the lower delta of the Paraná river (Argentina). 24th Session International Poplar Commission. Dehradun

Casaubon E (2013) Establecimiento de Sistemas Silvopastoriles: Efecto de la edad del material de multiplicación y manejo del pastoreo con bovinos. Tesis presentada para optar al título de Magister de la Universidad de Buenos Aires, Área Recursos Naturales

Casaubon E (2014a) Evaluación del comportamiento de guías de álamo, al séptimo año, en un sistema silvopastoril del delta del río Paraná. Cuarto Congreso Internacional de Salicáceas. La Plata

Casaubon E (2014b) Manejo silvicultural del sauce orientado a la producción de madera de calidad para usos sólidos en el delta del río Paraná. Cuarto Congreso Internacional de Salicáceas. La Plata

Casaubon E (2014c) Propuesta de una nueva silvicultura para los sistemas silvopastoriles de *Populus deltoides* en el bajo delta del río Paraná. Cuarto Congreso Internacional de Salicáceas. La Plata

Casaubon E, Gonzalez A (2008) Silvopastoral systems with poplar in the lower delta of the Paraná river (Argentina). 23° Sesión de la Comisión Internacional del Álamo. Beijing

Casaubon E, Gonzalez A (2009) Sistemas silvopastoriles en plantaciones de sauce en el bajo delta del río Paraná. XIII Congreso Forestal Mundial. Buenos Aires

Casaubón E, Spagarino C, Cueto G, González A (2001) Enraizamiento de guías de *Populus deltoides* en el Delta del Paraná. XVI Jornadas Forestales de Concordia, Entre Ríos

Casaubon E, Gurini L, Cueto G, Gomez L, Zanelli M, Berrondo G, Gonzalez A (2004) Caracterización estacional de una plantación de *Populus deltoides cv Catfish 2* en el Bajo Delta Bonaerense del Río Paraná (ARGENTINA). Investigación Agraria. Sistemas Recur Forestales 13(2):369–385

Casaubon E, Cueto G, González A, Spagarino C, Ortiz S (2005a) Resultados preliminares de dos ensayos orientativos de épocas de poda en *Populus deltoides cv Stoneville 67* en el bajo delta bonaerense del Río Paraná. III Congreso Forestal Argentino y Latinoamericano. Corrientes

Casaubón E, Gurini L, Cueto G, Arano A, Torrá E, Corvalán G, González A, Ortiz S (2005b) Evaluación del efecto de diferentes labores culturales en un sistema silvopastoril de álamo en el bajo delta bonaerense del Río Paraná. III Congreso Forestal Argentino y Latinoamericano. Corrientes

Casaubon E, Cueto G, González A (2006) Rebrote de ramas en *Salix babylonica x Salix alba* 'A 131/27' frente a dos intensidades de poda sistemática del fuste. Primeras Jornadas de Salicáceas. Buenos Aires

Casaubon E, Peri P, Cornaglia P, Cueto G, González A (2009a) Comportamiento de guías de álamo en estaquero para la obtención de material adaptado a la instalación de sistemas silvopastoriles en el delta del río Paraná. 1° Congreso Nacional de Sistemas Silvopastoriles. Posadas

Casaubon E, Peri P, Cornaglia P, Cueto G, Rossi C, González A (2009b) Establecimiento de sistemas silvopastoriles en el delta del río Paraná: Evaluación del daño animal. 1° Congreso Nacional de Sistemas Silvopastoriles. Posadas

Casaubon E, Peri P, Cornaglia P, González A, Fernández, M (2010) Establecimiento de sistemas silvopastoriles en el delta argentino: valor forrajero de pasturas y hojas de álamos. 1er Congreso Internacional Agroforestal Patagónico. Coyhaique

Casaubon E, Cueto G, Madoz G (2011) Comportamiento dasométrico e interacciones entre clones de sauces y ganado vacuno en un sistema silvopastoril del Delta del Río Paraná. 3er Congreso Internacional de Salicáceas. Neuquén

Casaubon E, Arano A, Tassano Viaña J (2012a) Manual de buenas prácticas silvopastoriles para obtención de madera de álamo de calidad para usos múltiples y de ganado de carne bovina en el delta inferior del río Paraná. Versión 1. EEA Delta del Paraná

Casaubón E, Peri P, Cornaglia P, Carou N, Cueto G (2012b) Valor forrajero de hojas de álamo en el bajo delta del Río Paraná. Proceedings of the 2nd National Congress of Silvopastoral Systems, INTA Editions, Santiago del Estero

Casaubon E, Gurini L, Cueto G, Gomez L, Gonzalez A (2013) Caracterización de sitios forestales para *Salix babylonica x Salix alba* "Ragonese 131/27 INTA" en el delta bonaerense del Río Paraná. Cuarto Congreso Forestal Argentino y Latinoamericano. Iguazú

Casaubon E, Peri P, Cornaglia P, Cueto G (2014) Establishment of silvopastoral systems in poplar in the delta of the Paraná River (Argentina). World Congress on Agroforestry, Delhi

Ceballos D, Frangi J, Jobbágy E (2012) Soil volume and carbon storage shifts in drained and afforested wetlands of the Paraná River Delta. Biogeochemistry. doi:10.1007/s10533-012-9731-2

Cerrillo T (2010) El sauce: Potencial forestal y aptitud tecnológica de la madera. Jornada técnica sobre el sauce. EEA Delta del Paraná

Clavijo M, Nordenstahl M, Gundel P, Jobbágy E (2005) Poplar afforestation effects on grassland structure and composition in the Flooding Pampas. Rangel Ecol Manag 58:474–479

Clavijo M, Cornaglia P, Gundel P, Nordenstahl M, Jobbágy E (2010) Poplar Litterfall limits tall fescue plant recruitment in silvopastoral systems in Flooding Pampa, Argentina. Agrofor Syst 80:275–282

Clavijo M, Cornaglia P, Rodríguez D, Signorelli A, Borodowski E (2012) Manejo del raleo y siembra de exóticas en sistemas silvopastoriles del delta del Paraná. Proceedings of the 2nd National Congress of Silvopastoral Systems, INTA Editions, Santiago del Estero, pp 110–115

Clavijo M, Cornaglia P, Rodríguez D, Signorelli A, Borodowski E (2014) Receptividad ganadera de pastizales naturales y de sitios bajo forestaciones de álamo en sistemas silvopastoriles del Delta del Paraná. Jornadas de Salicáceas. IV Congreso Internacional de Salicáceas en Argentina "Sauces y Álamos para el desarrollo regional", Universidad Nacional de La Plata

Cornaglia P, Zunino H, Monlezun S, Pincemin J, Clavijo M, Borodowski E (2009) Silvopastoral systems under poplars: effect of cover of fallen tree leaves on initial growth of cool temperate grasses. XIII Congreso Forestal Mundial (WFC), 19 al 23 de octubre. Buenos Aires, Argentina. Publicado en CD

Cornaglia PS, Gatti ML, Pizzio P, Guilligan M, Cabo F (2010) Respuesta morfogenética de gramíneas perennes templadas al sombreo por árboles y a la fre-

cuencia de defoliación. IV Reunión Binacional de Ecología, XXIV Reunión Argentina de Ecología. Buenos Aires, p 212
Cornaglia P, Clavijo M, Rodríguez D, Signorelli A, Borodowski E (2011a) Management practices that maximize the herbaceous productivity in temperate silvopastoral systems. IX Internacional Rangeland Congress, Rosario
Cornaglia PS, Gatti, ML, Guilligan M, Pizzio P, Cabo F (2011b) Structural responses to shade under trees and defoliation frequency in cool season grasses. IX International Rangeland Congress, Rosario
Darwin C (1978) El naturalista en el Plata. Buenos Aires. Centro Editor de América Latina
Eason W, Gill E, Roberts J (1996) Evaluation of anti-sheep tree-stem-protection products in silvopastoral agroforestry. Agrofor Syst 34:259–264
FAO (1957) Los Chopos en la producción de madera y la utilización de las tierras. Colección FAO N° 12, Roma, 525 pp
Feldhake CM (2001) Microclimate of a natural pasture under planted *Robinia pseudoacacia* in central Appalachia, WestVirginia. Agrofor Syst 53:297–303
Fernández Tschieder E, Fassola H, García Cortés M (2011a) Ecuación de volumen total para *Populus deltoides* de plantaciones del Bajo Delta del Paraná. Rev Investig Agropecuarias 37:172–179
Fernández Tschieder E, Borodowski E, García Cortés M, Signorelli A (2011b) Efecto de la intensidad de raleo sobre el crecimiento de *Populus deltoides*. 3er Congreso Internacional de Salicáceas, Neuquén
Fracassi N (2012) Diversidad de mamíferos y aves en forestaciones y pajonales de Salicáceas del Bajo Delta del río Paraná. Tesis Magister de la Universidad de Buenos Aires, Área Recursos Naturales
Fracassi N (2014) Las aves del bajo delta del Paraná. Aves Argentinas N&C. Año XVI
Galafassi G (1995) Aproximación al proceso histórico de asentamiento, colonización y producción en el Bajo Delta del Paraná. Estudios Sociales, n° 11. http://theomai.unq.edu.ar/artguido002.htm
Garden DL, Bolger TP (2001) Interaction of competition and management in regulating composition and sustainability of native pasture. In: Tow PG, Lazenby A (eds) Competition and succession in pastures. CAB International Publishing, Wallingford, pp 213–232
Gatti ML, Cornaglia PS (2011) Respuestas de gramíneas C3 perennes al sombreo por árboles y a la defoliación. XXXIV Congreso Argentino de Producción Animal. 1st Joint Meeting ASAS-AAPA, 04 al 07/10/2011, Mar del Plata, República Argentina. Revista Argentina de Producción Animal, Asociación Argentina de Producción Animal –AAPA-, Balcarce, República Argentina, vol 31(1):440
Gaute M, Mari N, Borodowski E, Di Bella C (2007) Elaboración de un Sistema de Información Geográfica para el monitoreo de pólderes en el Bajo delta Bonaerense durante el período 1985-2005. Teledetección. Hacia un mejor entendimiento de la dinámica global y regional. Ed. Martin, pp 465–470
González G, Rossi C, Torrá E, De Magistris A (2006) Caracterización de un Sistema Silvopastoril bajo Normativas de Producción Orgánica en el Delta de Paraná (Argentina). Resúmenes del IV Congreso Latinoamericano de Agroforestería para la Producción Pecuaria Sostenible. Varadero, Cuba, pp 95–103
González G, Rossi C, Pereyra A, De Magistris A, Lacarra H, Varela E (2008) Determinación de la calidad forrajera en un pastizal natural de la región del delta bonaerense argentino. Zootec Trop 26(3):223–225
Haene E, Pereyra J (2003) Fauna de Otamendi. Inventario de los animales vertebrados de la Reserva Natural Otamendi, Campana, Buenos Aires, Argentina. Temas naturaleza Conservación 3:1–192. Aves Argentinas/AOP, Buenos Aires
Jose S, Gillespie A, Pallardy S (2004) Intersspecific interactions in temperate agroforestry. Agrofor Syst 61:237–255
Kalesnik F (2010) Usos de la vegetación. Bienes y Servicios Ecosistémicos de los Humedales del Delta del Paraná. Wetlands International. Fundación humedales. 32 pp
Kandus P (1997) Análisis de patrones de vegetación a escala regional en el Bajo Delta Bonaerense del Río Paraná (Argentina). Tesis Doctoral. Facultad de Ciencias Exactas y Naturales, Universidad de Buenos Aires, Buenos Aires
Kandus P, Malvárez A (2004) Vegetation patterns and change analysis in the lower delta islands of the Paraná river (Argentina). Wetlands 4:620–632
Kemp PD, Matthew C, Lucas RJ (2000) Pastures species and cultivars. In: White J, Hodgson J (eds) New Zealand pasture and crop science. Oxford University Press, Auckland, pp 67–99
Kemp DR, King WMG (2001) Plant competition in pastures – implications for management. In: Tow PG, Lazenby A (eds) Competition and succession in pastures. CAB International Publishing, Wallingford, pp 85–102
Laclau P (2012) Consideraciones económicas y ambientales para la toma de decisiones en sistemas silvopastoriles. Proceedings of the 2nd National Congress of Silvopastoral Systems, INTA Editions, Santiago del Estero, pp 359–370
Lazenby A, Tow P (2001) Some concluding comments. In: Tow PG, Lazenby A (eds) Competition and succession in pastures. CAB International Publishing, Wallingford, pp 305–314
Lefroy E, Dann P, Wildin R, Wesley-Smith R, Mc Gowan A (1992) Agrofor Syst 20:117–139. Netherlands
Loponte D (2013) La Arqueología del Delta del Paraná, estado actual y perspectivas, En: Delta del Paraná: Historia, presente y futuro. Simposio Científico Académico Delta del Paraná. http://www.unesco.org.uy/mab
Luccerini S, Subovsky E, Borodowski E (2013) Sistemas Silvopastoriles: una alternativa productiva para nuestro país. Apuntes Agroeconómicos. Año 7. Nro. 8
MAGyP (2011) Superficie implantada con álamos y sauces en el delta del Paraná. Fuente área SIG. Dirección de Producción Forestal. Secretaría de

Agricultura, Ganadería y Pesca. www.minagri.gob.ar/forestación. Argentina

Malvárez I (1997) Las comunidades vegetales del Delta del Río Paraná. Su relación con factores ambientales y patrones de paisaje. Tesis doctoral. Universidad de Buenos Aires. Buenos Aires, 167 pp

Malvárez I (1999) El Delta del río Paraná como mosaico de humedales. In: Malvárez AI (ed) Tópicos sobre humedales subtropicales y templados de Sudamérica. MAB UNESCO, Montevideo

McAdam J (2003) An evaluation of tree protection methods against Scottish Blackface sheep in an upland agroforestry system. For Ecol Manag 45:119–125

Mead D (2009) Biophysical interactions in silvopastoral systems: a New Zealand perspective. 1er Congreso Nacional de Sistemas Silvopastoriles. Posadas

Minotti P (2010) Biodiversidad de peces. Bienes y Servicios Ecosistémicos de los Humedales del Delta del Paraná. Wetlands International. Fundación humedales. 32 pp. http://lac.wetlands.org

Monlezun S, Cornaglia P, Pincemin J, Zunino H, Clavijo M, Borodowski E (2008) Cover of senesced tree leaves reduces herbaceous productivity under poplars in silvopastoral systems. 23° International Poplar Commission, China

Mosquera-Losada M, Fernández-Núñez E, Rigueiro-Rodríguez A (2006) Pasture, tree and soilevolution in silvopastoral systems of Atlantic Europe. For Ecol Manag 232:135–145

Neiff J (1999) El régimen de pulsos en ríos y grandes humedales de Sudamérica. In: Malvárez AI (ed) Tópicos sobre Humedales Subtropicales y Templados de Sudamérica. UNESCO-ORCYT-MAB, Montevideo, pp 97–146

Neiff J, Malvárez A (2004) Grandes humedales fluviales. In: Malvárez AI, Bó RF (eds) Documentos del Curso Taller Bases ecológicas para la clasificación e inventario de humedales en Argentina, Buenos Aires

Nordenstahl M, Gundel P, Clavijo M, Jobbágy E (2011) Forage production in natural and afforested grasslands of the Pampas: ecological complementarity and management opportunities. Agrofor Syst 83:201–211

Peri PL (2012) Implementación, manejo y producción en sistemas silvopastoriles: enfoque de escala en la aplicación del conocimiento Proceedings of the 2nd National Congress of Silvopastoral Systems, INTA Editions, Santiago del Estero, pp 8–21

Peri PL, Moot D, Jarvis P, Mcneil D, Lucas R (2007) Morphological, anatomical, and physiological changes of orchardgrass leaves grown under fluctuating light regimes. Agron J 99:1502–1513

Pincemin J, Monlezun S, Zunino H, Cornaglia PS, Borodowski E (2007) Sistemas Silvopastoriles en el Delta del Río Paraná: Producción de materia seca y estructura de gramíneas templadas bajo álamos. Actas de la XX Reunión Latinoamericana de Producción Animal, V Congreso Internacional de Ganadería de Doble Propósito, Cuzco

PTR (2009) Plan Tecnológico Regional de la Estación Experimental Agropecuaria Delta del Paraná. Centro Regional Buenos Aires Norte. Instituto Nacional de Tecnología Agropecuaria. Ministerio de Agricultura, Ganadería y Pesca. Argentina

Quintana R, Bó R (2010) Fauna silvestre. Bienes y Servicios Ecosistémicos de los Humedales del Delta del Paraná. Wetlands International. Fundación humedales. 32 pp. http://lac.wetlands.org

Rawnsley RP, Donaghy DJ, Fulkerson WJ, Lane PA (2002) Changes in the physiology and feed quality of cooksfoot (*Dactilys glomerata*) during regrowth. Grass Forage Sci 57(3):203–207

Robson MJ, Ryle GJA, Woledge J (1988) The grass plant – its form and function. In: Jones MB, Lazenby A (eds) The grass crop. The physiological basis for production. Chapman and Hall, London, pp 22–83

Rossi C (2010) Forraje para la ganadería. Bienes y Servicios Ecosistémicos de los Humedales del Delta del Paraná. Wetlands International. Fundación humedales. 32 pp. http://lac.wetlands.org

Rossi C, Torrá E, González G, Lacarra H, Pereyra A (2005) Evaluación de las hojas de álamo y sauce como forraje en un sistema silvopastoril del Delta del Paraná. XIX Reunión de la Asociación Latinoamericana de Producción animal, Tamaulipas

Rossi C, De Magistris A, Torrá E, Medina J, González G (2006) Evaluación de la presencia de especies forrajeras y malezas para el aprovechamiento silvopastoril de parcelas con Salicáceas en el Delta del Paraná (Pcia. de Buenos Aires). Jornadas de Salicáceas 2006

Rossi C, González G, De Magistris A, Torrá E (2009) Composición botánica del pastizal natural en un sistema silvopastoril del Delta del Paraná: clasificación forrajera. Proceedings of the 1st National Congress of Silvopastoral Systems, INTA Editions, Misiones

Rossi C, Gonzalez G, De Magistris A, Carou N, De Loof E (2012) Contenido proteico en forrajeras en un sistema silvopastoril del bajo delta del Paraná. Proceedings of the 2nd National Congress of Silvopastoral Systems, INTA Editions, Santiago del Estero, Argentina

Sanhueza A (1998) Cultivo del álamo. (*Populus spp*.). Parte 2. Corporación Nacional Forestal. Santiago de Chile, Chile, 214 pp

Sarmiento D (1974) El Carapachay. Eudeba, Buenos Aires

Signorelli A, Borodowski E, Cornaglia P, Fernandez Tschieder E (2011) Efecto de la intensidad de raleo sobre la transmisión de luz en plantación de *Populus deltoides*. 3er Congreso Internacional de Salicáceas, Neuquén

Simón M, Ibrahim M, Finegan B, Pezo D (1998) Efectos del pastoreo bovino sobre la regeneración de tres especies arbóreas comerciales del Chaco Argentino: un método de protección. Agroforestería Américas 5:17–18

Sinclair K, Fulkerson WJ, Morris SG (2006) Influence of regrowth time on forage quality of prairie grass, perennial ryegrass and tall fescue under non-limiting soil nutrients and moisture conditions. Aust J Exp Agric 46:45–51

Somarriba E (1997) Pastoreo bajo plantaciones forestales. Agroforestería Américas 4(15):26–28

Taranaki Regional Council (2001) Poplar and Willow varieties available from Taranaki Regional Council. Sustainable Land Management

Thomas E (2011) Producción de biomasa de hojas de álamo para uso forrajero en valles irrigados de Patagonia Norte. Tercer Congreso Internacional de Salicáceas en Argentina. Neuquén

Zoffoli ML, Kandus P, Madanes N, Calvo D (2008) Seasonal and interannual analysis of wetlands in South America using NOAA-AVHRR NDVI time series: the case of the Parana Delta Region. Landsc Ecol 23:833–848

# Silvopastoral Systems in the Western Chaco Region, Argentina

4

Carlos Kunst, Marcelo Navall, Roxana Ledesma, Juan Silberman, Analía Anríquez, Darío Coria, Sandra Bravo, Adriana Gómez, Ada Albanesi, Daniel Grasso, José A. Dominguez Nuñez, Andrés González, Pablo Tomsic, and José Godoy

**Abstract**

The Chaco region is a vast plain extending over northwestern Argentina, Bolivia and Paraguay. Over the years, the woody component has increased in all vegetation communities of the Chaco. This poses a major problem for grassland productivity because the woody plants compete for resources with grass, impact advanced tree regeneration, hamper livestock and personnel movements, and restrict accessibility for forage due to high stem density and thorns. These disadvantages far outweigh the advantages offered by such trees and shrubs in terms of providing shade, foliage and fruits, and improving soil conditions. To address this issue, a silvopastoral system called "Low intensity roller-chopping" (RBI) was tried, based on theoretical approaches and field research, in the western Chaco from 1990 to 2014. A distinctive feature of the system is that trees that are already present in the Chaco system are mechanically disturbed by properly adjusting the size of the tractor and the roller-chopper and regulating the frequency and severity of the number of passes over the standing vegetation. The operation is based on an appropriate soil and vegetation inventory of the area at a scale >1:50,000. By using RBI, the estimated improvement of stocking rate is 3–5 ha*animal units$^{-1}$ (Animal unit (AU): forage of average quality consumed by a cow of 350 kg liveweight weaning one calf a year) when native species are considered, and if *Panicum* species are used there is average increase of 1–2 ha*AU$^{-1}$. Effects on soil quality as assessed by soil organic matter (SOM); and soil nitrogen (N) mag-

C. Kunst (✉) • M. Navall • R. Ledesma • D. Coria • A. Gómez • A. González • P. Tomsic • J. Godoy
Santiago del Estero Experimental Station, Instituto Nacional de Tecnología Agropecuaria (INTA), Jujuy 850, G4200CQR Santiago del Estero, Argentina
e-mail: kunst.carlos@inta.gob.ar

J. Silberman • A. Anríquez • S. Bravo • A. Albanesi
Universidad Nacional de Santiago del Estero (UNSE), Av. Belgrano (S) 1239, Santiago del Estero, Argentina

D. Grasso
INTA, Soil Research Institute, National Center of Agriculture Research, Castelar, Argentina

J.A.D. Nuñez
Escuela Superior de Montes, Universidad Politécnica de Madrid, Madrid, Spain


P.L. Peri et al. (eds.), *Silvopastoral Systems in Southern South America*, Advances in Agroforestry 11, DOI 10.1007/978-3-319-24109-8_4

nitudes, tree regeneration, vegetation and bird diversity, and brush control were also studied. Results indicate that RBI is a promising approach to managing native vegetation of the Chaco in a sustainable manner.

**Keywords**
Disturbance • Forest management • Range management

## 4.1 Introduction

The Chaco region is a vast plain that extends into northwestern Argentina and surrounding countries (Fig. 4.1). *Chaco* is a *quichua* word meaning "a place for hunting" or "a place where I am self-sufficient" (Metraux 1946; Berton 2014, personal communication). The native vegetation of the Chaco region is complex, composed of plant communities where either woody species

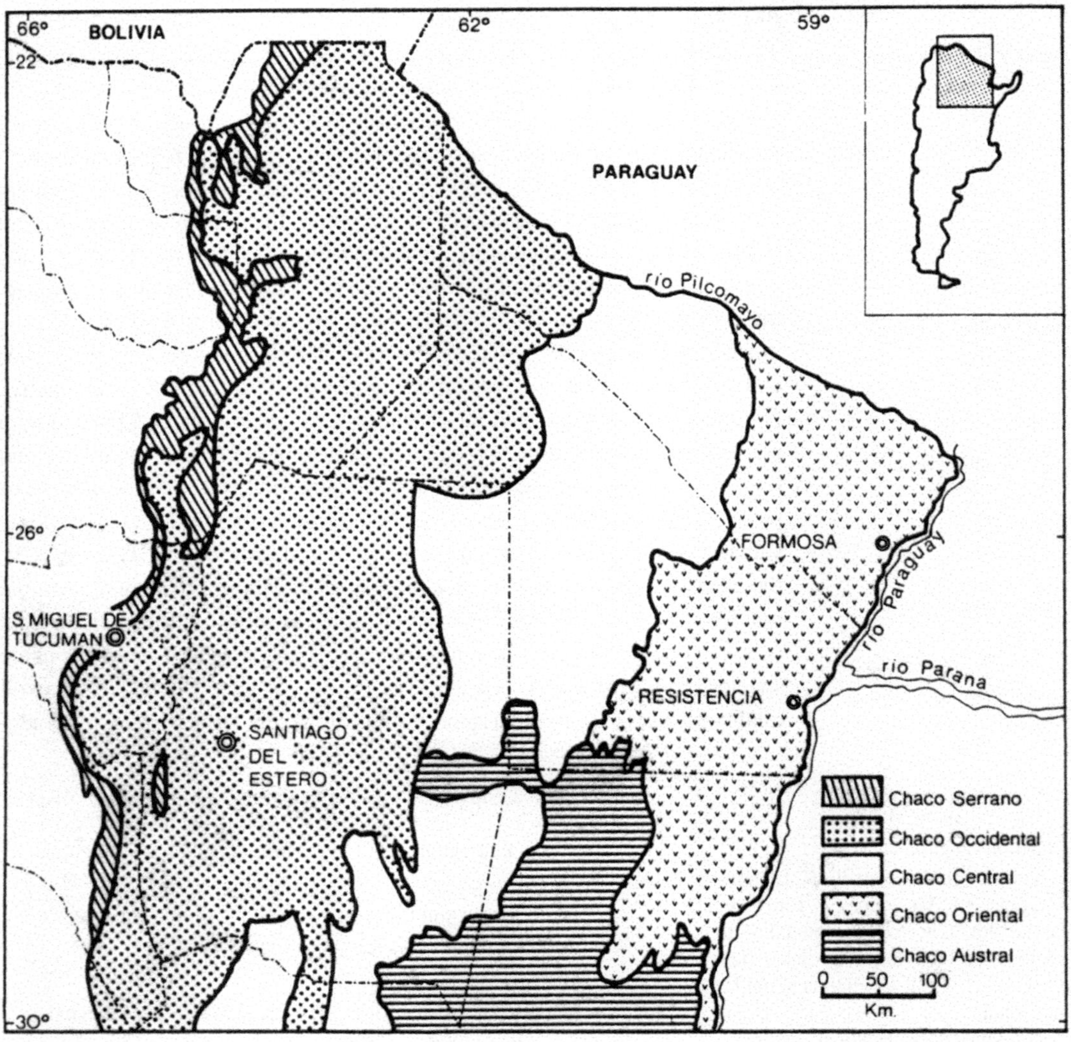


**Fig. 4.1** The Chaco region in Argentina (Source: Morello 1968). References: Chaco serrano: Hilly Chaco; Chaco occidental: Western Chaco, arid-semiarid climate; Chaco central: central Chaco, subhumid climate; Chaco oriental: Eastern Chaco, humid climate; Chaco austral: southern Chaco, temperate climate

(≈forests, shrub thickets) or grasses (≈grasslands, savannas) dominate. Throughout the years, the woody component has increased in all vegetation communities and has homogenized the landscape, decreasing diversity and the suitability of the ecosystem for providing goods and services. These problems are shared with other arid and subhumid regions of the world, creating complex management situations since there are production and conservation issues mixed together (Gifford and Howden 2001; Burrows 2002; Van Auken 2009). Silvopastoral systems are an option for solving these conflicts that is worth exploring. Concerns about the sustainability of the Chaco ecosystem favor reclamation procedures keeping most of the native tree and shrub individuals, a vegetation clearing system locally called "*silvopastoral*" (Provincia de Santiago del Estero Law 6841 and further regulations 2007).

In this paper, we will describe a silvopastoral system proposed by INTA and other institutions, called "Low intensity roller-chopping" (RBI for "*rolado de baja intensidad*" in spanish) based on theoretical approaches and results from field research conducted in the western part of the Chaco from 1990 to 2014. These findings were used to develop management recommendations to provide good and services (meat, wood, diversity and wildlife) from Chaco ecosystems in a sustainability framework.

## 4.2 The Chaco Ecosystem: Climate, Soils and Vegetation

The Chaco summers are hot and wet, while winters are dry and cold (Boletta 1988). Freezing temperatures as low as −10°/−15 °C are common during winter, while temperature can reach 38–40 °C in a spring-summer afternoon. Rain is quite variable within and between seasons and tends to decrease from 1000 to 1200 mm in the eastern Chaco to 300–400 mm toward the southern tip of the western Chaco (Boletta 1988). Average air temperatures follow the same pattern, balancing in some way the diminishing rainfall.

The western Chaco is a vast plain, with a slight slope toward the southeast and crossed by several rivers, such as the Salado and Dulce. Toward the south, as rainfall decreases, rivers become semipermanent streams. Sediments brought by rivers and streams from the mountains and sierras located in the western side are the original material for the soils. In the semiarid-subhumid subregions of the eastern Chaco, soils are Mollisols, and in the semiarid-arid subregions they are Aridisols and Entisols (western Chaco). There are also vast areas with saline-sodic soils. Mollisols, entisols and alfisols occupy 38 %, 28 % and 16.5 % of the Chaco region, respectively (INTA-SAGyP 1990). Mollisols have a better productive capacity than the other two, while Entisols are less evolved and shallow, presenting an ochric epipedon (Morello et al. 2012). The resilience and buffering capacity of these soils is limited when faced with poor land management such as overgrazing (Sanzano et al. 2005).

At the local scale, the vegetation of the Chaco region is a mosaic of hardwood forests, savannas and grasslands (Bucher 1982; Morello and Adámoli 1974; Bordón 1983). These vegetation types correspond to different soil and drainage features related to the geomorphological processes associated with water runoff (e. g. levees and flats, Bucher 1982). At a scale of ~1:20,000, soil and vegetation types within the Chaco are located along a catena from the 'upland' to the 'lowland'. A vegetation model, using the concept of ecological site, was provided by Kunst et al. (2006b, Fig. 4.2): hardwood forests occupy the upland sites, while the woodlands and savannas are located on the intermediate and lowland sites, respectively. The tree species *Schinopsis lorentzii* (Griseb.) Engl.[1] and *Aspidosperma quebracho-blanco* Schldtl. dominate in upper layers of the dry Chaco forests (c. 15 m tall). *Prosopis nigra* (Griseb.) Hieron. (algarrobo negro), *Ziziphus mistol* Griseb. and *Cercidium praecox* (Ruiz and Pav. ex Hook.) Harms are

[1] Botanical nomenclature follows Catálogo de Plantas Vasculares de la República Argentina, Instituto de Botánica Darwinion, 2011 [www.darwin.edu.ar/Proyectos/FloraArgentina].

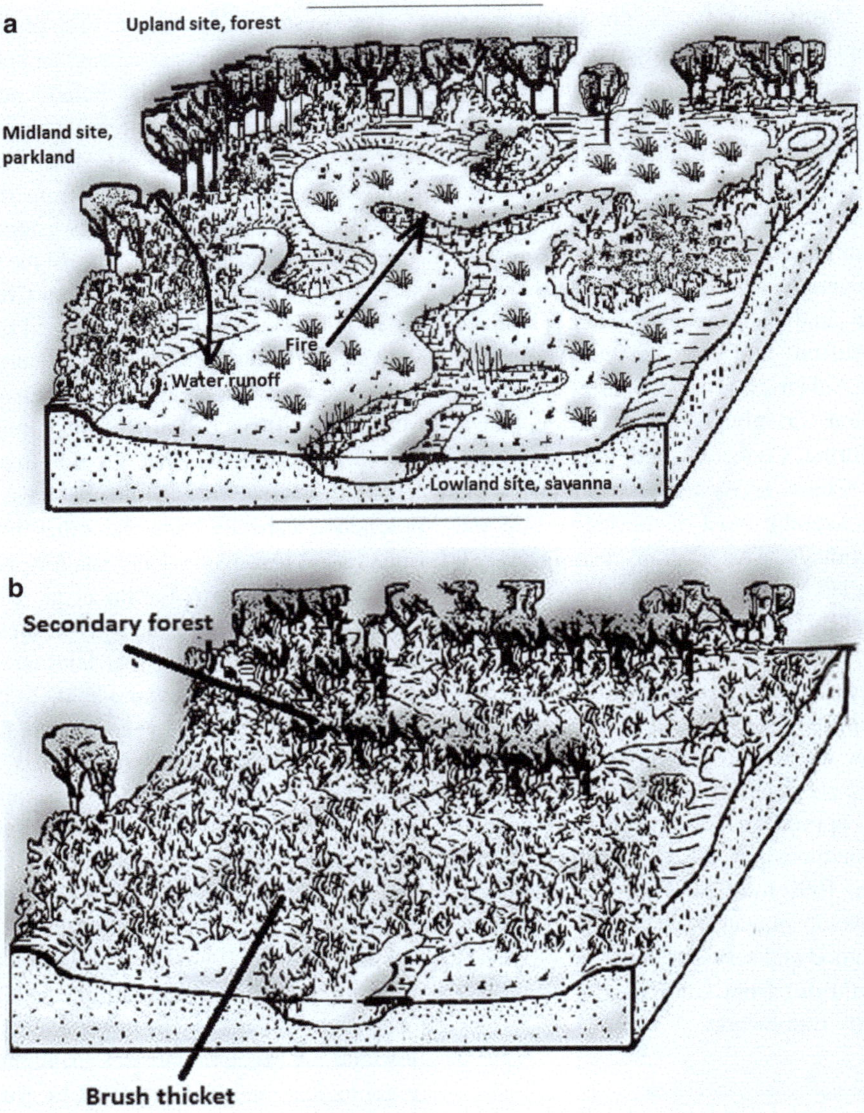


**Fig. 4.2** Ecological sites and associated vegetation types in the western Chaco at a cartographic scale ~1:20,000. (**a**) Original landscape and common disturbances; (**b**) current landscape resulting from overgrazing, indiscriminate logging and lack of fire (Modified from Kunst et al. 2006)

common species in the medium strata. The shrub layer is composed by *Acacia furcatispina* Burkart, *Capparis atamisquea* Kuntze, *Condalia microphylla* Cav. and *Celtis pallida* Torr. In the upland site, wildfire creates pyric grasslands dominated by *Trichloris crinita* (Lag). Parodi, *T. pluriflora* E. Fourn., *Gouinia paraguayensis* (Kuntze) Parodi, *G. latifolia* (Griseb.) Vasey and *Setaria leiantha* Hack (Morello and Saravia Toledo 1959). The midland and lowland sites are occupied by parklands and savannas, respectively. Parklands are dominated by the grass species *Pappophorum pappiferum* (Lam.) Kuntze, and *P. mucronulatum* Nees (Herrera et al. 2003), while savannas dominated by *Elionurus muticus* (Spreng.) Kuntze, *Heteropogon contortus* (L.) P. Beauv. ex Roem. and Schult., *Schyzachirium* spp., *Paspalum* spp. and *Botriochloa* spp. Woody species frequent in these ecological sites are *Prosopis nigra* (Griseb.) Hieron., *P. kuntzei* Harms, *Geoffroea decorticans* (Gillies ex Hook. and Arn.) Burkart, *Acacia aroma* Gillies ex. Hook and Arn, and *Schinus* spp. Trees and shrubs are scattered across the savannas in small patches or isolated trees (Kunst et al. 2003).

## 4.3 The Woody Species 'Problem' in the Chaco

Chronicles of European conquerors and scientists traveling along the vast territories of Chaco region described open areas dominated by grasses, with smaller regions of dense forest, where indigenous people carried out their subsistence activities such as hunting, fishing and harvesting of fruits (Caamaño 1955; Metraux 1946). Maps of the Chaco from the mid nineteenth to the twentieth centuries as well as the names of small cities and towns suggest that until well the mid twentieth century, grasslands/savannas and forests shared the Chaco landscape in an even proportion (Moussy 1873; Frenguelli 1940). Fire was probably responsible for maintaining that vegetation distribution, since the natives generated many fires for hunting and war purposes (Morello and Saravia Toledo 1959; Bravo et al. 2001).

At the beginning of the twentieth century, the vegetation of the Chaco region was suitable for livestock raising and/or timber operations. Nowadays, dense shrub thickets and overstocked secondary forests are widespread in the region, a fact that has homogenized the landscape (Adamoli et al. 1972). These changes in landscape and vegetation physiognomy have been caused by livestock overgrazing, indiscriminate logging, changes in the fire regime, and fencing (Adamoli et al. 1972; Kunst et al. 2006). Woody plants may reach a density of 10,000 stems*ha$^{-1}$ in some areas (Kunst et al. 2008).

The 'problem' of woody plants has many facets. High density of woody plants decreases the 'suitability' of vegetation types for the livestock industry. Specifically, the problem of woody species is twofold: (a) competition for resources with grass species (sunlight and water) and (b) hampering of livestock and personnel movements and low/nil forage accessibility due to the high stem density and their thorns (Nai Bregaglio et al. 1999). The current stocking rate of native vegetation in poor range condition is negligible if compared with vegetation in good condition: 30 ha*AU$^{-1}$ versus 3–5 ha*AU$^{-1}$, respectively. This fact negatively influences the economy of ranching operations in areas dominated by with 'woody states' (Fumagalli et al. 1997; Kunst et al. 2006). The restoration of the original savannas and grasslands would be an unrealistic proposal due to high costs involved.

From the point of view of forest management, the excess of shrubs and young tree individuals (advance regeneration) decrease the availability of resources for tree growth, thus the community requires thinning.

On the other hand, depending on the circumstances, tree and shrub species can be viewed favorably by ranchers and shepherds. In semiarid-arid areas like the Chaco a paddock that possess trees and shrubs has several advantages over a treeless pasture: (a) the environment under the shade of woody plants is more comfortable, animals do not endure intense heat neither cold conditions (Ledesma 2006; Obispo et al. 2008); (b) most of the native tree and shrub species provide foliage and fruits with high nutritive value (Bordón 1983); (c) grass forage has a better nutritional quality under trees than in open areas (Ludwig et al. 2008) and (d) in arid and semiarid ecosystems trees and shrubs can improve subcanopy soil conditions by increasing microbiological activity, thus generating more nutrient availability (Belsky et al. 1993; Durr and Rangel 2000, 2002; Smit 2004). Moreover, soil physical properties, such as structure, porosity and water storage capacity are improved (Ledesma 2006). They also are a source of biodiversity, habitat for wildlife, and carbon storage (Gifford and Howden 2001; Ansley and Rasmussen 2005; McIvor 2006; Navall 2008; Le Brocque et al. 2009).

## 4.4 Silvopastoral Systems in the Chaco: The Santiago del Estero Experimental Station (EEASE) Approach

### 4.4.1 Theoretical Approach

A silvopastoral system may be defined as a combination of two activities occurring in the same place: livestock grazing and timber operations. In the Chaco region both activities take place in a paddock, and in an ideal setting, both should be complementary and sustainable. Around the world, many silvopastoral systems are created by planting trees; in the Chaco the practical challenge is to develop a silvopastoral system based on already existing native tree and shrub species. In this situation, the manager/operator cannot manage the initial tree densities by planting. The individuals already exist in a paddock, and its origin may be associated to previous mismanagement of the area.

From a range/grass management standpoint, whenever a threshold in woody encroachment is attained, the limitations of grass standing crop cannot be surmounted unless a major disturbance aimed at reducing the dominance of woody plants is applied (Cabral et al. 2003; Ansley et al. 2006; McIvor 2006; Díaz 2007; Ansley and Wiedemann 2008; Smit 2005). Traditional 'active approaches' (McIver and Starr 2001) such as chemical products, prescribed fire and mechanical treatments have been assessed experimentally in the Chaco to disrupt woody states and to transform them into grassy states. Research reports a yield increase >500 % of native grass species in comparison with untreated areas (Rodriguez Rey et al. 1983; Alessandria et al. 1987; Galera 1990; Passera et al. 1996; Boetto et al. 1999; Kunst et al. 2001a). Nowadays, mechanical treatments are widely used in the Chaco region. Reasons are a lack of fine fuel for carrying out a prescribed fire, high cost of chemical products and ecological concerns about the sustainability of the ecosystem. However, we acknowledge that the fossil fuels used for mechanical treatments are also considered unsustainable.

However, all these practices harm the woody plants (especially trees) in one way or the other, thus endangering one of the key components of a silvopastoral system (Araujo 2003). The technical staff of the EEASE surveyed existing paddocks reclaimed by mechanical treatments in the Chaco during the late 1990 and early 2000 and conducted field trials in the 'La María' Experimental Ranch of the INTA Santiago del Estero Research Station (Kunst et al. 2001a, b). In these surveys and experiments, spatial and temporal variation was partitioned in ecological sites, treatment and years since initial treatment. Ecological site was considered as a classification factor representing the combined effects of topographical position, water runoff-runon phenomenon, soil texture, and introduced the spatial variation at cartographic scale larger than 1:20,000. Treatment represented the mechanical treatment proper and seeding effects, and years since initial treatment represented the mixed effect of several factors: grazing and weather features during a specific year -especially rainfall amount and distribution-, and the spatial variation resulting from the changing locations of sampling transects every year. Based on this research, the silvopastoral system was developed using the following criteria:

1. Agronomic and forestry practices such as tree thinning, herbivory, mechanical and herbicide applications, etc. could be viewed as disturbances. A disturbance is a discrete event in time that disrupts ecosystem structure, composition and/or processes by altering its physical environment and/or resources, causing a destruction of plant biomass (Seidl et al. 2011). In a ranching or a timber operation context, these practices are usually repeated throughout time to increase grass standing crop, or aimed at reducing competition so that a timber yield is sustained (Heitschmidt and Pieper 1983; Seidl et al. 2011). A disturbance could be characterized by its intensity, severity and frequency, the latter defined by the longevity (~return interval) of treatments (Kunst et al. 2012a). Based on the Prygogyne dissipation equation, Naveh (1990, 2004), stated that the main issue of a management program is to establish the regime of disturbance. Therefore, to create or manage a silvopastoral system, the research question is to ascertain the appropriate features of a series of disturbance practices

with a specific, well defined regime in order to sustainability be maintained.

2. The intensity and severity of the disturbance(s) should be such as to create/maintain positive interactions such as facilitation between woody plants and grasses (Ong and Leakey 1999). Surveys indicate that mechanical treatments that are too severe reset the succession to zero, and so called pioneer species, that are well adapted to disturbances, dominate. In the Chaco region, that means that woody species of the genera *Acacia* and *Prosopis* thrive (Kunst et al. 2003) and this may not be a desirable effect.
3. Native woody plants return because most are sprouters: mechanical treatments and/or fire rarely kill them (Bond and Midgley 2001; Van Auken 2009; Bravo 2008, 2010; Bravo et al. 2011, 2012). An important aspect in integrated woody plant management programs is their longevity, referred in the literature as persistence, duration of the treatment, treatment life, and thickening (Heitchmidt and Pieper 1983; Noble et al. 2005; Gifford and Howden 2001). It estimates how long the disruption caused by a disturbance will last. Thresholds should be defined in order to define an appropriate timing for a treatment, since concerns are both economical (treatment amortization, Teague et al. 2008) and ecological: a disturbance cannot have a return interval that may endanger the sustainability of the population of a desirable woody or wildlife species (Navall 2008).

### 4.4.2 Practical Approach: Low Intensity Roller-Chopping (RBI)

The most refined and precise disturbance that could be used in a silvopastoral system to thin woody plants in the Chaco is hacking/axing by hand. In that way, selection of individual plants is stressed and the manager could manage tree regeneration, a key issue in forest management. However, hand thinning is costly, labor intensive and time consuming. Therefore, a mechanized alternative should be used. In the Chaco region the widely used mechanical treatment is roller-chopping. A 'roller-chopper' is an iron drum with diameter = 1.4 m and width = 2.5 m armed with blades. It may be filled with 3000 kg of water, and it is usually pulled either by a small tractor, a four-wheel articulated tractor, or by a D4 Caterpillar bulldozer.

Roller-chopping serves several purposes: (a) to create site availability by specifically reducing the aboveground structure of shrubs and decreasing their competence for resources (also called thinning or clearing in literature); (b) to increase sunlight availability, (c) to facilitate forage accessibility and livestock movements, and (d) to thin the woody plant population to an acceptable density. Roller chopping should be aimed at controlling shrub plants, since they are the main source of problems instead of trees. The top soil horizon is slightly disturbed so to seed germination is enhanced.

Roller chopping research at the Santiago del Estero Experimental Station is aimed at creating a park-like vegetation structure, which is more suitable for livestock operations than the current dense, homogeneous 'woody state' (Kunst et al. 2003, 2012a). Woody individuals – especially trees- with DBH >10–15 cm are left standing in different patterns and densities, while at the same time shrubs, are crushed and left as debris that can degrade naturally, harvested as firewood, or be burnt. The latter practice should be carefully implemented because fire is extremely soil degrading due to the high wood density of the Chaco species. After a roller-chopping treatment, forage yields and livestock stocking rates usually increase due either to the recuperation of the native grass species; or by seeding exotic species of the genus *Panicum* and *Cenchrus* (Poaceae) simultaneously with the roller chopping (Ledesma 2006). *Panicum* species are summer growth perennials which possess the $C_4$ syndrome, and are adapted to shadow. The species of choice is usually *Panicum maximum* Jacq.

By handling properly the size of the tractor and the roller-chopper, especially its length, and the number of passes over the standing vegetation, intensity and severity of the mechanical disturbance could be appropriately planned (Figs. 4.3 and 4.4). This is the main principle of the 'Low Intensity roller-chopping' (RBI) the technique developed by research personnel of the

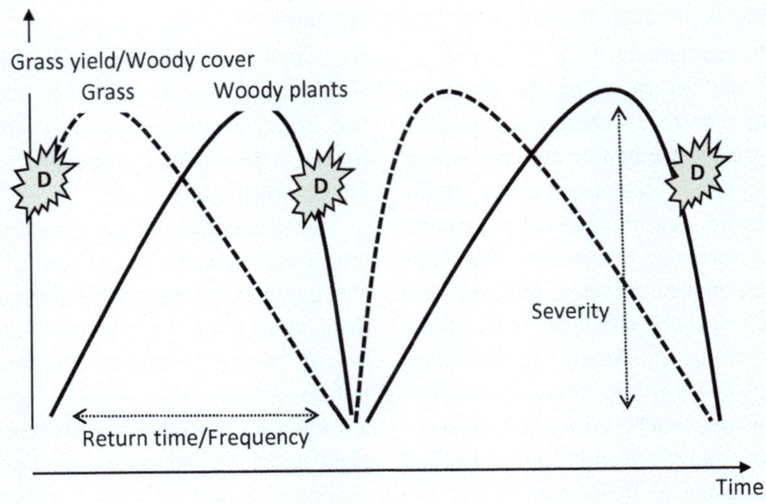


**Fig. 4.3** Theoretical basis of the low intensity roller-chopping (RBI): dynamics throughout time of the grass yield (*broken line*) and its interaction with woody cover (*continuous line*), severity and frequency of the mechanical disturbance. D stands for the periodical mechanical disturbance needed for maintaining the woody cover checked, and thus allowing enough grass yield and accessibility for livestock operations. The art and science of RBI is based in scheduling an appropriate disturbance regime characterized by these two descriptors: magnitude of severity (*vertical arrow*) and length of the return time (*horizontal arrow*)

Santiago del Estero Experimental Station to establish a silvopastoral system in the Chaco region. Intensity could be associated by the size of the machine used for the treatment and the number of passes. In RBI, the roller-chopper is passed twice, the second pass in 45° angle with the first. Severity is assessed by: (a) density and DAP of the woody individuals left standing, and (b) the amount of woody biomass crushed and left as a residue. The assessment and characterization of disturbances in forestry has been developed by Roberts (2007) and Seidl et al. (2011). The use of carefully planned disturbances in a range management context has been already proposed by Fulbright (2004).

### 4.4.3 Ecological Sites and RBI

Ecosystems are a mix of vegetation patches produced by climate, landform, soil and disturbance regimes. Although the definition of ecosystem is often arbitrary and not restricted to any particular spatial unit or scale, the grouping of their biological and physical components in homogenous ecological portions (e.g. ecological sites) in order to separate the ecosystem into ecologically meaningful and more tractable units is a practice well established in the science of vegetation management for several purposes (Kunst et al. 2006). Empirical observations as well as surveys conducted parallel to this research determined that in the Chaco ecological sites generate plant communities that differ in physiognomy, and botanical and growth form composition. The target plant community (reference or benchmark vegetation type) should be carefully defined, and the disturbance regime should be defined accordingly to meet that target. Although the restoration of the original forest and savannas may be impractical due to the high costs involved, the adequate disturbance regime should be tailored for each ecological site (ecosystem) and vegetation condition or state, that is its suitability for producing goods and services.

The implementation of a silvopastoral system should be based on an appropriate soil and vegetation type inventory of a paddock, ranch or region. The choice of method of ecosystem mapping and classification; and the level of perception at which the spatial variation should be perceived are key practical questions: research has shown that the concept of ecological site is

**Fig. 4.4** Upland site forest treated with Low intensity roller chopping (RBI) and seeded with *Panicum maximum* cv Gatton. (**a**) Before treatment, December 2006. (**b**) One rainfall season after treatment, April 2007. INTA Santiago del Estero, La María Experimental ranch

quite sound and should be used as another basis for the development and implementation of silvopastoral systems in the Chaco. The cartographic scale at which the ecological sites should be mapped should be >1:50,000.

## 4.5 RBI and Resources

### 4.5.1 Sunlight Availability

In experiments conducted from 1996 to 2002, mean sunlight availability in RBI plots was ~40 % to ~55 % greater than in untreated, control plots (930.83 $\mu mol.s^{-1}.m^{-1}$ versus 631 $\mu mol.s^{-1}.m^{-1}$), respectively ($p>|t|<0.0001$) (Kunst et al. 2012a). Differences among ecological sites were not significant. During 1998–1999, treatment and seasonal effects were significant ($p>F=0.0001$). The mean sunlight availability was 461 $\mu mol.s^{-1}.m^{-1}$ and 329 $\mu mol.s^{-1}.m^{-1}$ ($p>|t|<0.0002$) in the treated and control plots respectively (Kunst et al. 2012a).

### 4.5.2 RBI, Soil Processes and Soil Quality

Soil is an essential part of any production system. The homogenization of the vegetation into shrub thickets does not necessarily mean that soils are also degraded, a fact that could confuse managers. In this section, we will address the effect of silvopastoral systems created by RBI on soil quality, e.g. dynamics of resources such as water availability, organic matter, and the structure and functions of soil microbial community. Each ecological site responds to a specific disturbance differentially; soil quality being essential to quantify in the stability and function of the site (Albanesi et al. 2014).

**Water infiltration**. Woody canopies intercept rainwater, decreasing the kinetic energy of the raindrop that facilitates water infiltration into the soil (Haworth and McPherson 1994; Binkley and Giardina 1998; Serrato and Díaz 1998; Wainwright et al. 1999). The rainfall intercepted by the canopy of forest and shrub thickets of the Chaco may amount to 20 % of the annual rainfall (Acuña and Juarez 2001). If a part of the woody canopy is crushed by RBI mechanical treatments, the soil infiltration rate proper becomes a key issue in the local hydrologic cycle. Kunst et al. (2003) assessed soil water infiltration in RBI treated plots. Results suggested that the litter of woody species generated by the roller-chopper significantly affected infiltration ($p=F<0.05$, Fig. 4.5). This result suggests the importance of keeping trees and shrubs in the Chaco, giving support for the development of silvopastoral systems.

**Soil moisture dynamics**. In RBI treated plots, soil moisture followed the seasonal rainfall pattern. There was an increase of the mean soil moisture (%) of the upper soil horizon immediately after the treatment when compared to control plots and deep soil horizons (Kunst et al. 2012a). However, this increase was not statistically significant, and only lasted ~2 weeks after the treatments were applied. Afterwards, the temporal pattern of soil moisture was similar for both controls and treated plots (Kunst et al. 2012a). The effect of ecological site on soil moisture was statistically significant in experiments conducted from 1996 to 2002 ($p>F=0.0946$) and from 2002 to 2006 ($p>F=0.0414$), respectively, despite the large temporal and spatial variation considered. Lowland sites presented the highest mean soil moisture when compared to the other sites, irrespective of treatment (Kunst et al. 2012a). This short term effect of roller-chopping on soil moisture has been also reported by Galera (1990) in the arid of the Chaco.

**RBI and soil processes**. The effect of RBI on soil processes depends on the ecological site (Anriquez et al. 2005). Here we present and discuss effects on upland and midland sites:

In secondary forests in upland sites, soil bulk density was not affected 5 years after being treated with RBI and rotational grazing (Table 4.1, Albanesi et al. 2013). This result suggests that water and air movement in the soil were not reduced. In the same plots, the soil total organic

**Fig. 4.5** Mean basic soil infiltration rates ($mm^*h^{-1}$) in Low intensity roller chopping (RBI) plots according to type of litter estimated by the Kostiakov equation. INTA Santiago del Estero, La María Experimental Ranch. Means within sites followed by the same letter are not significantly different, Duncan test, $\alpha = 0.05$ (Kunst et al. 2003)

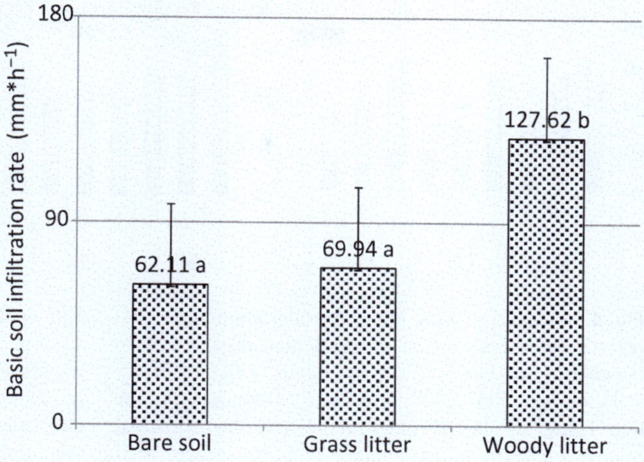


**Table 4.1** Mean soil bulk density, total organic carbon (TOC), particulate organic carbon (POC), total nitrogen (TN), microbial biomass carbon (MBC), ratio MBC:TOC and metabolic quotient ($qCO_2$) at a soil depth 0–15 cm, in a "Low intensity roller chopping" (RBI) experiment (2006–2012). INTA Santiago del Estero Experimental Research Station, La María Experimental Ranch (Albanesi et al. 2013)

| | BA (gr soil $cm^{-3}$) | TOC (g C $kg^{-1}$ soil) | POC (g C $kg^{-1}$ soil) | TN (g N $kg^{-1}$ soil) | MBC (ug C g $suelo^{-1}$) | MBC:TOC (%) | $qCO_2$ |
|---|---|---|---|---|---|---|---|
| Control | 0.84 a[1] | 30.10 b | 25.37 b | 2.24 a | 226.07 b | 0.58 a | 0.26 ab |
| RBI | 0.84 a | 28.77 b | 24.14 b | 2.72 a | 167.63 b | 0.55 a | 0.28 ab |
| RBI plus cattle grazing | 0.87 a | 22.94 a | 18.91 a | 2.11 a | 91.84 a | 0.37 a | 0.46 b |

[1]$\alpha = 0.05$

carbon (TOC), particulate organic carbon (POC), microbial carbon biomass (MBC), and the ratio MBC:TOC were not affected by treatments (Table 4.1). These results were attributed to the woody residues crushed by the roller chopper and left for decomposition, mainly composed by branches and leaves. If *Panicum* grass was seeded, its roots were probably quickly degraded and incorporated into the soil.

However, rotational cattle grazing, (grazing in winter and resting in summer) decreased TOC, POC and the ratio MBC:TOC, likely due to the loss of plants tissues being consumed by the animals (Taddese et al. 2002). On the other hand, the level of total nitrogen (TN) increased, a fact attributed to addition of Nitrogen and Phosphorus via faeces, as it has been reported for silvopastoral systems in other world ecosystems (Nyakatawa et al. 2012).

Plant litter plays a significant role in the dynamics of soil carbon in silvopastoral systems, since the soil retains more C than the plants (Kumada et al. 2008). In RBI plots, woody cover, especially that of trees, had a significant effect. TOC and POC decreased as the distance to the main bole of a tree increased, probably due as response to gradient of litter accumulation (Fig. 4.6). This result suggests that in a silvopastoral system is essential that tree canopies overlap, because this cover provides litter rich in carbon (Ng et al. 2014). This effect depends on the species: under *Z. mistol*, TOC content was higher than under the canopy of *A. quebracho blanco* and *S. lorentzii* . *Z. mistol* produce between 16 and 64 % more litter than the other two species (Fig. 4.6).

The microbial metabolic quotient ($qCO_2$) increased, a fact suggesting some stress that

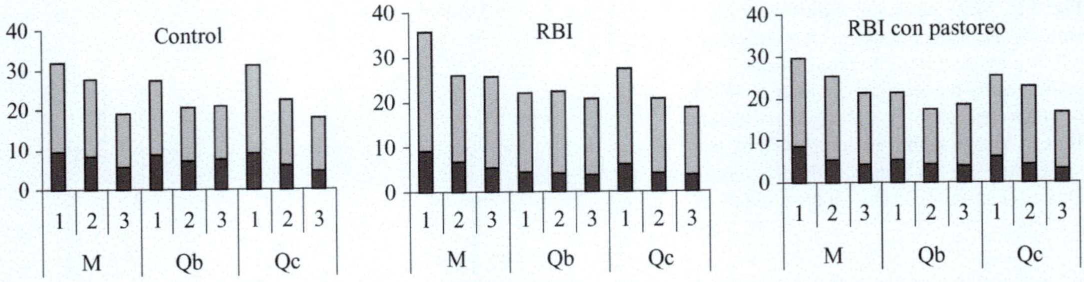


**Fig. 4.6** Effects of Low intensity roller chopping (RBI) on total soil C (*entire bar*) and particulate organic carbon (*shaded in grey*) at 0–15 cm soil depth (g C kg$^{-1}$ of dry soil), average 2007–2012, at different distances from the tree trunk. INTA Santiago del Estero Experimental Station, La María Experimental Ranch. References: RBI con pastoreo = RBI plus seeding of *Panicum* and grazing; *RBI* = RBI with seeding of *Panicum* but no grazing; *Control*: no treatments. *M*: *Zyziphus mistol*; *Qb*: *Aspidosperma quebracho blanco*; *Qc*: *Schinopsis lorentzii*. Distances: *1* = 0.50 m from the tree trunk; *2* = under tree canopy; and *3* = tree canopy border (Adapted from Albanesi et al. 2013)

**Table 4.2** Mean potential mineralizable Carbon ($C_0$), carbon mineralization rate (kc), and dehidrogenase activity (Dh-asa) at 0–15 cm soil depth observed in a "Low intensity roller chopping" (RBI) experiment observed between 1999 and 2003

| | $C_0$ (mg C kg$^{-1}$ soil) | | | kc (mg C kg$^{-1}$ soil day$^{-1}$) | | | Dh-asa (mg TPF kg$^{-1}$ de suelo h$^{-1}$) | | |
|---|---|---|---|---|---|---|---|---|---|
| | Sitios ecológicos | | | | | | | | |
| | UP | MD | LL | UP | MD | LL | UP | MD | LL |
| Control | 384.4 | 344.2 | 338.7 | 0.040 | 0.054 | 0.019 | 96.8 | 103.9 | 119.3 |
| RBI | 454.1 | 290.1 | 386.7 | 0.027 | 0.064 | 0.025 | 96.1 | 95.2 | 134.2 |

References: Ecological sites: *UP* upland, *MD* Midland, *LL* lowland. Treatments: Control, no treatment; RBI: low intensity roller chopping with seeding of *Panicum* (Albanesi et al. 2013)

decreases the water use efficiency of soil microorganisms (Table 4.1).

RBI keeps the majority of the tree individuals standing. This fact likely maintains a higher magnitude of the fractions of soil carbon of fast mineralization, increases the soil mineralization rate (kc), as well as the activity of the microbial population involved in the initial degradation of SOM. This finding is supported by the high activity showed by the dehydrogenase enzyme (Anriquez et al. 2005).

In midland sites, RBI and the seeding of *Panicum* decreased the carbon stock available for mineralization in the short term ($C_0$) while the mineralization rate (kc) increased (Table 4.2). The midland site has usually less TOC and POC of fast mineralization than the upland site. These facts suggest that the grazing management of exotic species in a silvopastoral system in this site should be carefully planned because a small magnitude of $C_0$ coupled with a fast mineralization rate could increase the loss of SOC.

**Genetic diversity in microbial communities**. Research conducted during 5 years both in the central (humid) and western Chaco (semi-arid) suggests that climate and soil type rather than silvopastoral management influence the structure of soil bacterial communities. In the western Chaco, results indicate that RBI, by modifying their habitat, alters the species composition of soil bacteria 1 year post disturbance. However, initial species composition was re-established 5 years after the disturbance (Albanesi et al. 2013). In the central Chaco, trends were similar, although the difference in bacterial genetic diversity (distance of the fingerprint profiles) between RBI treated plots and controls was lower than in the western Chaco (Fig. 4.7a). These results agree with studies reporting that changes in bacterial communities are driven mainly by modification of the plant communities (Lupatini et al. 2013). The addition of plant residues resulting from the initial crushing of the

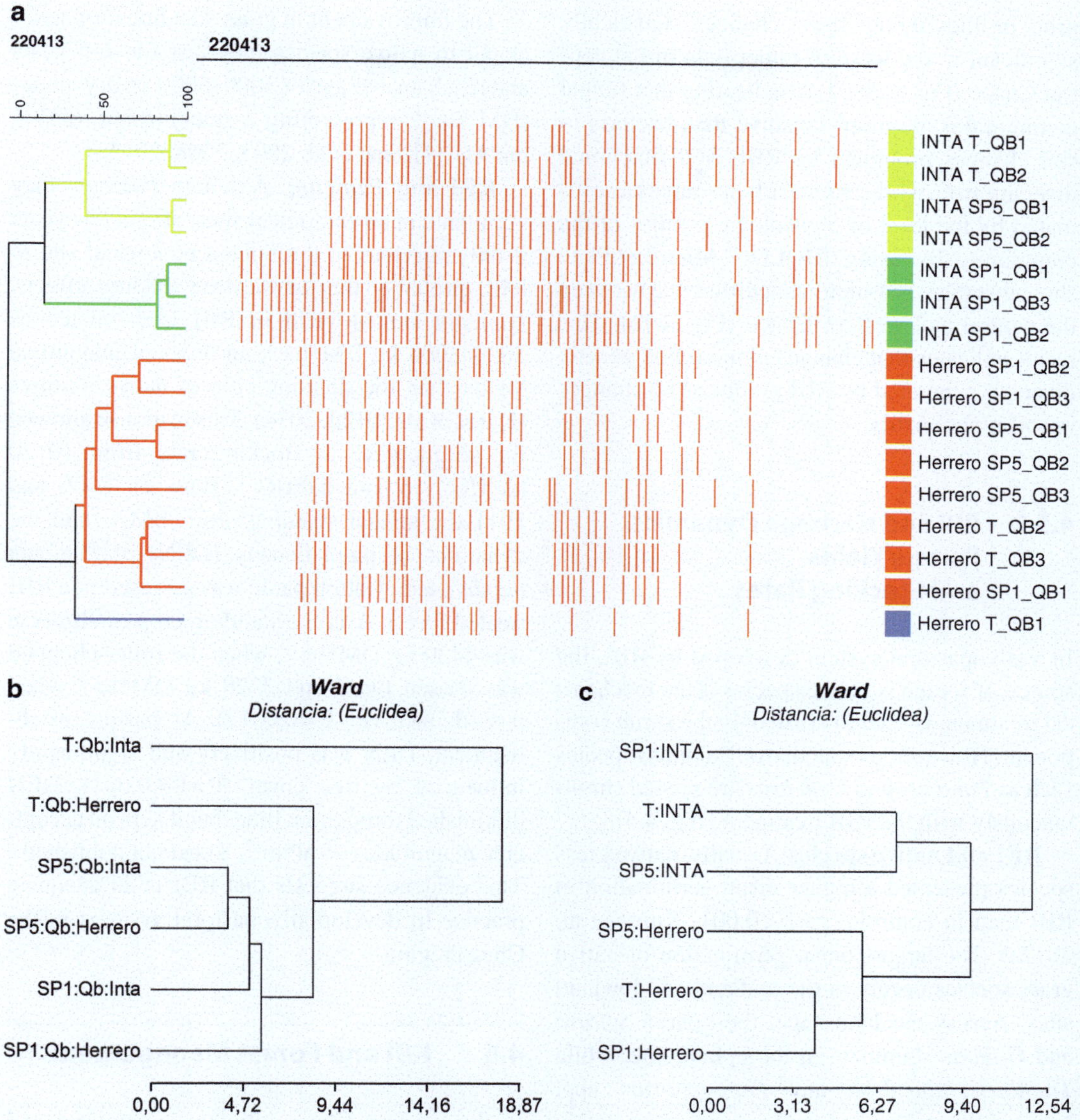


**Fig. 4.7** Dendrograms using ward distance of (**a**) ADNr 16S-DGGE profiles using Bionumerics software, (**b**) using ADNr 18S-TRFLP profiles, and (**c**) community-level physiological profiles (BIOLOG) for different sites and treatments. References: *T*: control, *SP1*: Low intensity roller chopping (RBI) after 1 year of the initial treatment, *SP5*: RBI after 5 years of initial treatment, *QB*: *Aspidosperma quebracho blanco*; *Herrero*: central Chaco, subhumid environment; *INTA*: semiarid environment, western Chaco

woody plant represents a significant C input to the soil, and 1 year after application of RBI bacterial communities were still affected by this fact. According to Ng et al. (2014) there is a direct relationship between C composition and the structure of soil bacterial communities. These results suggest that native bacterial populations have a great tolerance to change in environmental conditions, and also that RBI has, in fact, low impact on them. In conclusion, RBI does not significantly affect the diversity of soil bacteria.

Fungal communities, as revealed by rDNA 18S T-RFLP technique, are also distributed according to climate and soil type. However, RBI promotes fungal communities different from controls as measured by the distance of fingerprint profiles, being these changes statistically significant in the western rather than in the central Chaco (Fig. 4.7b), i. e. indicating that fungal communities are more sensitive than bacteria to soil changes promoted by RBI. Soil functional diversity from 31 different carbon sources as estimated by the level of physiologic profiles of the community by using BIOLOG MicroEcoPlate show that there is a significant difference between the central and western Chaco (Fig. 4.7c). This result indicates that changes in microbial genetic diversity generated by RBI produced no changes in soil functionality.

### 4.5.3 RBI and Herbage Dynamics, Forage Yields and Stocking Rates

In a silvopastoral system developed by RBI, the source of forage is native species if the mechanical treatment is used to crush only the shrub component. However, as stated, tropical grass species such as *Panicum* and *Cenchrus* are seeded simultaneously with the RBI treatment (Fig. 4.4).

**RBI and native species**. Usually, native grass species presented a higher mean germination in RBI than in controls ($p>F<0.001$, Kunst et al. 2012a). The largest mean germination of native grass species occurs in the midland and lowland sites. Across the landscape, the genera *Setaria* and *Gouinia* dominate in the upland site, while *Trichloris pluriflora* and *Pappophorum* spp. occur in the midland and upland site, respectively (Kunst et al. 2008). Germination of broadleaf herbaceous and woody species was also higher for the RBI plots than for the controls ($p>F<0.001$ and $p>F=0.0312$, respectively). Nevertheless, germination of both life forms may be observed only in the first season after application of roller-chopping.

The recruitment of new individuals by germination was relevant only in the case of native grasses: forb species changed the botanical composition for a while, but after two growing seasons, the herbaceous stratum was dominated by grasses. Few small plants of woody species survived up to the second season.

The improvement in grass standing crop associated to native species may last for 4–5 years and reach an average of 3000–5000 kg dry matter (DM)*ha$^{-1}$, representing a stocking rate of 3–5 ha*AU$^{-1}$ (Kunst et al. 2003, 2008, 2012a).

**RBI and *Panicum***. Although *Panicum* may germinate in all ecological sites, after a few years it only survives in the upland ecological site in RBI, probably due to the tree shading effects. *Panicum* annual yield in RBI is estimated as 5000–8000 kg DM*ha$^{-1}$, an order of magnitude higher than the standing crop of native pastures (Kunst et al. 2012d, 2014). This response allowed for an increase in stocking rate from 10–20 ha*UG$^{-1}$ to 1–2 ha*UG$^{-1}$. Between 2006 and 2011 the annual standing crop (BMA) and the dynamics of plant density (DEN) of *Panicum maximum* cv Gatton panic was assessed in a RBI treated forest. Average standing crop of *Panicum* was 4430 kg DM*ha$^{-1}$, when the roller-chopped was passed twice and 3329 kg DM*ha$^{-1}$ when passed once ($P>F=0{,}0012$). At pasture establishment, DEN was positively and significantly influenced by tree cover ($P>F=0{,}001$). DEN diminished throughout time, but it kept an acceptable magnitude (~6 pl*m$^{-2}$, Kunst not published). This evidence suggests that RBI is an adequate practice to develop silvopastoral systems in the Chaco region.

## 4.6 RBI and Forest Management

In the western Chaco, climate is semiarid and forests are uneven-aged (Araujo 2003). Provincial laws state that forest harvest should be regulated based using single tree selection: therefore DBh classes should resemble an inverse-J curve, and an appropriate q-value[2] should be used in order to obtain a 'balanced uneven aged forest'. This management approach requires meeting some minimum criteria for sustainability, as follows:

[2] Q-value is a constant associated to the BDq approach for managing forest structure and refers to the distribution of growing stock Large q values retain more trees than smaller (BC 2003).

(a) intensity and frequency of logging should be regulated in order to be lower than the forest growth rate and maintain ecosystem services,
(b) forest structure should be balanced, applying selection criteria in all size classes, with a q-value greater than 1.3 (for 5 cm DBH classes), and
(c) regeneration events should be promoted, at least three times in the life-span of target species selected for management.

In the western Chaco, tree growth rates range from 1 to 1.5 tn*ha$^{-1}$.year$^{-1}$ (Obst et al. 2007). Harvesting should be applied with frequencies of 15–20 years. Research in treated plots showed that trees under RBI and a harvesting program presented a relative growth 58 % greater than the observed in control plots (Navall 2012b, Fig. 4.8).

Other important interactions occur when disturbances are combined in a silvopastoral system. The return of RBI is needed to keep brush strata under an appropriate threshold so livestock transit and forage accessibility are not hampered (disturbance regime). In order to minimize negative interactions among the disturbances applied for managing a silvopastoral system based on RBI, the following plan was developed:

(a) Roller-chopping intensity, especially the initial treatment was reduced compared with the traditional magnitude, and its frequency increased, in order to minimize its effect on the forest structure. Significant increases in forage yield could be addressed by RBI. Research suggest that tree damage by RBI represents less than 8 % of the basal area of trees damaged compared with controls, mainly in smaller dbh classes (Gómez and Navall 2008). This research found that only 17.4 % of trees from 0.25 to 1.3 m height remain without damage after roller-chopping. However, the density of remaining individuals in this class was four times the required to keep a balanced structure (Fig. 4.9).

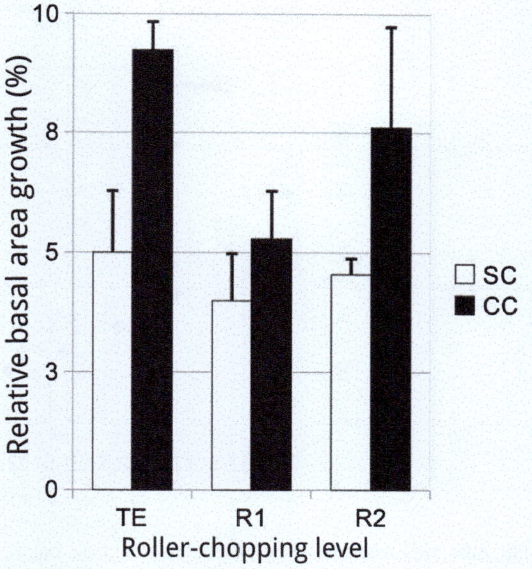


**Fig. 4.8** Effects of Low intensity roller chopping (RBI) in trees: mean basal area growth expressed in percent of initial, pre-RBI basal area (%) after 4-years of treatment (2006–2010). References: Treatments: *R1*: RBI with one pass of roller-chopper; *R2*: RBI with two passes of roller-chopper; *TE*: controls, *CC*: with timber harvesting (30 % of initial basal area) and *SC*: without timber harvesting. INTA Santiago del Estero Experimental Station, La María Experimental Ranch

(b) Improvement of visibility and accessibility generated by RBI, are synergic with forestry operations, like tree selection, marking, harvesting and extraction (Navall and Cassino 2009; Navall et al. 2013a, b and c).
(c) Grazing exclosures were incorporated after each forest harvesting (every 20 years), to promote a natural regeneration pulse, taking advantage of available resources generated by tree thinning (Navall 2011). Regeneration was monitored prior to subsequent RBI treatments.
(d) Maintenance of tree structure was able to sustain key environmental services, such as wildlife habitat (Coria et al. 2012) and litter fall. Low impact roller-chopping plus 30 % reduction of basal area by harvesting reduced litter fall from 2440 to 1607 Kg DM*ha$^{-1}$ compared with the control. More intensive harvesting treatments could endanger these environmental services (Navall 2012b).

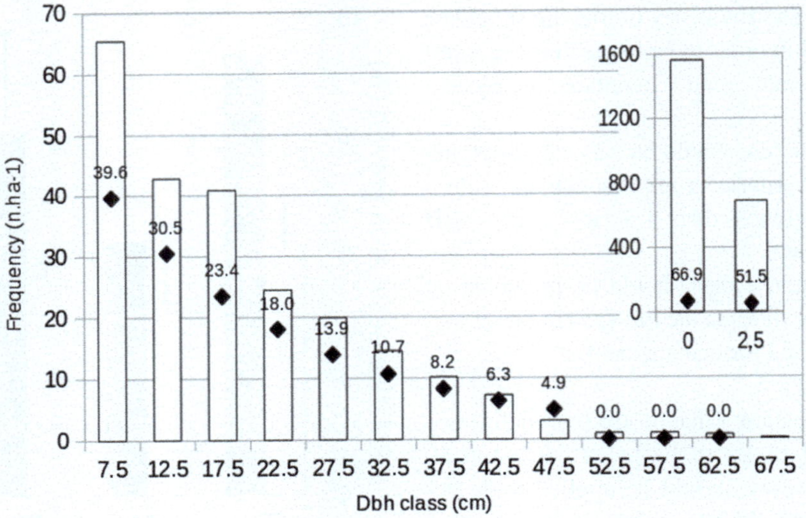


**Fig. 4.9** Tree frequencies by DBH classes in Low intensity roller chopping (RBI) treated plots: pre-harvesting (*bars*) and designed post-harvesting (*diamonds*), for a q value = 1.3 and G = 8.9 and a final DBH = 40 cm. Harvesting intensity 30 % of initial BA. Class "0" represents trees from 0.25 to 1.3 m height. INTA EEA Santiago del Estero Experimental Station, La Maria Experimental Ranch

(e) An appropriate rotation of forest, grazing and RBI treatments over time and space was designed (Navall 2011); and
(f) The added value to forest products improved revenue. The use of branches of low quality wood as firewood was an appropriate measure (Navall et al. 2013c; Cassino et al. 2013).

## 4.7 Shrub Control in a Silvopastoral Systems Created by RBI

An important aspect in silvopastoral systems is their longevity related to the shrub stratum (Heitchmidt and Pieper 1983; Noble et al. 2005; Gifford and Howden 2001), since shrub regrowth severely hampers the transit of livestock and personnel. Longevity estimates how long the disruption caused by the disturbance used to create the silvopastoral system will last. Cattlemen in the Chaco have reported that the average longevity of a traditional, severe roller chopper treatment is 3 years (Casas et al. 1978), an interval at odds with the conservation of forest species, especially with their advance regeneration (Gómez and Navall 2008; Gómez et al. 2009).

Accessibility was assessed at the end of two RBI experiments conducted between 1996 and 2002 as the area of an imaginary plane of 10 $m^2$ intercepted by woody plants (Kunst et al. 2012a). Treatment effect was somewhat inconclusive, since it was statistically significant only in one experiment ($p > F = 0.009$) while not significant in the other ($p > F = 0.21$). Mean accessibility was higher when prescribed fire was used; but there was an ecological site effect: in the upland and midland site, where the majority of large trees were left standing, mean accessibility was high. This suggests a competition effect of large trees on shrubs. The longevity of the treatment is directly associated to accessibility: between 2006 and 2012, RBI treated plots showed a longevity of at least 6 years, a more appropriate return interval for a sustainable silvopastoral system (Kunst et al. 2012a, e, Fig. 4.10).

In a RBI experiment assessed between 2006 and 2011 the volume growth of three native brush

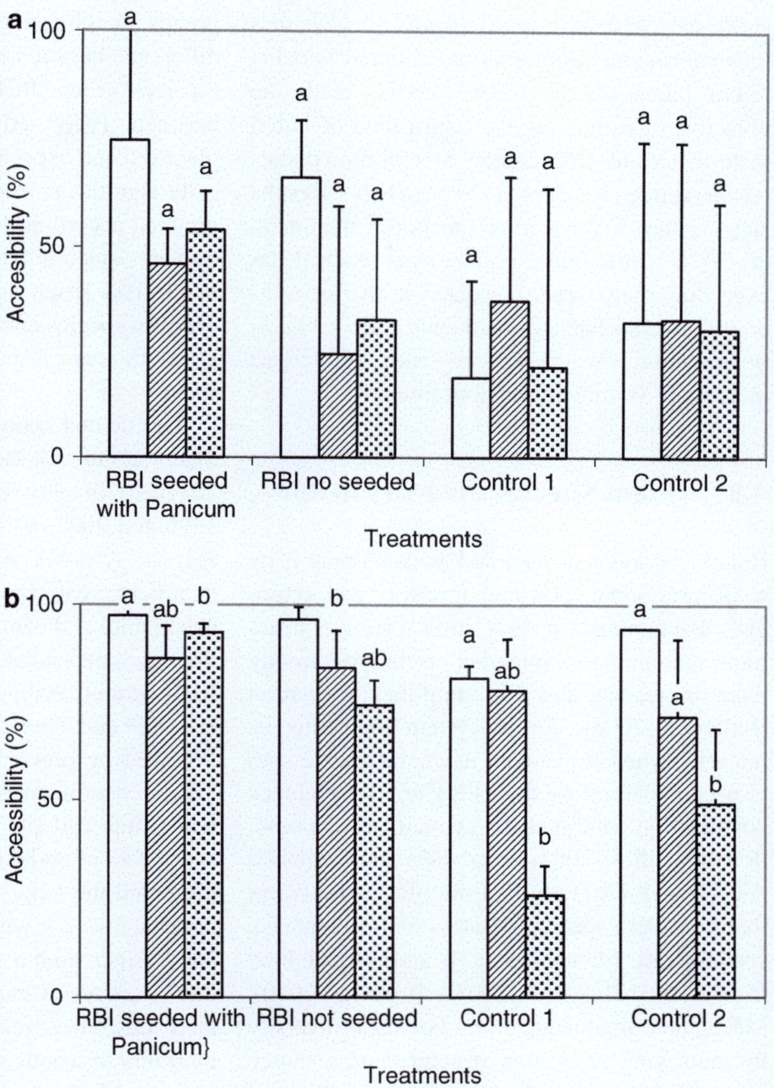


**Fig. 4.10** Mean accessibility (%) estimated at the end of Low intensity roller chopping (RBI) experiments, sorted by treatment and range sites. (**a**) Exp. 1, 1996–2002; (**b**) Exp. 2. *White bar* = upland site; *shaded bar* = midland site; *dotted bar* = lowland site. Within a treatment, different *letters* indicate significant differences, Duncan test, $\alpha = 0.05$. *Error bars* represent spatial variation within a treatment, five sampling stations by ecological site. INTA EEA Santiago del Estero Experimental Station, La Maria Experimental Ranch (Adapted from Kunst et al. 2012a)

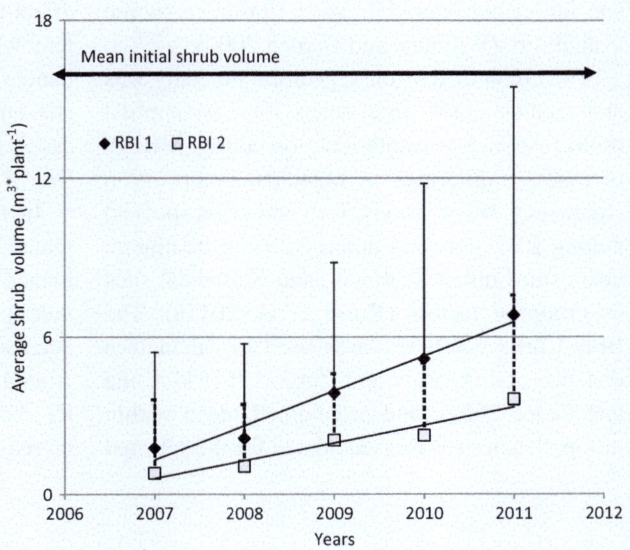


**Fig. 4.11** Volume dynamics ($m^3$*$plant^{-1}$) of shrub species (six plots, six plants per plot) from 2006 to 2011 in a Low intensity roller chopping (RBI) experiment. References: *RBI 1*: roller chopper passed once, *RBI 2*: roller chopper passed twice. INTA EEA Santiago del Estero Experimental Station, La Maria Experimental Ranch (Kunst, unpublished results)

species: *Acacia furcaptispina*, *Capparis atamisquea* and *Celtis spinosa* treated by two different intensities of roller-chopping with the objective of estimating the return time of initial disturbance and characterize the optimum disturbance regime (Kunst et al. 2009, 2012c). Results suggest that 5 years after the initial treatment, 30–50 % of the initial volume was reached, far exceeding the 3 years presented in the local literature for mechanical treatments (Fig. 4.11). *A. furcatispina* was the species with the largest increase in volume throughout time.

## 4.8 Plant Species Diversity in RBI

Roller chopping is regarded with suspicion by some traditional, classical foresters and ecologists that contend that it is a non-selective, structural manipulation intended to reduce woody plant frequency, and that simplifies vegetation (Fulbright 2004). An ecosystem maintains its integrity when it presents its characteristic species composition (~ diversity) or its 'reference composition', and in that way maintains its functionality (SER 2004; Ruiz Jaén and Mitchell Aide 2005). Diversity is a multifaceted concept that considers species richness and dominance, spatial scale (diversities α, β and γ) and time (Sheil and Burslem 2003; Brudvig 2010). Mechanical treatments affect botanical diversity by modifying vegetation structure and resource availability (Nolte et al. 1994). Besides, *Panicum* species are considered invasive, limiting resource availability (Williams and Baruch 2000).

The effect of RBI on vegetation diversity was analyzed using two approaches. First, we applied multi-response permutation procedure (MRPP) to assess similarities in botanical composition (frequency, %, of woody, forb and grass species) among RBI plots and controls, using treatment, years since initial treatment, and ecological sites as grouping factors (Kunst et al. 2012a). The Bray-Curtis coefficient was used to calculate a distance matrix (Bray and Curtis 1957; McCune and Grace 2002), and a weighted mean within group distance (δ) was calculated for the selected groups. A third step was the calculation of the difference between the observed and expected δ for each group. In RBI experiments conducted between 1996 and 2002, differences among observed and expected δ were larger and statistically significant when species frequencies were grouped according to ecological site rather than by years since initial treatment (time) and treatment, This result supports the idea that differences in species composition for shrub, forb and grass strata are strongly influenced by ecological site.

The second approach was to estimate H′, the Shannon index at the α and γ levels (Kunst et al. 2012b, c). In two experiments conducted between 1996 and 2002, H′ was used as dependent variable in ANOVA with repeated measures and treatment, years since initial treatment (time) as covariable, and ecological site and their interactions as independent variables. Treatments were: RBI without seeding (RR); RBI with seeding of *Panicum maximum* cv green panic (RS); and RBI followed by prescribed fire (RF), and controls. Treatment effect was not significant at the γ level, while time and site*treatment showed a significant effect of on H′ ($p>F=0.05$).The upland site presented the largest α diversity, while the lowland the lowest α when compared to controls. RR and RF increased α diversity in the upland, while there were no differences in the lowland and midland sites. These results are attributed to the low mortality of woody species and to the increase of grass and forbs as a result of RBI. RBI does not affect the plant γ diversity of the Chaco ecosystem when intensity and severity of the disturbances are medium, but the interaction between site and treatment should be fully explored to assess the effect of roller-chopping on α diversity (Fig. 4.12).

In the upland site, forest richness was estimated by the number of species present and dominance (evenness) by the index D = (1 – Simpson index). Species richness increased and D decreased in RR and RF compared to Controls. Abundance of grass species increased in RR and RF, but forb abundance diminished. In RS, D increased, likely because of the seeding of Green

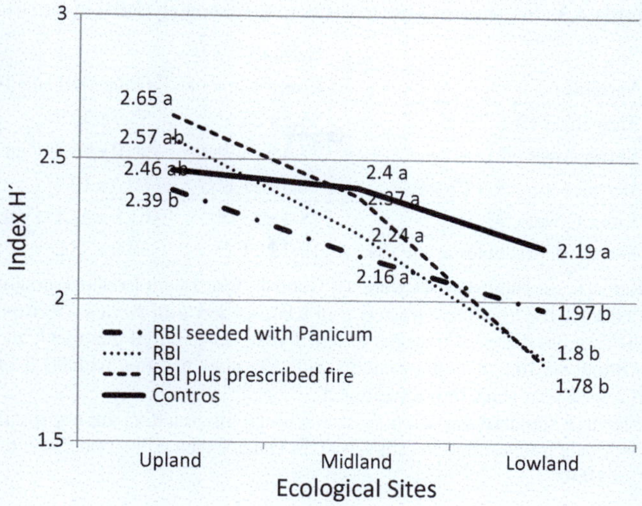


**Fig. 4.12** Effect of Low intensity roller chopping (RBI) in plant species diversity, as estimated by the Shannon index (H′) in three ecological sites of the western Chaco region. INTA Santiago del Estero Experimental Research Station, La María Experimental Ranch. Mean H′ calculated from botanical composition data collected from 1996 to 2002. Previous to the application of RBI, the original vegetation types were: upland site had a forest+shrub physiognomy; the midland site a low height forest+shrub and the lowland site a shrubland. Means within sites followed by the same letter are not significantly different, Duncan test, $\alpha=0.05$ (Kunst et al. 2012a, b, c)

panic. These effects could be considered positive for livestock operations and neutral for the timber operations.

## 4.9 RBI and Wildlife Habitats

Wildlife diversity in terrestrial neotropical habitats such as those of the western Chaco is determined by local factors such as vegetation structure (cover of dead woody material standing and on soil surface, litter, herbaceous, shrub and tree species) and floristic composition (Ojasti 2000). Wildlife assessments in the upland ecological site indicate that RBI retains much of these attributes. Habitat evolution from the time when the disturbance is applied is synthesized as follows (Kunst et al. 2008, 2012b, c, Table 4.3):

- The main effect of RBI is the generation of open woody communities/parklands that tend to close their canopy over time, mainly due to the recovery of the shrub stratum. The structural complexity is preserved (i.e. no cover is reduced).
- Some strata may reduce initially their cover, especially the shrub and herbaceous strata. However, they show a good rate of recovery: there is no difference between RBI treated plots and controls within 7 years.
- Litter cover is reduced relatively very little.
- A considerable amount of woody debris is generated, but by removing the medium and high diameter materials for use as firewood, the original levels are restored within a maximum period of 7 years.
- The botanical composition of woody species does not change. However, *Panicum* becomes increasingly dominant where it was planted (>37 %).

These changes may be assessed as globally "mild", and from a wildlife management standpoint, it can be predicted that RBI treated paddocks would have a high capacity to support a diversity of wildlife. This specific hypothesis was tested by assessing diversity of birds, reptiles and amphibians in the same forests treated with RBI (see Table 4.3). Results were consistent with the stated hypothesis, and in no case was a loss of diversity detected (Table 4.4). There was even an increase of bird diversity in the oldest treatment. However, habitat assessments in large areas treated with RBI are required to support the hypothesis more conclusively. Additional studies are also required to assess the consequences of the introduction of exotic pasture. However, results are supported in part by the regional study of Mastrangelo and Gavin (2012) who reported that low intensity silvopastoral systems retained most of the bird species of the forest.

**Table 4.3** Influence of Low intensity roller chopping (RBI) in the structure of a forest of the western Chaco

| | Habitats | | |
|---|---|---|---|
| Variables | Control (n=14) | Older RBI (n=10) | Recent RBI (n=14) |
| Tree cover (%) | 55.5±25.1[a] | 53.7±35.9[a] | 44.1±27.3[b] (Var. – 20.5 %) |
| Shrub cover (%) | 77.6±13.7[a] | 56.0±16.3[b] (Var. – 27.8 %) | 26.7±10.8[c] (Var. – 65.6 %) |
| Herbaceous cover (%) | 16.5±21.8[a] | 10.3±10.6[a] | 6.12±11.6[b] (Var. – 62.9 %) |
| Litter Cover (%) | 78.4±33.6[a] | 61.6±37.1[b] (Var. – 21.4 %) | 68.7±35.9[b] (Var. – 12.4 %) |
| Woody debris volume ($m^3$*$ha^{-1}$) | 19.7±21.7[a] | 38.0±78.2[a] | 428.4±262.4[b] (Var. + 2074.6 %) |

References: Habitat types were: (i) Control: forest with livestock grazing, without logging in the last three decades; (ii) Older RBI: 32 ha forest treated with RBI, seeded with *Panicum maximum cv Gatton panic*, with livestock grazing and 30 % of basal area of trees harvested, 7 years since initial treatment; and Recent RBI: 100 ha of forest treated with RBI, without seeding of exotic pastures, with livestock grazing, with 30 % of the basal area of trees felled but extracted only 8.2 %, 1 year since initial treatment

Means ± standard deviation. Means with common letters are not significantly different. Parametric one-way ANOVA, with data transformed to ranks, p=0.5. Date of data collection: September 2013

[a–c]$\alpha = 0.05$

**Table 4.4** Influence of Low intensity roller chopping (RBI) in the richness and diversity (Shannon index expressed by Weber) of communities of bird, reptiles and amphibians species by breeding season, in western Chaco forests. Habitat types similar to Table 4.3

| Wildlife classes | Index | Control (n=7) | Older RBI (n=5) | Recent RBI (n=7) | Overall richness |
|---|---|---|---|---|---|
| Birds | Richness | 38 | 44 | 44 | 54 |
| | Shannon weaver index | 2.7±0,3 a | 3.1±0,2 b | 2.8±0,1 a | |
| Reptiles and amphibians | Richness | 6 | Not assessed | 6 | 9 |
| | Shannon weaver index | 1.3±0,4 a | | 1.1±0,3 a | |

Means ± standard deviation. Means with common letters are not significantly different. Parametric one-way ANOVA, p=0.1. Date of data collection: October–December 2013

## 4.10 RBI and Energy Concerns

Roldan Bernhard (2012) has reported that the gross energy requirements between a silvopastoral system developed by RBI and a traditional, intensive pastoral system differ. Although most of the energy inputs of both systems are made up of fossil fuels (27 % versus 36 %, respectively) and wire for fencing (36 % and 44 %, respectively), the RBI silvopastoral system remunerated 8.54 times each energy unit invested, while the traditional, intensive pastoral returned only 0.99 energy units. A comparison of the power flow per unit area of the power dissipated per unit of accumulated biomass shows a high degree of sustainability of the RBI silvopastoral system, since the remuneration remains very close to the relationship shown by natural systems around the world.

## 4.11 Conclusions and Management Implications

Results presented here suggest that the implementation of silvopastoral systems in the western Chaco should be based on: (a) an appropriate map of the ecological sites and the current suitability (condition) of the their vegetation for producing goods and services, and (b) the definition of an appropriate disturbance regime, stating intensity, severity and frequency of treatments and taking into account livestock, wildlife and timber production requirements. Success in "ecosystem restoration" for production of good and services as well as for conservation is not clearly established neither in the Chaco region nor worldwide (Zedler 2007). However, silvopastoral systems seem to be an acceptable option for the

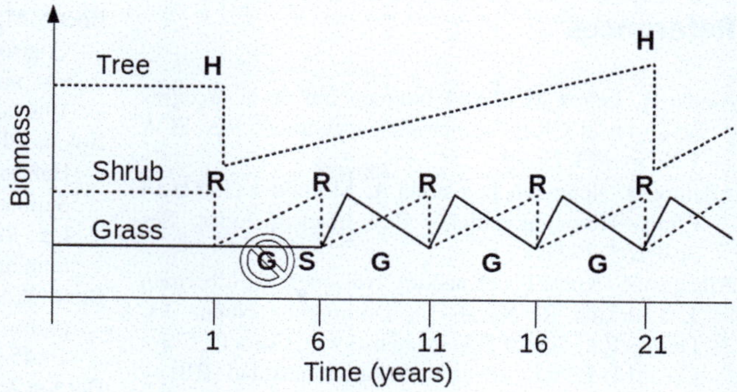


**Fig. 4.13** Volume dynamics ($m^3$*$plant^{-1}$) of shrub species (6 plots, 6 plants per plot) in RBi experiment from 2006 to 2011. References: RBI 1: roller chopper passed once, RBI 2: roller chopper passed twice

management of the current native vegetation of the western Chaco, where native woody plant communities (secondary forests and shrub thickets) dominate, implying that the removal of some individuals should be performed.

Results shown here suggest that the theoretical development of such as system should be based on disturbance theory, but applied in a context of production-conservation rather than exclusively for the latter. An appropriate disturbance regime should be proposed (Fig. 4.13). Intensity and severity of the initial and maintenance disturbance should be of medium magnitude. Although machinery has to be used, research shown here suggest that the initial and later interventions should not be severe or extremely severe and be carefully planned. Training of operators as well as managers is often key to success. Global positioning of treatment areas based on resource inventory may aid in the accuracy of RBI implementation in the future.

Since the development of a silvopastoral system means that an appropriate disturbance regime should defined, a key input for implementing the system in the field is a detailed ecosystem inventory, so that intensity, severity and frequency are specifically tailored to meet spatial demands.

**Acknowledgements** *Funding for RBI research was provided by the following research projects*:

INTA, Proyecto Integrado 1104081 *Manejo sustentable de sistemas de bosque nativo*, 2013–2016.

INTA, Regional Project *Intensificación de la Producción de Carne Bovina en el NOA*, 1996–2005.

Universidad Nacional de Santiago del Estero, Secretaría de Ciencia y Técnica, Proyecto Interinstitucional con Transferencia *Modificación de ecosistemas degradados para uso ganadero y caprino*, 1998–1999.

Project 1126074 *Manejo de la vegetación natural para fines ganaderos* 2013–2016.

INTA, Centro Regional Tucumán-Santiago del Estero PT 420108 *Relevamiento y evaluación de la receptividad ganadera bovina de la vegetación natural de Santiago del Estero*, 1991–1996.

INTA, Centro Regional Tucumán-Santiago del Estero, Proyecto Ganadero Regional *Intensificación de la Producción de Carne Bovina en el NOA*, 1996–2005.

INTA, Programa Carne, Proyecto Específico PNCAR 1502 *Incremento de la productividad primaria de pastizales*, 2003–2009.

INTA, Programa Carne, Proyecto Específico PNCAR 1503 Interacción de los Pastizales con la Producción Animal, 2003–2009.

INTA, Programa Forestal, Proyecto Específico, PNFOR 4322: Instalación y Manejo del Componente Forestal en Sistemas Silvopastoriles, 2003–2009.

INTA, Programa Forestal, Proyecto Específico, PNFOR 4321 *Interacciones Ecológicas en Sistemas Silvopastoriles*, 2003–2009.

INTA – Unión Europea – Instituto Superior de Agronomía (Portugal). Proyecto Fire-Paradox 018505 'An innovative approach of integrated wildfire management problem by regulating the wise use of fire: solving the fire paradox'. European Union, 6th Framework.

Gobierno de la Pcia de Santiago del Estero, Secretaría de Ciencia y Técnica: Proyecto de Investigación: *Compatibilización de objetivos ganaderos y forestales en la recuperación de áreas con vegetación degradada*. 2003–2007.

Secretaría de Agricultura, Pesca, Ganadería y Alimentación de la Nación, Proyecto PROINDER *Adaptación y prueba de herramientas para recuperación de oferta de forraje en ecosistemas degradados de Sgo del Estero* 2002–2004.

## References

Acuña R, Juárez M (2001) Intercepción de la precipitación en el Chaco semiárido (Argentina). Rev FCA UNCuyo 23:75–79

Adámoli J, Neumann R, Colina A, Morello J (1972) El Chaco aluvional salteño. Rev Investig Agrop Ser 3(9):165–237

Albanesi A, Kunst C (ex aequo), Anriquez A, Silberman J, Ledesma R, Navall M, Dominguez Nuñez JA, Duffau RA, Suarez RA, Werenistky D, Coria D, Coria O (2013) Rolado selectivo de baja intensidad (RBI) y sistemas silvopastoriles de la región chaqueña. In: Albanesi A, Paz R, Sobrero MT, Helman S, Rodriguez S (eds) Hacia la Construcción del Desarrollo Agropecuario y Agroindustrial de la FAyA al NOA. Ediciones Magna Publicaciones, Tucumán

Albanesi A, Anríquez A, Silberman J (2014) Microbiología agrícola. Un aporte de la Investigación en la Argentina. Ed. Magna, Santiago del Estero

Alessandria E, Karlin U, Casermeiro J, Ferreyra R (1987) Repuesta de algunos caracteres estructurales de la vegetación resultante de un rolado ante diferentes presiones de pastoreo. Memoria 1eras Jornadas Nacionales de Zonas Áridas y Semiáridas, pp 339–341

Anriquez A, Albanesi A, Kunst C, Ledesma R, López C, Rodriguez Torresi A, Godoy J (2005) Rolado de fachinales y calidad de suelos en el Chaco occidental, Argentina. Cienc del Suelo 23:145–157

Ansley J, Rasmussen G (2005) Managing native invasive juniper species using fire. Weed Technol 19:517–522

Ansley J, Wiedemann H (2008) Reversing the woodland steady state: vegetation responses during restoration of *Juniperus*-dominated grasslands with chaining and fire. In: Van Auken OW (ed) Western North American Juniperus communities: a dynamic vegetation type. Springer, Berlin, pp 272–290

Ansley J, Wiedemann H, Castellano M, Slosser J (2006) Herbaceous restoration of Juniper dominated grasslands with chaining and fire. Rangel Ecol Manag 59:171–178

Araujo PA (2003) Bases para la gestión sostenible de bosques en regeneración del chaco semiárido. Tesis doctoral de Madrid, Spain

BC-British Columbia. Ministry of Forests. Forest Practices Branch (2003) Silvicultural systems handbook for British Columbia. British Columbia Ministry of Forests, Forest Practices Branch, Victoria. http://www.for.gov.bc.ca/HFP/Pubssilvsystems.htm

Belmonte Serrato F, Romero Díaz A (1998) A simple technique for measuring rainfall interception by small shrub: 'interception flow collection box'. Hydrol Process 12:471–481

Belsky A, Mwonga S, Duxbury J (1993) Effects of widely spaced trees and livestock grazing on understory environments in tropical savannas. Agrofor Syst 24:1–20

Binkley D, Giardina C (1998) Why do tree species affect soils? The warp and woof of tree-soil interactions. Biogeochemistry 42:89–106

Boetto M, Alessandría E, Maccagno P, Díaz M (1999) Control químico de *Acacia furcaptispina* Burkart en un sistema ganadero. Rev Argent Prod Anim 19:411–423

Boletta PE (1988) Clima. In: Casas RR (ed) Desmonte y Habilitación de Tierras en la Región Chaqueña Semiárida. Red de Cooperación Técnica en el uso de los Recursos Naturales en la Región Chaqueña Semiárida. FAO, Santiago, pp 7–21

Bond W, Midgley J (2001) Ecology of sprouting in woody plants: the persistence niche. Trends Ecol Evol 16:45–51

Bordón A (1983) Comentarios e ideogramas sobre la vegetación de la Provincia del Chaco emergentes de una muestra de descripciones de vegetación en relación a series de suelo. Bol. No 86. INTA EERA, Saénz Peña

Bray J, Curtis J (1957) An ordination of the upland forest communities of southern Wisconsin. Ecol Monogr 27:325–349

Bravo S (2008) Manejo de rolados: caracteres estructurales de leñosas que influencian la producción de rebrotes frente tratamientos mecánicos. In: Kunst C, Ledesma R, Navall M (eds) Rolado Selectivo de Baja Intensidad. Ediciones INTA, Santiago del Estero, pp 126–131

Bravo S (2010) Anatomical changes induced by fire-damaged cambium in two native woody species of argentine Chaco region. IAWA J 31(3):283–292

Bravo S, Kunst C, Giménez A, Moglia G (2001) Fire regime of *Elionurus muticus* Spreng. savanna, Western Chaco region, Argentina. Int J Wildland Fire 10:65–72

Bravo S, Kunst C, Basualdo M, Leiva M, Ledesma R (2011) Assessment of the aerial bud bank of native tree and shrub species of the Chaco region, Argentina. International Rangeland Congress, Rosario

Bravo S, Kunst C, Leiva M, Ledesma R (2012) Espesor de corteza y tolerancia la fuego en especies nativas del Chaco en estadio juvenil, Proceedings of the 2nd National Congress of Silvopastoral Systems. INTA Editions, Santiago del Estero

Brudvig L (2010) Woody encroachment removal from midwestern oak savannas alters understory diversity across space and time. Restor Ecol 18:74–84

Bucher E (1982) Chaco and Caatinga. South American arid savannas, woodlands and thickets. In: Huntley B, Walker B (eds) Ecology of tropical savannas. Ecological studies, vol 42. Springer, Berlín, pp 48–79

Burrows W (2002) Seeing the wood(land) for the trees – an individual perspective of Queensland woodland studies (1965–2005). Trop Grassl 36:202–217

Caamaño J (1955) Noticia del Gran Chaco (1778). In: Furlong G (ed) Escritores Coloniales Rioplatenses VIII. Librería del Plata S.R.L., Buenos Aires, 186 pp

Cabral A, De Miguel J, Rescai A, Schmitz M, Pineda F (2003) Shrub encroachment in Argentinean savannas. J Veg Sci 14:145–152

Casas R, Irurtia C, Michelena R (1978) Desmonte y habilitación de tierras para la producción agropecuaria en la República Argentina. INTA-CIRN, 114 p

Cassino W, Navall M, Carignano P, D'Angelo P (2013) Agregado de valor a la leña y su efecto sobre la rentabilidad. IUFROLAT – 3er Congreso Forestal Latinoamericano de IUFRO. San José de Costa Rica

Coria R, Coria O, Kunst C (2012) Rolado de baja intensidad (RBI) y aves de un bosque del Chaco semiárido. Actas 2do Congreso Nacional de Sistemas Silvopastoriles, 371–376. ISBN 978-987-679-123-6

de Moussy M (1873) Description geographique et statistique de la Confederation Argentine, 2nd edn. Paris Librairie de Fermin Didot Freres, Atlas

Díaz M (2007) Utilización de pastizales naturales. Encuentro Grupo Editor, Córdoba, Argentina, 456 p

Durr F, Rangel J (2000) The response of *Panicum maximum* to a simulated subcanopy environment. I. Soil x shade interaction. Trop Grassl 34:110–177

Durr F, Rangel J (2002) Enhanced forage production under *Samanea saman* in a subhumid tropical grassland. Agrofor Syst 54:99–102

Frenguelli J (1940) Rasgos Principales de Fitogeografía Argentina, vol 2, Publicacion periódica No 2. Museo de La Plata, La Plata

Fulbright T (2004) Disturbance effects on species richness of herbaceous plants in a semiarid habitat. J Arid Environ 58:119–133

Fumagalli A, Kunst C, Pérez H (1997) Intensificación de la producción de carne en el NOA. In: Carrozzino J, Rearte D (eds) Memoria del 1er. Congreso Nacional de Producción Intensiva de Carne. INTA – Forum Argentino de Forrajes, Buenos Aires, pp 53–61, 261 p

Galera M (1990) Dinámica y manejo de bosques xerofíticos del chaco árido y comunidades sustitutivas posttala, rolado, con pastoreo en el Noroeste de la Provincia de Córdoba, Argentina. Memoria Taller interregional Africa/America Latina, MAB. Unesco, Chile

Gifford R, Howden M (2001) Vegetation thickening in an ecological perspective: significance to national greenhouse gas inventories. Environ Sci Policy 4:59–72

Gómez A, Navall M (2008) Manejo de rolados: efectos del rolado sobre la estructura del bosque. Implicancias para el manejo forestal. In: Kunst C, Ledesma R, Navall M (eds) Rolado Selectivo de Baja Intensidad. Ediciones INTA, Santiago del Estero, pp 117–125

Gómez A, Kunst C, Navall M (2009) Sistemas silvopastoriles en el Chaco semiárido II: Efecto del rolado sobre la regeneración natural de dos especies arbóreas nativas, Proceedings of the 1st National Congress of Silvopastoral Systems. INTA Editions, Misiones

Haworth K, McPherson G (1994) Effects of *Quercus emoryi* on herbaceous vegetation in a semiarid savanna. Vegetatio 112:153–159

Heitschmidt R, Pieper R (1983) Integrated brush control and grazing management. In: Mc Daniel K (ed) Proceedings brush management symposium. Society for Range Management, Albuquerque, pp 89–95

Herrera P, Torella S, Adámoli J (2003) Los incendios forestales como modeladores del paisaje en la región chaqueña. In: Kunst C, Bravo S, Panigatti J (eds) Fuego en los Ecosistemas Argentinos. Ediciones INTA, Santiago del Estero, pp 145–156

INTA-SAGyP. Instituto Nacional de Tecnología Agropecuaria-Secretaría de Agricultura, Ganadería y Pesca de La Nación (1990) Atlas de suelos de la República Argentina, Buenos Aires, 500 p

Kumada S, Kawanishi T, Hayashi Y, Ogomori K, Kobayashi Y, Takahashi N, Saito M, Hamano H, Kojima T, Yamada K (2008) Litter carbon dynamics analysis in forests in an arid ecosystem with a model incorporating the physical removal of litter. Ecol Model 215:190–199

Kunst C, Cornacchione M, Gelid L, Godoy J (2001a) Ganadería en el este santiagueño: Aumento de la oferta de forraje. Manejo desmonte y posdesmonte. Persistencia de pasturas subtropicales. INTA-UNSE, Gobierno de la Provincia de Santiago del Estero, 28 pp

Kunst C, Cornacchione M, Godoy J (2001b) Prácticas de control del fachinal I: Efectos sobre la oferta de forraje de herbáceas. Proceedings 1er Congreso Nacional sobre Manejo de Pasturas Naturales. Asociación Manejo de Pastizales Naturales, San Cristóbal

Kunst C, Ledesma R, Basan M, Angella G, Prieto D, Godoy J (2003) Rolado de fachinales e infiltración de agua en el suelo en el Chaco occidental argentino. Revista de Investigaciones Agropecuarias 32:105–122

Kunst C, Monti E, Perez H, Godoy J (2006) Assessment of rangelands of southwestern Santiago del Estero for management and research. J Environ Manag 80:248–265

Kunst C, Ledesma R, Navall (2008) Rolado Selectivo de Baja intensidad. INTA EEA, Santiago del Estero, 140 pp

Kunst C, Ledesma R, Bravo S, Godoy J, Navarrete V (2009) Sistemas silvopastoriles en el Chaco Semiárido III: efecto del rolado sobre la estructura de especies arbustivas nativas, Proceedings of the 1st National Congress of Silvopastoral Systems. INTA Editions, Misiones, pp 1–7

Kunst C, Ledesma R, Bravo S, Albanesi A, Anriquez A, van Meer H, Godoy J (2012a) Disrupting woody steady states in the Chaco region (Argentina): responses to combined disturbance treatments. Ecol Eng 42:42–53

Kunst C, Ledesma R, Bravo S, Albanesi A, Godoy J (2012b) Rolados y diversidad botánica I: ¿sitio ecológico, tiempo o perturbación? Proceedings of the 2nd National Congress of Silvopastoral Systems. INTA Editions, Santiago del Estero, pp 377–382

Kunst C, Ledesma R, Bravo S, Albanesi A, Godoy J (2012c) Rolados y diversidad botánica II: efecto en el sitio ecológico alto, Proceedings of the 2nd National Congress of Silvopastoral Systems. INTA Editions, Santiago del Estero, pp 388–393

Kunst C, Ledesma R, Godoy J, Navarrete V (2012d) Dinámica del volumen de arbustivas en rolados de baja intensidad, Proceedings of the 2nd National Congress of Silvopastoral Systems. INTA Editions, Santiago del Estero, pp 76–80

Kunst C, Ledesma R, Godoy J (2012e) Acumulación de biomasa aérea de *Panicum máximum* cv gatton panic en rolados, Proceedings of the 2nd National Congress of Silvopastoral Systems. INTA Editions, Santiago del Estero, pp 61–65

Kunst C, Ledesma R, Castañares M, Cornacchione M, van Meer H, Godoy J (2014) Yield and growth features of *Panicum maximum* (Jacq.) var Trichoglumecv Petrie (Green Panic) under woody cover, Chaco region, Argentina. Agrofor Syst 88:157–171

Le Brocque A, Goodhew K, Zammit C (2009) Overstorey tree density and understorey regrowth effects on plant composition, stand structure and floristic richness in grazed temperate woodlands in eastern Australia. Agric Ecosyst Environ 129:17–27

Ledesma R (2006) Desarrollo de sistemas ganaderos: una alternativa de manejo en ecosistemas degradados del Chaco semiárido. Universidad Nacional de Santiago del Estero, Santiago del Estero. Tesis de Maestría, 134 pp

Ludwig F, De Kroon H, Prins H (2008) Impacts of savanna trees on forage quality for a large African herbivore. Oecologia 155:487–496

Lupatini M, Suleiman A, Jacques R, Antoniolli Z, Kuramae E, de Oliveira Camargo F, WürdigRoesch L (2013) Soil-borne bacterial structure and diversity does not reflect community activity in Pampa biome. PLoS ONE 8(10):e76465

Mastrangelo M, Gavin M (2012) Trade-offs between cattle production and bird conservation in an agricultural frontier of the gran Chaco of Argentina. Conserv Biol 26:1040–1051

Mc Cune B, Grace J (2002) Analysis of ecological communities. MJM Software Design, Oregon

McIver J, Starr L (2001) Restoration of degraded lands in the interior Columbia River basin: passive vs. active approaches. For Ecol Manag 153:15–28

McIvor J (2006) Pasture management in semiarid tropical woodlands: effects on tree regrowth. Trop Grassl 40:14–23

Métraux A (1946) Ethnology of the Chaco. In: Steward J (ed) Handbook of South American Indians. Smithsonian Institution, Bull. The marginal tribes, US Government Printing Office 143(1), pp 197–379

Morello J (1968) Las grandes unidades de vegetación y ambiente del Chaco argentino. INTA Serie fitogeográfica No 10, Buenos Aires

Morello J, Adámoli J (1974) Las grandes unidades de vegetación y ambiente del Chaco argentino. 2da. parte: Vegetación y ambientes de la Pcia. del Chaco. INTA Serie Fitogeográfica No 13, Buenos Aires

Morello J, Saravia Toledo C (1959) El bosque chaqueño I y II. Rev Agron Noroeste Argent 3:209–258

Morello J, Matteucci S, Rodriguez A, Silva M (2012) Ecorregiones y complejos ecosistémicos argentinos. Orientación Gráfica Editora, Buenos Aires, 752 pp

Nai Bregaglio M, Karlin U, Coirini R (1999) Repuesta de especies arbustivas forrajeras a tratamientos de desmonte selectivo en el Chaco árido, Argentina. Multequina 8:101–109

Navall M (2008) Rolados y Manejo Forestal. In: Kunst C, Ledesma R, Navall M (eds) Rolado Selectivo de Baja Intensidad. Ediciones INTA, Santiago del Estero, pp 71–85

Navall M (2011) Silvopastoral management in the semiarid Chaco: low impact roller-chopping and forest harvesting combined effects on tree basal area growth. IX International Rangeland Congress, Rosario

Navall M (2012a) Aporte de hojarasca en un quebrachal semiárido santiagueño bajo manejo silvopastoril, Proceedings of the 2nd National Congress of Silvopastoral Systems. INTA Editions, Santiago del Estero

Navall M (2012b) Efectos del rolado y la corta sobre el crecimiento de un quebrachal semiárido santiagueño, Proceedings of the 2nd National Congress of Silvopastoral Systems. INTA Editions, Santiago del Estero

Navall M, Cassino W (2009) Sistemas silvopastoriles en el Chaco Semiárido IV: efecto del rolado sobre la eficiencia de tareas de manejo forestal. Resumen. "1er Congreso Nacional de Sistemas Silvopastoriles"

Naval M, Cassino W, Carignano P, D'Angelo P (2013a) Un nuevo método de marcación y control de cortas en bosques irregulares. IUFROLAT – 3er Congreso Forestal Latinoamericano de IUFRO. San José de Costa Rica

Navall M, Cassino W, Carignano P, D'Angelo P (2013b) SilvoINTA: una aplicación móvil para asistir la silvicultura de bosques irregulares. Proceedings IV Congreso Forestal Argentino y Latinoamericano. Puerto Iguazú

Navall M, Cassino W, Carignano P, D'Angelo P (2013c) Leña en blanco: una alternativa para el uso sustentable de bosques nativos. IUFROLAT – 3er Congreso Forestal Latinoamericano de IUFRO. San José de Costa Rica

Naveh Z (1990) Fire in the Mediterranean – a landscape ecological perspective. In: Goldammer J, Jenkins M (eds) Fire in ecosystem dynamics. SPB Academic Publishing, The Hague, pp 1–20

Naveh Z (2004) Multifunctional, self organizing biosphere landscapes and the future of our total human ecosystem. World Futur 60:469–503

Ng E, Patti A, Rose M, Schefe C, Wilkinson K, Smernik R, Cavagnaro T (2014) Does the chemical nature of soil carbon drive the structure and functioning of soil microbial communities? Soil Biol Biochem 70:54–61

Noble J, Muller W, MacLeod N, Boduluvic Z, Jones P, Wood J (2005) Integrated shrub management in semiarid woodlands of eastern Australia: ground and aerial application of defoliant to shrubs regenerating after disturbance. Rangel J 27:117–134

Nolte K, Gabor T, Hehman M, Asleson M, Fulbright T, Rutledge J (1994) Long term effects of brush management on vegetation diversity in ephemeral drainages. J Range Manag 47:457–459

Nyakatawa EZ, Mays D, Naka K, Bukenya J (2012) Carbon, nitrogen, and phosphorus dynamics in a loblolly pine-goat silvopasture system in the Southeast USA. Agrofor Syst 86:129–140

Obispo N, Espinoza Y, Gil J, Ovalles F, Rodriguez M (2008) Efecto del sombreado sobre la producción y calidad del pasto guinea (*Panicum maximum*) en un sistema silvopastoril. Zootec Trop 26:285–288

Obst K, Ortíz G, Soto G (2007) Manual para el manejo sustentable de los bosques nativos de la provincia del Chaco. UNIQUE Forest Consultants

Ojasti J (2000) Manejo de fauna silvestre neotropical. SI-MAB, Washington, DC

Ong K, Leakey R (1999) Why tree-crop interactions in agroforestry appear at odds with tree-grass interactions in tropical savannahs. Agrofor Syst 45:109–129

Passera C, Borsetto O, Allegreti L (1996) Short term effects of shrub control on two different plant communities in Argentina. J Arid Environ 34:415–420

Provincia de Santiago del Estero (2007). Ley 6841: Conservación y Uso Múltiple de las Áreas Forestales de la Provincia de Santiago del Estero. Santiago del Estero

Roberts M (2007) A conceptual model to characterize disturbance severity in forest harvests. For Ecol Manag 242:58–64

Rodríguez Rey J, Rovati A, Guzmán L, Juárez V (1983) Recuperación de pasturas naturales de monte mediante el uso de herbicidas hormonales y clausura. Pub. Misc. 75. Est. Agro-Industrial Obispo Colombres, Tucumán

Roldan Bernhard S (2012) Los sistemas silvopastoriles desde una perspectiva energética, Proceedings of the 2nd National Congress of Silvopastoral Systems. INTA Editions, Santiago del Estero, p 488

Ruiz Jaen M, Mitchell Aide T (2005) Vegetation structure, species diversity and ecosystem processes as measures of restoration success. For Ecol Manag 218:159–173

Sanzano GA, Corbella RD, García JR, Fadda GS (2005) Degradación física y química de un Haplustol típico bajo distintos sistemas de manejo de suelo. Cienc del Suelo 23:93–100

Seidl R, Fernandez P, Fonseca TF, Gillet F, Jönssong AM, Merganicová K, Netherer S, Arpaci A, Bontemps J-D, Bugmannl H, González-Olabarria JR, Laschn P, Meredieuo C, Moreira F, Schelhaas M-J, Mohren F (2011) Modelling natural disturbances in forest ecosystems: a review. Ecol Model 222:903–924

SER. Society for Ecological Restoration (2004) International primer on ecological restoration. In: http://www.ser.org/content/ecological_restoration_primer.asp

Smit G (2004) An approach to tree thinning to structure southern African savannas for long term restoration from bush encroachment. J Environ Manage 71:179–191

Sheil D, Burslem D (2003) Disturbing hypothesis in tropical forests. Trends Ecol Evol 18:18–26

Smit G (2005) Tree thinning as an option to increase herbaceous yield of an encroached semi-arid savanna in South Africa. BMC Ecol 5:1–15

Taddese G, Mohamed Saleem M, Abyie A, Wagnew A (2002) Impact of grazing on plant species richness, plant biomass, plant attributes, and soil physical and hydrological properties of Vertisol in East African highlands. J Environ Manag 29:279–289

Teague W, Grant U, Kreuter H, Diaz-Solis S, Dube M, Kothmann M, Pinchak W, Ansley R (2008) An ecological economic simulation model for assessing fire and grazing management effects on mesquite rangelands in Texas. Ecol Econ 64:611–624

Van Auken O (2009) Causes and consequences of woody plant encroachment into western North American grasslands. J Environ Manag 90:2931–2942

Wainwright J, Parsons A, Abrahams A (1999) Rainfall energy under creosotebush. J Arid Environ 43:111–120

Williams D, Baruch Z (2000) African grass invasion in the Americas: ecosystem consequences and the role of ecophysiology. Biol Invasions 2:123–140

Zedler J (2007) Success: an unclear, subjective descriptor of restoration outcomes. Ecol Restor 25:162–168

# Silvopastoral Systems Based on Natural Grassland and Ponderosa Pine in Northwestern Patagonia, Argentina

5

G. Caballé, M.E. Fernández, J. Gyenge, V. Lantschner, V. Rusch, F. Letourneau, and L. Borrelli

**Abstract**

This chapter summarizes the results of research conducted over the past 15 years and provides management guidelines for the development of silvopastoral systems based on ponderosa pine plantations established on natural grasslands in northwestern Patagonia, Argentina. The goals of the studies were to generate knowledge for improving environmental and production efficiency in that region. Ecological interactions among the different life forms which make up a silvopastoral system (grasses, trees and animals) were studied to determine what management system would optimize environmental resource sharing. In addition, aspects affecting environmental sustainability of the systems were analyzed, such as resource use (mainly water use) by silvopastoral components and biodiversity patterns. The results suggest that applying silvicultural practices leading to a tree canopy cover level equal to or below 50 % is compatible with proper production of natural grassland species, particularly the palatable *Festuca pallescens*. Animal rearing should include rotation between different areas in this type of system, where partial shade may limit grass re-growth compared to open grasslands. However, microclimatic benefits of the trees on the animals may be particularly significant, and deserve specific future research. Currently available information indicates that the

G. Caballé (✉) • V. Rusch
Grupo de Ecología Forestal, EEA Bariloche INTA cc 277, 8400 San Carlos de Bariloche, Río Negro, Argentina
e-mail: caballe.gonzalo@inta.gov.ar

M.E. Fernández • J. Gyenge
Consejo Nacional de Investigaciones Científicas y Técnicas, CONICET, AER Tandil, EEA Balcarce INTA, Buenos Aires, Argentina

V. Lantschner
Grupo de Ecología de Poblaciones de Insectos, Consejo Nacional de Investigaciones Científicas y Técnicas, CONICET, EEA Bariloche INTA, Bariloche, Argentina

F. Letourneau
Grupo de Silvicultura, EEA Bariloche INTA, Bariloche, Argentina

L. Borrelli
Laboratorio de Microhistología, EEA Bariloche INTA, Bariloche, Argentina



P.L. Peri et al. (eds.), *Silvopastoral Systems in Southern South America*, Advances in Agroforestry 11, DOI 10.1007/978-3-319-24109-8_5

studied silvopastoral systems constitute a biologically and environmentally sustainable activity in the fragile ecosystem of semiarid Patagonia.

**Keywords**
Festuca pallescens • Ecological interaction • Resource use by silvopastoral components

## 5.1 Introduction

Land-use activities in northwestern (NW) Patagonia, Argentina, South America, should aim – as everywhere else – at improving environmental and production efficiency. Traditional practices such as extensive livestock raising, based on continuous or rotational grazing with limited stocking management in the region, have resulted, however, in the loss of vegetation cover, changes in species abundance leading to reduced number of forage species, and an increased proportion of bare soil. This has favored erosion processes leading to losses of soil and organic matter (León and Aguiar 1985; Soriano and Movia 1986; Paruelo et al. 1993). The same husbandry practice in areas of native forest or forest-steppe ecotone threatens the natural regeneration of shrub and tree species as a result of sheep, goat or cattle browsing (Veblen and Lorenz 1988; Peri et al. 2006). The logging of native forests, another major production activity during modern human settlement in NW Patagonia at the beginning of twentieth century, has declined gradually due to the creation of National Parks, the pressure of a society concerned about environmental conservation, and the expansion of "green" tourism activity seeking pristine forests and landscapes (Laclau 1997). Coupled with population growth in recent decades, this has caused an increase in the demand for timber and timber products in the region, which cannot be satisfied with native forest products. Faced with this reality, national and provincial governments have promoted planting of fast-growing exotic forest species in Patagonia and other regions of Argentina. The aim has been to meet increasing timber demand from a fast-growing population, while diversifying regional production systems by using grassland or shrubland sites with different levels of degradation.

Despite these national and provincial efforts to promote forestry activity over the past four decades, however, only 82,000 ha have been afforested out of 800,000 ha of land that has relatively high forestry potential available in Patagonia (Laclau and Andenmatten 2005; Loguercio and Deccechis 2006). Thus, although forestry has been shown to be a profitable and competitive activity compared to extensive livestock raising (Laclau et al. 2002; Laclau and Andenmatten 2005), only 10 % of the regional potential has been achieved, which is far from the original goal. The causes underlying this situation are varied and complex. Usually, the blame is assigned to the prevailing cattle culture in the region. Another important factor is the lack of a sustained policy framework, with strong fluctuations in promotion systems, and the inherent long term rotations (about 35 years or more) of the forestry activity in this cold temperate region (Danklmaier 2004).

The improvement of environmental and production efficiency of Patagonian land-use activities is critical, and one option is the use of silvopastoral systems (SPS) based on plantations established on natural grassland in the forest-steppe ecotone. One potential advantage of SPS is that traditional extensive livestock raising compatible with the "new" forestry activity at the farm level. Simultaneously, this diversification brings quick returns for the forestry investment as it generates annual revenues from animal products. The low planting density typical of these systems reduces the risk of pest damage by *Sirex noctilio* or other insects (Villacide and Corley 2012), or from fires. Additionally, this agroforestry system has clear environmental advantages over high-density timber plantations. In terms of biodiversity, it increases the richness and diversity of flora and avifauna (Rusch et al. 2004; Lantschner et al. 2008). Regarding the use of water, a critical environmental resource for primary production in northern Patagonia, SPS consume less water than high-density plantations, with water use being similar to natural vegetation systems despite the higher biomass productivities (Gyenge et al. 2002, 2003; Licata et al. 2008).

The area where SPS could be developed corresponds to the Sub-Andean and Occidental districts (38° to 46° 30′S) of the Patagonian phyto-geographical region with altitudes between 300 and 1800 m.a.s.l. (Leon et al. 1998). This area includes the eastern foothills of the Andes Mountains, extra Andean Sierras, basaltic plateaus, alluvial and glacier valleys, and hills. As mentioned above, the climate is cold temperate, humid towards the Andes, with more than 1000 mm annual precipitation, and sub-humid at the eastern extreme (400 mm isohyet). Rainfall is concentrated in the autumn and winter, leading to dry summers with less than 150 mm precipitation during October to April (Paruelo et al. 1998). Natural meadows are dominated by three of the nine vegetated units defined by Paruelo et al. (2004): Grass Steppe and Grass Shrub Steppe in the periphery of relatively high topographic positions (NPP: 300–800 kg DM $ha^{-1}$ $year^{-1}$; CC: 0.5–1.5 PSUE $ha^{-1}$ $year^{-1}$), and Prairie in lower central areas (NPP: 1500–6000 kg DM $ha^{-1}$ $year^{-1}$, CC: 2.8–11.3 PSUE $ha^{-1}$ $year^{-1}$) [NPP: Net primary productivity; CC: Carrying capacity; PSUE: Patagonian sheep unit equivalent, the requirements of 530 kg DM $year^{-1}$ for 1 Corriedale ewe of 49 kg of live weight.]

Grass Steppes and Grass Shrub Steppes are dominated by perennial C3 tussock grasses, *Festuca pallescens* and *Pappostipa speciosa*, and the spaces among tussocks are dominated by exotic herbs such as *Taraxacum officinale*, native graminoids such as *Juncus balticus* and *Carex gayana*, and C3 grasses, especially the exotic *Poa pratensis*. The major shrub components are the native *Nassauvia* sp. and *Berberis* sp. Prairies are dominated by the same species found among tussocks described above, plus *Phleum pratense* and *Holcus lanatus*, two C3 exotic grasses and the introduced legume *Trifolium repens*.

Ponderosa pine (*Pinus ponderosa*) has been and is still the fastest growing conifer species most commonly used in afforestation in NW Patagonia. Developing SPS based on this species may be feasible if the net balance of biological interactions of competition and facilitation with natural grassland species is neutral or positive. Competition occurs when the component species of a system overlap in the use of growth resources to the point that the growth, survival and/or reproductive success of one or more of them are adversely affected (Harper 1990). Facilitation occurs when one species of the system changes the biophysical environment, creating one or more conditions favorable to the development of the remaining species (Holmgren et al. 1997). The net result of the interactions will also depend on the life stage of the species involved, their physiology and the intensity of the stress caused by abiotic factors (Callaway and Walker 1997).

Using the conceptual framework of ecological interactions that occur between system components, this chapter reviews research results and provides management guidelines for SPS based on ponderosa pine plantations established on natural grasslands of NW Patagonia. Suggested management guidelines are based on ecophysiological processes evaluated at individual plant and stand levels. On a larger scale, environmental sustainability advantages of this type of production system are compared with high-density plantations.

## 5.2 Understorey Component of Silvopastoral Systems in NW Patagonia

### 5.2.1 Growth of Patagonian Grasses under Pines

Silvopastoral systems may have greater yields compared to monocultures because of higher resource capture and/or facilitation effects of the trees on the pastures growing beneath or between them (Huang and Xu 1999). Many studies have reported that, as trees grow, pasture production decreases (Kellas et al. 1995; Ong and Leakey 1999) due to high competition for environmental resources such as water and radiation. However, other results have shown that understorey productivity can increase under trees under certain circumstances (e.g. Belsky 1994; Holmgren et al. 1997).

Trees may improve the water storage capacity of soils (Joffre and Rambal 1988) and nutrient availability (Belsky 1994). Moreover, by reducing radiation levels beneath them, trees decrease the evaporative demand for understorey species (Breshears et al. 1997; Holmgren et al. 1997). Trees also buffer extreme temperatures in winter and summer (e.g. Garnier and Roy 1988). The net result of these effects on understorey productivity will depend on species characteristics (Belsky 1994; Pugnaire et al. 2011), and the intensity of adverse environmental conditions (Callaway and Walker 1997; Kikvidze et al. 2012; Pugnaire et al. 2011).

In many temperate regions, radiation is the major limiting resource (Kho 2000). In these regions, opportunities for yield benefits by agroforestry technologies may exist if other factors, such as nutrient and water availability, are more limiting for understorey production than radiation is (Kho 2000). In the same way, in a given environment, water may be a more limiting resource for one species than for another. Thus, species more vulnerable to soil water deficits might be more able to take advantage of improvement in water conditions under trees than drought-tolerant species are.

Grasslands in NW Patagonia are commonly dominated by tussock grasses of the genera *Poa*, *Pappostipa* and *Festuca*. Some of them co-exist at several sites, but due to their different stress tolerance, sometimes they occupy different niches within an ecosystem. *Pappostipa speciosa* is a heliophilous drought-resistant species (Nicora 1978), whereas *Festuca pallescens* is more drought-sensitive (Nicora 1978; Fernández 2003). Thus, it is expected that these two species may respond differentially to the presence of tree canopy cover.

Studies carried out at the plant level, i.e., measuring periodically the number of tillers and the amount and length of green leaves, showed that, as hypothesized based on Kho's (2000) model of relative limiting resources, the relative growth of *P. speciosa* decreased rapidly and linearly with increasing tree canopy cover (Fernández et al. 2002), whereas the growth of *F. pallescens* is maintained until tree canopy cover level is around 60–70 %, and then decreases significantly (Fernández et al. 2006a; Caballé 2013). At those canopy cover levels, the radiation that reaches the understorey throughout the day during the growing season was around 45–55 % (Fig. 5.1) rela-

**Fig. 5.1** Silvopastoral system based on natural grassland and ponderosa pine with 350 trees $ha^{-1}$ at 21 years old. The radiation that reaches the understorey throughout the day during the growing season (October to April) was around 45–55 % relative to open grassland

**Fig. 5.2** Silvopastoral system based on natural grassland and ponderosa pine with 500 trees $ha^{-1}$ at 21 years old. The radiation that reached the understorey throughout the day during the growing season (October to April) was ≤30 % relative to open grassland

tive to open grassland, whereas it decreased to 30 % (Fig. 5.2) or less in the more dense situation, where grass growth was significantly affected (Fernández et al. 2006a; Caballé 2013).

In *P. speciosa*, the most negatively affected growth variable under shade was tillering. Maximum tiller production – recorded in early autumn- was statistically similar (but lower on average) to open areas up to a canopy cover level of around 45 % (Fernández et al. 2002). However, net tillering was negative (i.e. more tillers die at each time than new tillers appear) or nil during most of the growing season under tree canopy cover levels between 25 and 60 % (Fernández et al. 2002). On the other hand, despite its lower impact on whole plant growth, the mean number of green leaves per tiller also decreases under shade conditions in *P. speciosa* as well as in *F. pallescens*, whereas tiller height and maximum leaf length tend to increase under moderate shade levels in both species (Fernández et al. 2002, 2006a) probably due to etiolation. Thus, on sites where the two species normally coexist, the reduced growth of *P. speciosa* plants under SPS, results in the species being replaced in favor of *F. pallescens* (Fernández et al. 2005).

Ecologically therefore, the net effects of trees on *P. speciosa* growth is negative (net competition) even when some positive effects on its water status have being detected (Fernández et al. 2002, see below). Radiation is the most limiting growth resource for this drought tolerant species, and therefore light competition by the trees dominates this particular tree-grass interactions.

In contrast, the net balance of tree interactions on growth of the fairly shade-tolerant *F. pallescens* may be nil or positive up to quite high tree canopy cover levels (Fig. 5.3). This is most

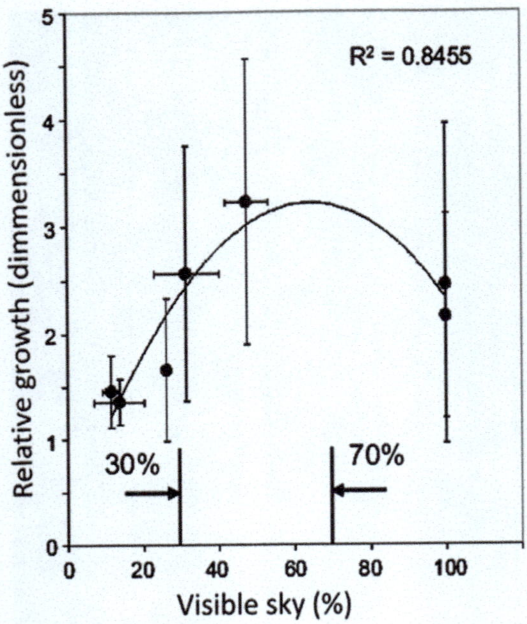


**Fig. 5.3** Relative growth (dimensionless value, mean and SE *vertical lines*) of *Festuca pallescens* as a function of the proportion of ponderosa pine canopy openness or visible sky (%, mean of a class in *horizontal lines*). The relative growth expresses the ratio between the growth at the peak production period and the initial value (beginning of the growing season, October). The values 30 and 70 % of visible sky indicate the range of the tree canopy cover where the growth of *F. pallescens* is reasonable (see text for references)

noticeable between tree crowns compared to beneath crown positions, where tree shade is quite high and competition for soil water between species is quite low (Fernández et al. 2006a, 2007; see next section). In addition, the positive net effects of trees on grasses are more pronounced in wet seasons or humid periods within the whole growing season (such as early spring) (Fernández et al. 2007; Caballé 2013).

These results suggest that *F. pallescens* could be a major species for the development of SPS with ponderosa pine in NW Patagonia. Further, this species often makes up over 20 % of most livestock diet and is widely distributed in the forest-steppe ecotone zone (Pelliza Sbriller et al. 1984; Bonino et al. 1986; Caballé et al. 2009; Bertiller and Defosse 1990; Somlo 1997).

### 5.2.2 Ecophysiology of Patagonian Grasses Growing under Ponderosa Pine

One of the main biological hypotheses for SPS is that trees ameliorate grass water-stress under their canopies due to reduced evaporative demand, and that this effect is higher than potential competition for soil water and reduced soil water content from increased rain interception. If this is true, even when soil water content may be lower under the tree canopy cover than in open grasslands, pre-dawn water potential should be higher (less negative) in grasses growing under or between trees than in the open areas: stomatal conductance should also be higher at times of high evaporative demand such as midday or early afternoon.

Pre-dawn water potential, a measure of plant water status, has been measured in grasses growing in a wide range of pine canopy cover and climatic conditions during several growing seasons. This parameter showed different trends depending on tree canopy cover level, position of the grass with respect to the tree, soil water availability and period during the growing season. However, even though there is less information available on *P. speciosa* than on *F. pallescens*, both species presented similar patterns of pre-dawn water status (Fernández 2003). Generally a neutral or a positive effect of trees on the pre-dawn water potential of grasses was detected, and when present, the positive effect increased with tree canopy cover. However, this positive effect was restricted to periods of relatively high soil water content (i.e. to early spring or in seasons with higher than the average rainfall) (Fernández et al. 2007; Caballé 2013). In dry seasons – compared to average climatic conditions – and during periods of prolonged low soil water content (below 13 % by vol) and high evaporative demand, pre-dawn water potential of grasses growing in SPS was lower than in the open grassland, reflecting a negative effect of the trees on grass water status (Fernández et al. 2007; Caballé 2013). These effects might be related to direct competition for limited soil water between trees and grasses (Fernández et al. 2008) with or without significant differences in soil water availability between SPS and grassland (Fernández et al. 2007; Caballé 2013). Studies with stable isotopes of H and O have indicated that *F. pallescens* tussocks extract >80 % of their water from the upper 20 cm of soil, despite them rooting to more than 100 cm, throughout the growing season, and P. ponderosa trees in SPS use a low, though significant fraction (about 20 % of its total water use) of water from this shallow soil level, particularly during the spring (Fernández et al. 2008). This leads to some competition for soil water during the most active growth period in both species.

A study of the relative spatial distribution of the trees and grasses found that at a mean canopy cover of 50–70 %, the radiation level and soil water competition differ for grasses under the tree crowns compared to the midpoint between tree crowns (Fernández et al. 2006a, 2007). Due to the high latitude of the Patagonian region, tree shadows are often displaced – particularly if trees are pruned and their crowns are mostly concentrated in the upper third of the tree – so that the between-tree area is more shaded than those beneath the vertical projection of the crown (Fernández 2003). Thus, for a fairly shade-tolerant and drought-sensitive species such as *F. pallescens*, the position between tree crowns is the most suitable for its growth. This is due to both reduced evaporative demand, as a result of higher shade levels, plus reduced direct competition for soil water with trees, particularly when this is a limiting resource in late spring or summer (Fernández et al. 2007).

During a sunny day in the growing season, the lower evaporative demand under trees was significantly reflected in leaf transpiration of the grasses. This was lower under high tree canopy cover than in the open grassland, with intermediate values in SPS with relatively low canopy cover (Fernández et al. 2006a). Stomatal conductance did not show any clear instantaneous response to radiation level (Fernández et al. 2006b), and therefore leaf transpiration was highly correlated with leaf water potential. Leaf water potential reached mean minimum values of about −2,8 MPa in the open grassland, −2,2 MPa in open SPS and −1,7 MPa under the deepest shade conditions (Fernández et al. 2006a). However, this better water status and conservative water use was associated with low radiation levels, which may limit photosynthesis. In general, plants acclimated to mean low radiation levels tend to reduce their maximum photosynthetic capacity, which reduces respiration costs, and therefore decreases the light compensation point (LCP. this the minimum radiation level needed for a positive C balance) (Lambers et al. 1998). Curves relating net photosynthesis to photosynthetically active radiation (PAR) in *P. speciosa* and *F. pallescens* leaves showed that both species have low photosynthetic plasticity. Thus, for these species the maximum photosynthetic rate per leaf area (Pmax), quantum yield and LCP were similar between plants growing in the open grassland and under different levels of tree canopy cover (Fernández et al. 2002, 2006a, b). In addition, the general form of the curve, with quite low curvature factors compared to other temperate grasses, reflects the heliophilous behavior of both species, which, however, is more marked in *P. speciosa* than in *F. pallescens* (curvature factor: 0.34 and 0.84, respectively; Fernández et al. 2002; Caballé et al. 2011a, b). The low photosynthetic plasticity largely precludes both species from acclimating to very low radiation levels. Nevertheless, another study conducted on *F. pallescens* did find a significantly lower Pmax of SPS grasses compared to those in open grassland after they had been growing for several years under high shade condition (Caballé et al. 2011b). This lower Pmax seems to be related to lower maximum stomatal conductance measured in shaded plants, which in turn may be related to their lower stomatal density compared to individuals growing in open grassland (Caballé 2013).

In contrast to its relatively low photosynthetic plasticity, high plasticity was observed in *F. pallescens* plants at morphological and architectural levels, leading to increased capacity for capturing light under shade conditions. Specific leaf area (SLA, leaf area per unit of leaf biomass) increased significantly under shade conditions (Caballé 2013). In addition, the proportion of biomass allocated to leaves relative to the whole plant biomass also increased from 27 to 37 % of total biomass at the expense of root biomass, leading to increased shoot: root biomass (Fernández et al. 2004). This magnitude of change in biomass partitioning is high compared to a mean change of 4 % (maximum 13 %) in leaf biomass fraction found in a survey of 130 herbaceous species growing in open vs. shade conditions (Poorter and Nagel 2000).

Another architectural change observed in *F. pallescens* plants growing under shade conditions was related to mean leaf angles, the leaves being more horizontally oriented in shaded plants than in plants in the open. Most of the leaves in open grasslands plants have an angle relative to the horizontal plane of 35–50°, whereas most of the leaves in plants where canopy cover >50 % in SPS are <10° (Fernández et al. 2004). Accounting for all these morphological and architectural changes in a simulation model indicated that *F. pallescens* plants growing in SPS may capture 40 % more radiation than that expected if they had the architecture of an open grassland plant (Fernández et al. 2004).

### 5.2.3 Festuca Pallescens Responses to Defoliation under Tree Canopy Cover

Rangeland forage harvesting, the primary objective of herbaceous production in silvopastoral systems, involves considering the interactive effect of defoliation and tree canopy cover. There is limited attention to this interaction in the literature, and it was studied in *F. pallescens* under field conditions. The research considered different intensity and frequency of defoliation in relation to the microenvironment under different canopy cover levels of *P. ponderosa* (Caballé 2013).

Defoliated tussocks growing under trees generally had higher relative growth than non-defoliated plants in the same situation. However, the response depended on the frequency of defoliation and the climatic conditions of the growing season; there was no significant difference between defoliation intensities and tree canopy cover levels. Under shade, defoliation once at any moment in the growing season had a higher re-growth response than where there was more frequent defoliation. For example, after a third defoliation under 45 % of canopy cover level or a second defoliation under 75 % of canopy cover relative growth was similar to that of non-defoliated plants. However, due to the prior defoliation they had much lower absolute growth.

In open grassland conditions, defoliated plants also had much higher relative growth than non-defoliated plants; the difference was higher than under shade conditions by 20–25 % when defoliation was applied once in the growing season. However, in contrast to plants under shade, the defoliation intensity did have an effect on grass re-growth. Removal of 70 % of the above-ground biomass led to a higher relative growth than did removal of 50 % of the foliage.

Comparing defoliation responses of grasses growing in grassland *vs*. SPS, the observed responses were varied over the three growing season studied, but there was no clear relationship with annual climatic conditions. In a dry year, low-frequency defoliated plants in SPS grew in similar magnitude to defoliated plants of the open grassland, whereas in the following humid year, they grew significantly more than in the open grassland, suggesting net facilitative interactions with the trees. However, in the third growing season (also humid), defoliated grasses under trees grew less than in open grasslands. Thus, it seems that the responses are variable, and that non-analyzed variables are influencing these complex relationships.

The changes in relative growth induced by defoliation treatments were not accompanied by clear changes in morpho-physiological parameters such as *Pmax*, water status or *SLA*. These parameters tended to vary more in response to tree canopy cover level or climatic conditions (discussed in the previous section) than in response to defoliation itself. Thus, defoliation did not affect the *SLA* of the plants in shade treatments. SLA decreased 50 % from early spring to late summer in plants with and without defoliation whether under tree canopies or in the open. Similarly, defoliation did not affect plant water status (pre-dawn water potential) in either cover situation. However, frequent defoliation did induce higher water stress in grasses growing in open grasslands or in 45 % canopy cover treatments on a few dates or particular growing seasons. Similarly, *Pmax* was the same in defoliated and non-defoliated plants, and decreased as tree canopy cover level increased. However, defoliation did significantly affect instantaneous gas exchange through differences in stomatal conductance. This latter parameter was higher in defoliated than in non-defoliated plants within each canopy cover level. In addition, compensatory photosynthesis, which is the difference between photosynthesis in defoliated *vs.* non-defoliated plants, was two to threefold higher in open grassland plants than in grasses growing under tree canopy cover. In spite of the negative effect of shade on photosynthesis compensatory capacity, defoliated plants of *F. pallescens* growing under trees had maximum levels of compensation between 31 and 44 % (difference between *Pmax* of defoliated *vs.* non-defoliated plants), a magnitude which is higher than that reported for other herbaceous species with no radiation limitation (*Agropyron spicatum:* 21 % (Nowak and Caldwell 1984); *Lolium multiflorum:* 15 % (Gifford and Marshal 1973), and contrasting to results in *Lolium perenne* which presented no compensatory photosynthesis at all when it was subjected to shade conditions (Woledge 1977).

The activation and deactivation of the photosynthetic processes from the variable light conditions when grasses are growing under tree canopy cover, altered the induction times and magnitude between defoliated and non-defoliated plants. Leaves of *F. pallescens* from non-defoliated plants, after a shade period of 60 min, had a 20 % lower *Pmax* than when they were subjected to saturating light levels, and they reached 93 % of *Pmax* after 15 min. In defoliated plants, the reduction in *Pmax* was about 45 %, and subsequent recovery only reached 73 % of *Pmax* after 15 min. These differences are important because they mean that defoliated plants under tree canopy cover need longer periods of light and/or shorter shade periods than non-defoliated plants in order to reach their maximum C fixation capacity when radiation is available (Caballé 2013).

All these results indicate that shade has to be properly managed in SPS when grasses are subjected to grazing. Although *F. pallescens* may maintain its growth capacity up to tree canopy cover levels of about 70 % (see Fig. 5.3), its lower resistance to defoliation under shade conditions moves this tree cover threshold to lower values (no more than 50 % tree canopy cover, Caballé 2013) and even in this case, defoliation frequency must be reduced compared to open grassland conditions. Rotational grazing-based livestock management, i.e. livestock being moved to different grazing areas over the year, must be planned for long-term maintenance of forage supply in SPS.

## 5.3 Tree Component in Silvopastoral Systems in NW Patagonia

### 5.3.1 Ponderosa Pine Growth, Stand Productivity and Wood Quality

As the trees of SPS grow the ecological relationships between grasses and trees change in magnitude and direction. Immediately after tree planting, grasses and shrubs compete with the young trees for a few growing seasons (three to five, depending the availability of resources). In the first year after planting pines growing with good weed control had lower levels of drought stress and produced more and larger needles (75 % longer) than pines growing with grasses (Letourneau and Andenmatten 2007). In this experiment pines without competition were 2.1 m height at age 7 years compared to the 1.6 m in the control plots (Letourneau and Andenmatten 2007). In this experiment survival was unaffected. Similar results have been found in another experiment comparing weeding treatments in Patagonia (Davel et al. 2001). These latter experiments which also included fertilizer, found that the application of N-P-K 10-3-5 (10 gr $plant^{-1}$) did not increase pine growth or survival (Davel et al. 2001). As pines become larger, competition with grasses decreases or becomes neutral. As an example, weeded pines at an age of 5 years (85 cm height) had similar growth rates and drought stress levels to pines growing with grasses (Schlichter 2002). Based on the dynamics of water availability during the growing season (see below), it seems that grass competition becomes low or even negligible when the root system of the pines reaches the soil layers beneath those explored by grasses. Weeding treatments at planting can be used to increase the initial growth of pines and allow animals to grazed earlier with minimal damage to the trees, depending on the resilience of the grasses to re-growth after pine establishment.

It is well known that trees have higher diameter growth rates at low stockings compared to higher plantation densities. In the particular case of ponderosa pine plantations in NW Patagonia, on average, trees growing in SPS showed an annual stem growth 2.5–3 times higher than those in forestry stands (diameter growth at 1.3 m height (DBH): 18 and 6 mm year$^{-1}$, respectively, Gyenge et al. 2010; DBH growth in another study: 12 and 5 mm year$^{-1}$ in dominant trees of SPS and forestry stands, respectively, Gyenge et al. 2012). It is important to note that the relative growth rate (annual rate normalized by the diameter at the beginning of the growth season) and the annual volume growth were also higher in trees growing at low than at high density plantation (about four times higher, Gyenge et al. 2002). The differences in relative growth rates could means changes in carbon allocation or in the efficiency of resource use, that deserve more studies (Fernández and Gyenge 2009; Gyenge et al. 2012, 2014). Differences in annual growth rate between pines growing at different plantation densities in NW Patagonia could be observed early during the stand development, at age (at breast height) 8–10 (Martínez Meier et al. 2013), indicating the importance of thinning over annual aboveground productivity. Comparing growth rates of trees growing under same field conditions, it seems that ponderosa pines growing at lower densities show a longer period with positive increments, added to the higher rates mentioned above (Fernández et al. 2012).

Thinned SPS stand at age 15 years with 350 and 500 stems ha$^{-1}$ had current annual increments (CAIs) of 14 and 19 m$^3$ ha$^{-1}$ year$^{-1}$, respectively (Gyenge et al. 2010). An adjacent unthinned plantation with 1300 stems ha$^{-1}$ had a lower current annual increment of 13 m$^3$ ha$^{-1}$ year$^{-1}$; possibly as a product of that this plantation had already passed the time of peak CAI. These productivity values were similar to the average values published for commercial plantation in NW Patagonia (10–18.5 m$^3$ ha$^{-1}$ year$^{-1}$; SAGPyA 2001; Andenmatten et al. 2002).

At age 15 years, the annual growth-ring width of ponderosa pines in SPS showed higher variability between years than trees growing at higher plantation density in a site with a relatively positive water balance (Gyenge et al. 2010). Thus annual ring growth width varied by about 10 mm between years in the SPS pines (stands with 350 and 500 trees ha$^{-1}$), while the variation was about 2 mm in the high density stand (around 1300 pines ha$^{-1}$, Gyenge et al. 2010). In a more stressful environment, the opposite pattern was observed, i.e., the trees growing at higher plantation densities had higher growth variability than those growing at low plantation density (1.5–0.4 mm year$^{-1}$, respectively; Gyenge et al. 2012). It is important to note that at both sites, the minimum annual growth rate of the SPS trees was always greater than the maximum growth values of the trees growing in high plantation densities. So, even when the growth variability (i.e. the sensitivity to environmental conditions) could be higher in SPS trees than in high density stands, that variability is around a higher average value.

We have pointed out the high rates of individual stem annual growth of ponderosa pines growing at low plantation density. But what kind of physical properties does the wood have when it is produced under these particular growing conditions in Patagonia? Several studies were conducted to determine wood density and the functional (hydraulic) role of the wood formed in SPS. Wood density studies found that average density of the stem does not change with ponderosa pines growing at different plantation densities, despite differences in annual growth rates (Martínez Meier et al. 2013). Nevertheless, it was possible to determine statistical differences between the earlywood and latewood portions of the annual growth-ring (Martínez Meier et al. 2013). Regarding wood hydraulic functionality, measurements in trees planted at different initial stocking density showed that mean specific hydraulic conductivity of the growth-ring was similar between different stand densities and tree sizes (dominant vs. suppressed trees in each stand) (Fernández et al. 2012). This was in agreement with similar anatomical characteristics (tracheid lumen diameter and inter-tracheid pit dimensions) of the wood formed in the different treatments (Fernández et al. 2012). However, extrapolation of these results as a response to thinning treatment must be specifically tested in future experiments.

The abovementioned studies were carried out on growth rings where wood density had attained stabilized values (20- to 25-year-old trees). However, from the point of view of bole wood quality, taking into account that rotation periods are shorter in SPS due to higher growth rates and that transition from juvenile to mature wood in ponderosa pine in NW Patagonia is about 20 years, the proportion of juvenile wood in SPS logs could be substantially greater than in similarly sized logs from traditional plantations. Depending on the industrial use of the wood, this would mean a decrease in the proportion of high quality wood in SPS logs compared to those from traditional forest managements (Letourneau et al. 2014).

In order to analyze stem productivity of stands of different tree densities growing at sites with different water availability, simulations were performed of annual growth of ponderosa stands using $3PG_{PJS}$ (Sands 2004), a free excel spreadsheet version of the 3-PG model (Landsberg and Waring 1997). Some original parameters were changed based on Patagonian ecophysiological measurements and validated with growth data of different stands (Andenmatten and Letorneau 2003, Fig. 5.4). The simulations considered different conditions of water balance (dry and humid sites, with low and high soil depths in each case), different initial plantation densities (1200, 750 and 500 trees per ha.), and different management schedules. A control (no management at all) and a managed condition were simulated at each site x stand density combination. The management consisted of applying a thinning treatment each time the stand reached a threshold stand density index (SDI), which took a different value in the "traditional" schedule (SDI threshold = 800, initial plant density: 1200 trees $ha^{-1}$), than in the SPS scheme (SDI threshold = 600, initial plant density: 750 and 500 trees $ha^{-1}$). The target SDI in each thinning was 512 and 400 in traditional vs. SPS stands, respectively. The traditional management schedule was that recommended for Patagonian ponderosa pine plantations (Gonda 2001).

The predictions of $3PG_{PJS}$ partially agreed with the patterns of wood volume accumulation observed under field conditions (Andenmatten and Letorneau 2003), underestimating values of stands with high wood volume accumulation. This underestimation could be mathematically corrected (for example, by changing the self-thinning threshold), but the results shown in the following paragraphs were obtained without any correction. Considering the most contrasting situations, pines growing at under SPS, with a thinning schedule based on relative density index and planted in humid sites, reached a final size of average DBH = 40 cm at about 25–30 years of age, compared to 45–60 years in the high density unmanaged stands growing in the drier sites (Fig. 5.5a). This means that the SPS stands could be harvested around two times during the period in which the trees are growing in the high density stand to reach the target size of DBH = 40 cm. In contrast, the amount of harvested wood volume and mean annual increment followed the opposite pattern (Fig. 5.5b, c). By adding up the amount of wood harvested in the low density stands, it is possible to reach a similar amount to the total wood harvested in the high density stand. If we consider now that the stands must be harvested when trees are 30 years old, total wood harvested and mean annual productivity attain similar values, but most importantly, trees from

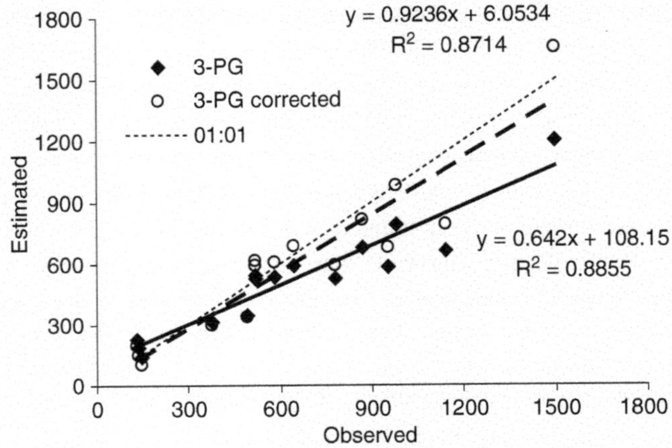


**Fig. 5.4** Observed (Andenmatten and Letourneau 2003) and estimated stand wood volume ($m^3\ ha^{-1}$) of ponderosa pine stands using 3-$PG_{PJS}$ model parameterized with ecophysiological data published by Gyenge (2005). Corrected data were obtained by multiplying the original simulated data by the difference between observed and estimated values/observed value

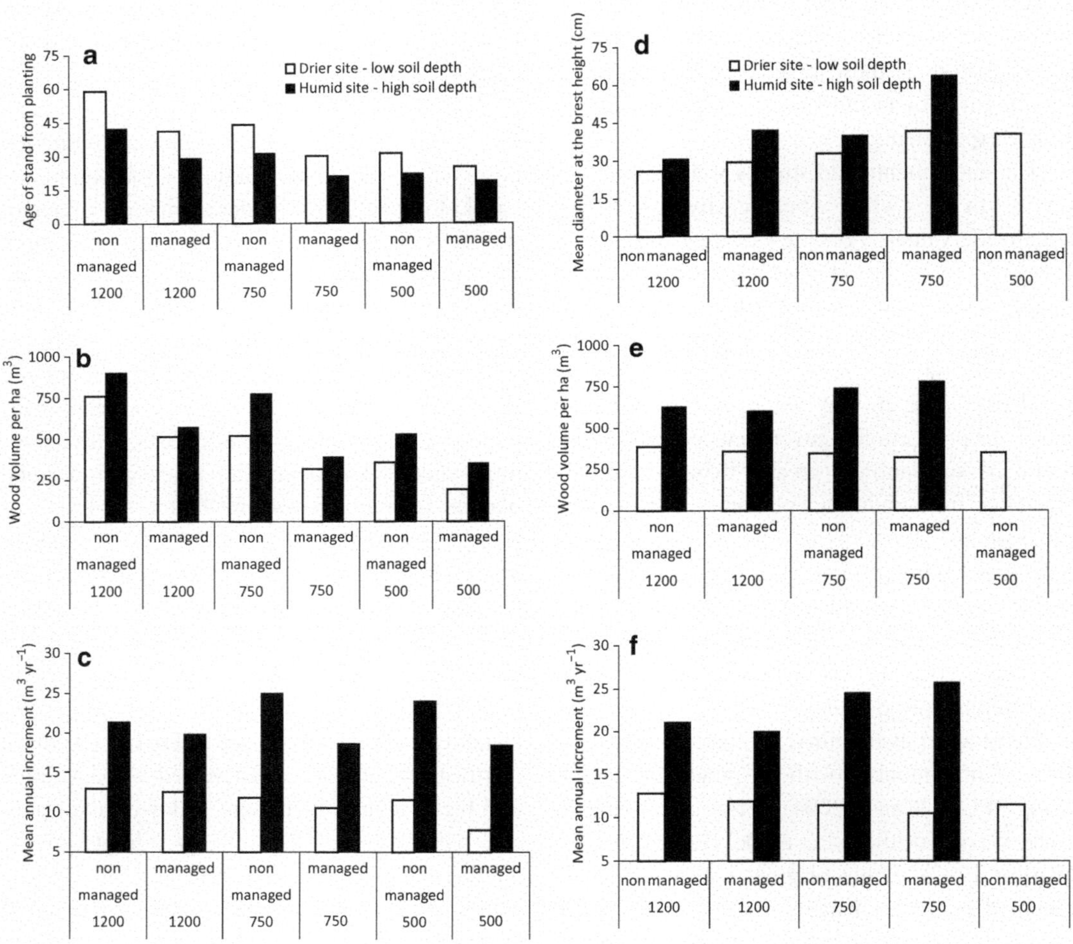


**Fig. 5.5** Simulation of ponderosa pine stand productivity using 3-$PG_{PJS}$. *Left panels* represent the age, total wood volume per ha and mean annual increment (**a**, **b** and **c**, respectively) when the mean diameter at breast height (DBH) of the stand reaches 40 cm. *Right panels* represent the mean DBH, total wood volume per ha and mean annual increment (**d**, **e** and **f**, respectively) when the stand reaches 30 years of age. The simulated situations shown in the figure are the most contrasting: dry site with low soil depth (negative water balance), humid site with high soil depth (positive water balance); managed and non-managed plantations with initial 1200, 750 and 500 pines per hectare. See text for details about the simulated management practices

low-density plantations attain higher diameter at breast height (Fig. 5.5d–f).

### 5.3.2 Silviculture of Ponderosa Pine in Silvopastoral Systems

Ecophysiological studies of grasses growing in the understorey of ponderosa pine plantations indicate that the critical threshold of tree canopy cover for which the net balance between positive and negative interactions are neutral or positive, is close to 70 % (Caballé 2013). Above this level, the facilitation effect generated by the trees on the water status of the grasses is lost due to strong competition for radiation.

Silvicultural management, thinning and pruning can regulate the levels of canopy cover in order not to exceed this critical threshold. Therefore, easy-to-measure indicators are required in order to apply these practices at the right time. At the same time, it is essential to know how canopy cover changes as trees grow. This can be achieved through the use of forest growth and yield models. In the following paragraphs a summary of another simulation study is presented, based on a different modeling approach from the one described in the previous section.

"Piltriquitron" (Andenmatten et al. 2007) is an empirical model developed by the National Institute of Agricultural Technology (INTA, Argentina) to specifically predict ponderosa pine growth and yield under NW Patagonian conditions. Through the analysis of hemispherical photographs taken at 63 ponderosa pine stands distributed between 36°.5 S and 42°.5 S (most of the potential plantation area of the species in Patagonia), parameters of tree canopy cover were accurately estimated, and the relationship between structural parameters of the stand and each tree canopy cover was established by forest inventory. The best fit, taking into account correlation coefficients, the distribution of errors in a validation sample and some practical parameters related to the application of the model, was found between canopy openness (or % visible sky) and Curtis relative density index (Curtis 1982). The Curtis relative density is based on the relationship between tree size and the number of trees per hectare (Letourneau et al. 2010). This relationship was used in the Piltriquitron model, and by considering critical thresholds, thinning schedules were simulated to maintain adequate levels of tree canopy cover.

Simulations showed that for a stand planted at 1111 trees ha$^{-1}$ on a site of intermediate site quality, it was necessary to have two thinnings in order not to exceed the critical threshold of 70 % of tree canopy cover (Table 5.1). At the final harvest at age 33 years the stand had 75 trees ha$^{-1}$ with a mean DBH of 45 cm. Including thinnings the total production was 280 m$^3$ ha$^{-1}$ (8.4 m$^3$ ha$^{-1}$ year$^{-1}$) of which 190 m$^3$ ha$^{-1}$ was harvestable. Clearly, in order to maintain an open canopy typical of SPS throughout the rotation, it is necessary to forego wood production.

On the other hand, studies of pruning effects on ponderosa pine in NW Patagonia have shown that pruning reduces stem productivity in SPS stand, and this is magnified in the trees growing at higher densities (Gyenge et al. 2010). This implies that the pruning threshold (i.e. the percentage of green canopy pruned) depends inversely on tree density and resource availability.

Green pruning also be changed branch growth. There was an increase in the diameter of branches located at the middle and bottom of the canopy when pruning was to height of 4.5 m than where trees were pruned to 3 m (Gyenge et al. 2009).

The decrease in productivity could be a result of the water stress induced by pruning or from reduced leaf area. Pines with a high level of pruning showed lower stomatal conductance and water potential at pre-dawn and during the day (Gyenge et al. 2009). Other results have shown quite low capacity of ponderosa pine to compensate water use in the short term after the loss of leaf area. Pruning the branches at the bottom of the canopy produced a decrease in whole tree transpiration. However, a difference of only 10 % in water use was measured in trees that differed by 40 % in foliar area as a result of pruning (Gyenge et al. 2009).

**Table 5.1** Simulation of silvicultural management for ponderosa pine stands in three different site qualities of NW Patagonia, Argentina to maintain the tree canopy cover level throughout the cycle below 70 %

| Site Index $IS_{(20)}$ (m) | 1st thinning | 2nd thinning | Final harvest | Commercial production ($m^3ha^{-1}$) >15 cm thin tip diameter | Gross production ($m^3ha^{-1}$) including thinning |
|---|---|---|---|---|---|
| 18.6 | 13 years | 24 years | 29 years | 196 | 290 |
| | 1111–290 trees $ha^{-1}$ | 290–77 trees $ha^{-1}$ | 77 trees $ha^{-1}$ | | |
| | Dg = 13.9 cm | Dg = 34.0 cm | Dg = 45.6 cm | | |
| 16.4 | 15 years | 29 years | 33 years | 195 | 287 |
| | 1001–248 trees $ha^{-1}$ | 248–68 trees $ha^{-1}$ | 68 trees $ha^{-1}$ | | |
| | Dg = 15.3 cm | Dg = 36.8 cm | Dg = 44.8 cm | | |
| 13 | 18 years | 36 years | 45 years | 185 | 276 |
| | 1083–290 trees $ha^{-1}$ | 290–78 trees $ha^{-1}$ | 78 trees $ha^{-1}$ | | |
| | Dg = 13.9 cm | Dg = 33.6 cm | Dg = 45.3 cm | | |

Site Index ($IS_{20}$): indicates the height (m) of dominant trees at 20 years old. Dg: diameter (cm) of the average tree basal area

Ecophysiological studies have demonstrated that leaf and hydroactive xylem area are functionally correlated in ponderosa pine, the value of the ratio being higher at sites with a more positive water balance (Callaway et al. 1994). This implies that leaf area of a branch or the canopy as a whole show a lineal relationship with the hydroactive xylem of the branch or the stem, respectively.

Based on this relationship, it seems that for ponderosa pine growing in a broad range of situations in NW Patagonia, the leaf area of each branch could be estimated by using the diameter at the base of the branch as predictor variable, at least for branches with basal diameter under 47 mm (Gyenge et al. 2009). It is important to note that for branches greater than this diameter and for trees growing at sites with a more negative water balance (drier sites), the equation may over-estimate leaf area. On the contrary, at the tree scale, a specific site equation is necessary to estimate canopy leaf area from hydroactive xylem area in stem. The higher values of this ratio, which reaches 0.20 $m^2$ $cm^{-2}$, were found in the more humid sites, compared to lower ratios in the drier sites (Callaway et al. 1994; Gyenge et al. 2010). This ratio also changed in response to tree density, being higher in trees growing at low density than at high density, even growing under similar general ecological conditions, or in response to drought in a particular year showing high plasticity in this character (Gyenge et al. 2012).

Another variable to be considered when silvicultural practices are oriented to canopy cover management is the longitude of the crowns (i.e. the distance between the top and the bottom of the canopy). If no pruning is applied, shade tolerance of ponderosa pine leaves determines that the relative amount of solar radiation that reaches the base of the canopy (the lower branch of the crown with green needles) is always about 11 % of the total radiation at the top of the crown, regardless of the plantation density (Gyenge et al. 2012). This determines, of course, different crown lengths in different plantation densities, as well as different canopy openness. In this regard, the amount of available solar radiation to each individual crown determines that the annual basal area increment of branches was higher at the top or middle than at the bottom of the canopy (Gyenge et al. 2009).

## 5.4 Animal Component in Silvopastoral Systems in NW Patagonia

### 5.4.1 Animal Production in Silvopastoral Systems in NW Patagonia

The forest-steppe ecotone zone in NW Patagonia where SPS can be conducted is a narrow strip 50 km wide extending along the province of

Neuquén, Río Negro and Chubut, between 36°30′ and 42°30′S. Different animal-based production systems are common in the different provinces and the steppe and the more humid sites in the mountain valleys. These include goats in the Northern portion of the region, extensive sheep raising in the steppe plateaus and cattle breeding in humid sites next to or within natural forests. Unfortunately, the available information about animal production under SPS schedules is very scarce in this region, with most of the studies coming from the northern area (north of Neuquén Province), where the traditional production system is transhumance livestock based on "local criollo goat". This activity involves about 1700 smallholders who use highlands of the Andes Mountains as summer grazing sites (known locally as "veranada" and which are highly productive lands used during the summer drought period) and steppes in Eastern Andes which are used as wintering sites (known locally as "invernadas").

The development of SPS to include the transhumance livestock of the northern province of Neuquén may be possible only in highland areas since the 300 mm rainfall of the steppe areas limits ponderosa pine growth. The period of "veranada" usually begins the first day of December (late spring in Southern hemisphere) and runs for 120 days until early April, (Fig. 5.6). Assuming that goats have a daily intake of 1.2 kgDM and eat 50 % of the biomass present in the grassland, the carrying capacity of the SPS in these areas would be 2–4 goats $ha^{-1}$ over this 4-month period.

Live weight and body condition score of goats throughout the summer period were not significantly different between animals on natural grasslands (conventional livestock production) and grazing under trees in SPS (Table 5.2). However, the daily live weight gains showed a favorable trend toward SPS. Young goats with "milk teeth" (main product of these production systems), doubled their weight during the "veranada" period reaching live weights of 23–26 kg by April. Postpartum mother goats recovered between 5 and 10 kg during the summer in the SPS. This is similar to traditional farming systems in the region. Apparently, regardless of initial weight, the animals can leave the "veranada" sites with

**Fig. 5.6** Silvopastoral system based on natural grassland, ponderosa pine and "local criollo goat" in northern of Neuquén Province in Patagonia Argentina

**Table 5.2** Live weight growth (kg) of "local criollo goats" of different ages and body condition (scale from 1 to 5) during the summer period of transhumance livestock in northern Neuquén Province, NW Patagonia, Argentina

| | | Silvopastoral systems | | Conventional Livestock | |
|---|---|---|---|---|---|
| | | December | April | December | April |
| Milk teeth | Liveweight (kg) | 12.1 (1.24) | 24.6 (1.42) | 13.9 (0.25) | 21.7 (0.41) |
| | Body condition | 2.4 (0.39) | 2.6 (0.25) | 2.3 (0.35) | 2.5 (0.2) |
| Two-toother | Liveweight (kg) | 26.1 (1.65) | 33.6 (1.31) | 27.1 (0.17) | 31.8 (0.88) |
| | Body condition | 2.3 (0.29) | 2.4 (0.13) | 2.5 (0.2) | 2.8 (0.11) |
| Four-toother | Liveweight (kg) | 30.2 (2.28) | 38.2 (1.06) | 30.2 (0.07) | 33.4 (4.11) |
| | Body condition | 2 (0.27) | 2.9 (0.08) | 2.2 (0.08) | 2.4 (0.44) |
| Six-toother | Liveweight (kg) | 31.6 (1.53) | 38.2 (0.24) | 32.6 (0.07) | 35.1 (3.69) |
| | Body condition | 1.8 (0.14) | 2.4 (0.05) | 2 (0.03) | 2.5 (0.48) |

The values indicate the mean (and SE) ≥15 goats per age class during three different summer periods (2009–2012)

similar weight and the body condition which is above the minimum threshold (BC: 2) required for breeding.

### 5.4.2 Diet Selection, Tree Damage and Time at Which Animals Are Introduced into Ponderosa Pine Silvopastoral Systems

Browsing and trampling by animals can cause serious damage to trees during early stages of SPS, and in some cases result in tree mortality. The voluntary intake of tree foliage, branches or bark is due to the animal's digestive physical capacity, diet preference, energy demands, forage quality and relative availability of these components (Minson 1990). Regarding diet preference, it has been suggested that goats are less selective than sheep and cattle, and their diet generally includes more woody species (Animut and Goetsch 2008). Therefore, goats may potentially cause more damage to young trees in SPS systems than other domestic animal species.

The botanical composition of the goat's diet when they were placed in young unpruned ponderosa pine SPS, found a higher proportion of trees and shrubs (47 %, SE=1.5) than grasses (30 %, SE=2.0), herbs (17.4 %, SE=1.65) and graminoids (5.3 %, SE=0.35). The tree and shrub component was higher in the younger animals (milk teeth and two-toother goats), while the grass component was higher in older animals.

Eighty percent of the tree and shrub component in the diet was represented by the following species, in order of importance: *Pinus ponderosa* (13 %), *Berberis* sp. (7 %), *Nothofagus antarctica* (7 %), *Chuquiraga* sp. (5.5 %), *Ephedra* sp. (4 %), *Adesmia* sp. (3 %) and *Gaultheria* sp. (3 %). The botanical composition for 75–95 % of the grass component was represented by four species. In order of importance, based on their average contribution, the main species were: *Festuca pallescens* (11 %), *Poa* sp. (6 %), *Rytidosperma* sp. (6 %) and *Bromus setifolius* (3 %). The herb component was almost the single genus, *Acaena*. On average, during the "veranada" period, 37 % (SE=1.4) of the goats' diet was composed of *F. pallescens*, *P. ponderosa* and *Acaena sp.*

The diet preference, which is the relationship between diet contribution and availability of each species, followed the order: (1) the native tree *Nothofagus antartica* (ñire) and the legume shrub *Anarthrophyllum rigidum*, locally called "mata guanaco", (2) the herb *Acaena splendens*, (3) the natural grasses *Poa lanuginosa*, *P ligularis* and *Festuca pallescens*, locally called "coirones", and (4) ponderosa pine needles.

While the pine needles were in fourth place in the diet preference of goats, their average contribution to the diet was greater than 13 % with a peak in

February of 18 %, at which time it was the plant species most consumed. This was because of their homogeneous spatial distribution and high availability (756 kg DM $ha^{-1}$) of this component due to their low crowns in the first 2–3 years of grazing. This amount of dry matter is about 70 % of the total dry matter of well-conserved natural grasslands and more than twice the dry matter present in degraded grasslands (Caballé et al. 2011).

The worst browsing damage associated with pine needle consumption was usually in trees shorter than 1.5 m. Trees taller than 1.5 m exhibited partial or total defoliation of the basal branches but they did not show any damage to the trunks after 120 days of continuous goat grazing. The frequency of damage to the pines increased in poor grasslands conditions, where there was the herb *Acaena splendens*, the C3 tussock *Pappostipa speciosa* and 30 % of bare soil. Thus, in SPS with ponderosa pine, goats should be introduced once the trees exceed 1.5 m in height. Also, special attention should be paid to the state of grassland conservation. Depending on the site quality, the period of grazing exclusion necessary for trees exceeding 1.5 m in height can be from 2 to 6 years.

### 5.4.3 Forage Quality and Availability of Ponderosa Pine Needles

Natural grasslands of NW Patagonia, outside areas of high productivity meadows, have a marked gradient in production, mainly defined by the west-east rainfall distribution and the highly seasonality of rainfall which is mainly concentrated in winter and early spring. Aboveground primary production is controlled during the winter by the low temperatures and during the late spring and summer by water availability (Jobbagy and Sala 2000). These climatic conditions cause the nutritional value of the main forage species to decrease markedly as the growth season progresses and to be enough only to supply the maintenance requirements of livestock (Somlo et al. 1985).

The average value of crude protein in ponderosa pine green needles of 9.2 % (was considerably higher than the average 5.7 % concentration foundin perennial grasses of the natural grassland (*Poa* sp., *Agrostis* sp., *Festuca* sp., *Pappostipa* sp.,) but lower than the leaves of shrubs (12 %), as most of these are legumes (Table 5.3; Caballé et al. 2009). Further, unlike grasses, where the crude protein concentration falls from 7.2 to 4.6 % as the growing season progresses, the crude protein in green pine needles remained the same or even increased towards the end of the season (Caballé et al. 2010).

The differences found between the crude protein concentration of pine needles and main grasslands forage species suggest that the contribution of pine to goat diet in the north of the region is important, especially at the end of the dry season when grasses contain only half their concentration. Nevertheless, except for the legumes, none of the understorey species present in SPS or the natural grasslands has the minimum 12 % of crude protein required for animals post-weaning or the 9 % for proper development and animal maintenance (Huston et al. 1971).

However, the digestibility of pine needle dry matter is very low and is similar to the digestibility of the poorest native grasses such as *Pappostipa speciosa* (Table 5.3; Somlo et al. 1985). This is directly related to the high lignin concentration of the pine needles (Table 5.3); lignification increased with increasing water deficit at the end of the growing season (Caballé et al. 2010). The low digestibility and the negative effect of secondary compounds present in the needles on rumen microflora probably prevent the animals from increasing their needle intake (Pfister et al. 1992).

Tannins and some of the main secondary compounds, generate astringency and depress preference levels and intake. However, a complementary study found an average of 3 % total tannins in the leaves (González et al. 2013). This may also explain the relatively low preference of goats for pine needles. Nevertheless, in concentration similar to those found in this study, tannins have been

**Table 5.3** Average forage quality of ponderosa pine needles and leaves of grasses, herbs and shrubs present in natural grassland and SPS in NW Patagonia

| Species | Site | ADL (%) | CR (%) | DMD (%) |
|---|---|---|---|---|
| Agrostis sp. | Open grassland | 3.8 | 6.8 | 67.5 |
| | Silvopastoral systems | 4.3 | 7.0 | 65.2 |
| Festuca sp. | Open grassland | 7.1 | 5.4 | 56.5 |
| | Silvopastoral systems | 7.8 | 5.2 | 54.7 |
| Poa sp. | Open grassland | 4.3 | 6.1 | 63.6 |
| | Silvopastoral systems | 4.6 | 6.7 | 65.9 |
| Poppostipa sp. | Open grassland | 7.3 | 4.6 | 54.7 |
| | Silvopastoral systems | 8.2 | 5.5 | 53.1 |
| Acaena splendens | Open grassland | 6.2 | 6.4 | 68.3 |
| | Silvopastoral systems | 8.4 | 7.5 | 61.5 |
| Anarthrophylum rigidum | Open grassland | 19.3 | 12.0 | 52.5 |
| | Silvopastoral systems | 18.4 | 12.5 | 56.6 |
| Pinus ponderosa | Silvopastoral systems | 15.6 | 9.2 | 53.3 |
| Nothofagus antartica | Open grassland | 12.2 | 8.3 | 59.8 |

ADL (%) is acid detergent lignin, CR (%) is crude protein and DMD (%) is dry matter digestibility (Source: Caballé et al. 2009, 2010)

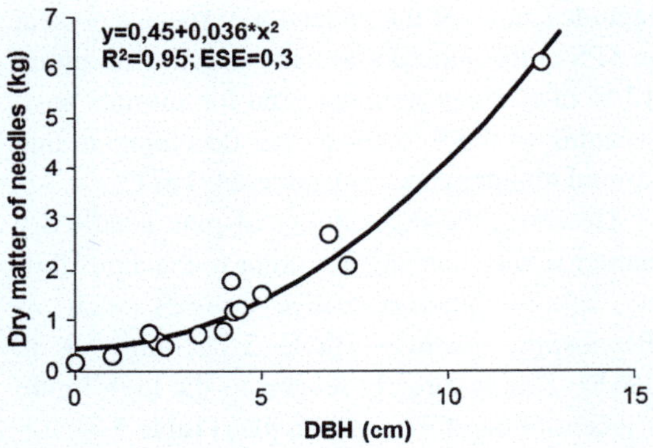


**Fig. 5.7** Relationship between total dry matter (kg) of green needles per tree (all biomass present below 1.3 m in height) and stem diameter al breast height (DBH, cm) for ponderosa pine in SPS in NW Patagonia, Argentina

found to be beneficial. Thus, some studies have shown that 2–3 % of tannins increased the production of meat, milk and wool by 10–15 % (Waghorn et al. 1990). This positive effect was attributed to increased protein by-pass reaching the duodenum, although it may been caused by other factors in the intake and digestion processes.

Regarding needle quantity, there is a good non-linear relationship between total dry mater of green needles per tree (all biomass present below 1.3 m in height) and stem DBH (Fig. 5.7; Caballé et al. 2010). This can be used to estimate needle supply in young SPS.

## 5.5 How Silvopastoral Systems Contribute to the Sustainability of Forestry in NW Patagonia

### 5.5.1 Conservation of Biodiversity

Land-use change typically result in the modification of vegetation cover and this can, in turn, lead to loss of habitat for many species (Lindenmayer and Fischer 2006). Human-modified lands usually have less habitat diversity and complexity than natural systems, and offer poorer food and shelter sources for many native species (Brockerhoff et al. 2008). Nevertheless, silvopastoral systems compare favorably with many other productive land uses. Across the scale of management intensity and conservation value, SPS have a higher conservation value than intensive forestry, range management or agriculture with introduced pastures. They may enhance biodiversity due to their higher structural heterogeneity than found in exclusively pastoral or forestry systems (Rois et al. 2006). Promoting SPS rather than agriculture or other intensive land uses may be important for conserving biodiversity, especially in landscapes where the natural habitat has been highly fragmented (Schroth et al. 2004).

As mentioned above, the forest – steppe ecotone, where pine plantations are commonly established in NW Patagonia, is a narrow transition zone along the Andes Mountains and is dominated by a gramineous steppe with sparse shrubs and trees. Since the early twentieth century, these systems have been used for extensive sheep and cattle production (Soriano 1983). Some pine plantations have been established at high densities but most were never managed, so resulting in dense, highly shaded environments. There has been great public concern in Patagonia because of potential negative environmental impacts of these plantations, particularly on biodiversity.

However, several studies in the region have suggested that changes in biodiversity in forest plantations depend on landscape design, stand structural characteristics at the site level and the taxa considered. At the site level SPS have much lower densities (sparse) than normal pine plantations, and with a more open canopy of between 30 and 70 % coverage. Such plantations have been shown to have a lower impact on biodiversity than dense plantations. Rusch et al. (2004) found that in sparse plantations with 52 % tree canopy cover there was higher shrub and herbaceous cover and richness than in dense plantations with 87 % tree canopy cover (Fig. 5.8).

However, such open plantations had lower richness than native grassland vegetation. Importantly, the new structural element, the trees, resulted in a new resource for fauna (Lantschner et al. 2008). Thus, when SPS are introduced in extended native grasslands, they increase β diversity (landscape heterogeneity) and structural diversity, while maintaining a high level of original biodiversity.

The way species change when *P. ponderosa* plantations are established also depends on the composition of the original community. Thus, two steppe communities may react differently. Steppes dominated by *P. speciosa* and *Mulinum spinosum* (neneo shrub) showed a strong tendency for *M. spinosum* mortality; while *P. speciosa* was less affected. Communities dominated by *F. pallescens*, increased in coverage when compared with open grasslands suggesting an improved site productivity.

Bird species abundance and richness have been shown to decrease in dense plantations compared to native vegetation, whereas they remained similar in sparse plantations (Fig. 5.9a; Lantschner et al. 2008). Moreover, changes in bird assemblage and composition also depended on the original vegetation type that was replaced. Though small changes in species composition resulted when plantations replaced native forest having a similar vegetation structure, the changes were more marked when plantations replaced steppe habitats, or where open plantation like SPS were developed (Lantschner et al. 2008). Regardless of the type of vegetation replaced, the abundance of many bird species (e.g., *Scelorchilus rubecula, Aphrastura spinicauda, Pteroptochos tarnii, Leptasthenura aegithaloides, Scelorchilus rubecula, Colorhamphus parvirostris*) in pine plantations has been observed to be positively associated with the amount of native vegetation in the understorey (Lantschner and Rusch 2007; Lantschner et al. 2008). Thus, sparse plantations planted for silvopastoral activities are better habitats for more species than are dense plantations.

Small mammals have also been shown to be sensitive to understorey vegetation cover. The decrease in abundance and richness of small mammals in pine plantations compared to native vegetation, and the presence of the most abundant species (e.g. *Abrothrix longipilis*), are best explained by herbaceous cover and shrub richness (Lantschner et al. 2011). The decline in small mammals in pine plantations is strongly associated with decreased understorey cover and plant species richness leading to the reduction in shelter and food resources (Lantschner et al. 2011). Therefore, the presence of higher understorey cover in SPS than in traditional forestry plantations may also enhance the abundance and diversity of small mammals.

Carnivore mammal species also use pine plantations depending on their vegetation structure. Sparse plantations favor the presence of two species, the culpeo fox (*Lycalopex culpaeus*) and Molina's hog-nosed skunk (*Conepatus chinga*), relative to traditional dense plantations (Fig. 5.9c; Lantschner et al. 2012). Thus, landscapes with greater areas of sparse plantations could have

**Fig. 5.8** Characteristic of vegetation structure and composition of native vegetation, sparse plantations (suitable for SPS) and dense plantations in steppe area (Rusch et al. 2004). *Letters* indicate statistical differences between vegetation types (α<0.05)

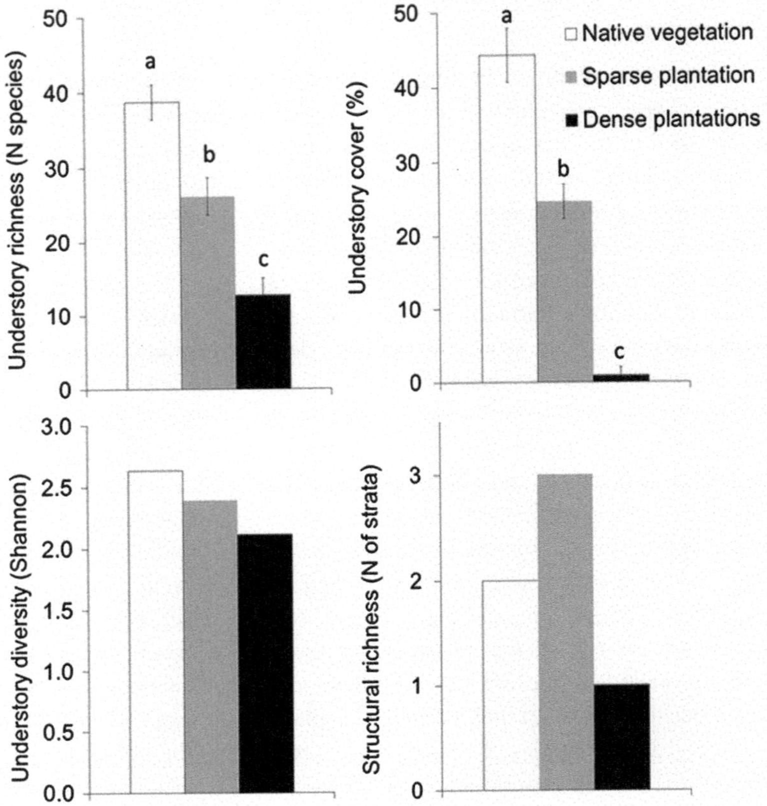


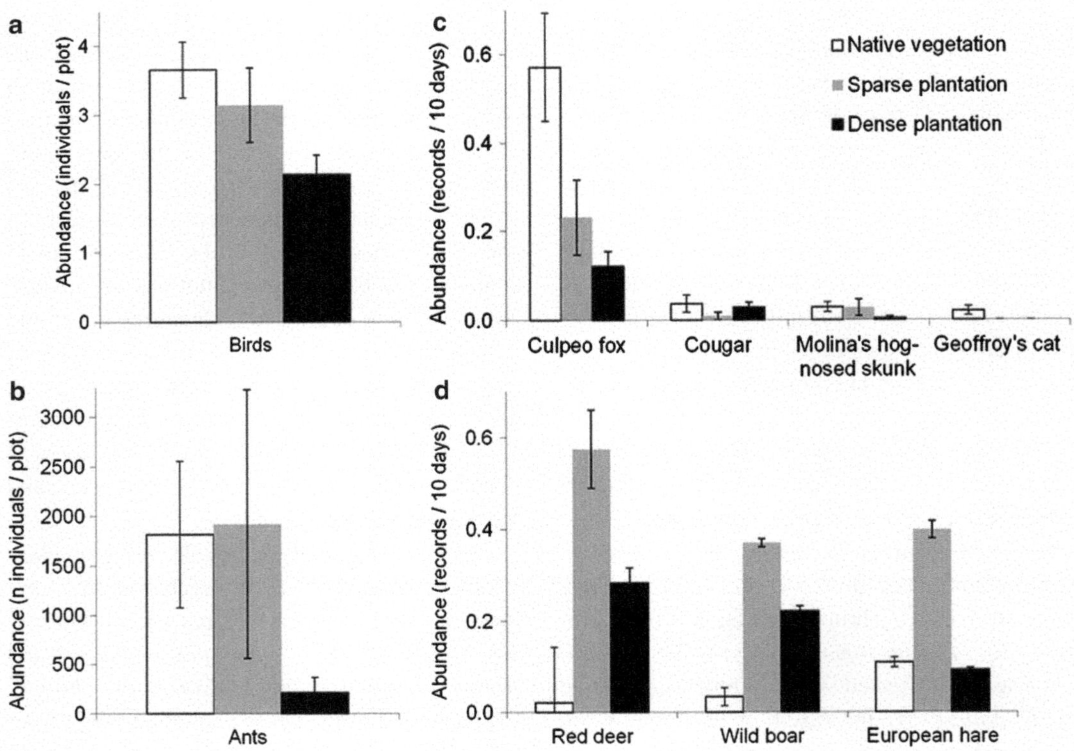


**Fig. 5.9** Abundance of wildlife species in native vegetation, sparse pine plantation, and dense pine plantations of NW Patagonia. (**a**) birds (Lantschner et al. 2008), (**b**) ants (Corley et al. 2006), (**c**) mammal carnivores (Lantschner et al. 2012), (**d**) mammal herbivores (Lantschner et al. 2013)

strong positive effects on these two species. Geoffroy's cat (*Leopardus geoffroyi*), on the other hand, was detected almost exclusively in native vegetation, and did not use either SPS or normal pine plantations. In contrast, the cougar (*Puma concolor*) was detected in dense and sparse plantations as frequently as in continuous native vegetation (Lantschner et al. 2012).

Structural differences within vegetation types did not appear to have a significant effect on habitat selection by wild herbivore mammals (Lantschner et al. 2013). Red deer (*Cervus elaphus*) and wild boar (*Sus scrofa*) were more abundant in pine plantations than in native vegetation, but no difference was observed in the use of dense or sparse plantations (Fig. 5.9d). This suggests that habitat selection by those species is shaped by the presence of an overstorey stratum, but not by internal differences in vegetation structure within forested and non-forested habitats. The European hare (*Lepus europaeus*) was much more abundant in firebreaks than in pine plantations, suggesting its preference for open habitats (Lantschner et al. 2013). It should be noted that these three herbivore species are introduced, and no native herbivore species was found in the forested landscapes (Lantschner et al. 2013).

Ant abundance within plantations is lower than in native steppe vegetation and species composition is impoverished (Corley et al. 2006). Moreover, the impact of plantations on ant abundance and richness varies with tree canopy cover. Dense plantations support the lowest ant abundance and species richness. In contrast, ants are less affected by sparser woods (Fig. 5.9b). Afforestation at lower tree density sustains very similar overall ant abundance and species composition to the natural steppe vegetation (Corley et al. 2006; Sackmann et al. 2008). The impact of sparse plantations on beetles resulted in a slight increase in species diversity. In dense plantations there were major changes in beetle communities, with an increased abundance of a few species, decreased number of species, and significant change the composition of the assemblage (Sackmann et al. 2008).

In conclusion, the presence of native vegetation in the understorey is one of the most important factors determining the use of plantations by native wildlife species. Understorey vegetation may provide escape cover against predators, safe nesting sites, and food resources for birds. Thus the presence of understorey vegetation in SPS seems to be a key factor to favor the presence of most native bird species. SPS may also enhance biodiversity at landscape scale in these highly modified lands, by increasing connection within native vegetation remnants for most sensitive species. With planted trees, the appearance of a new vertical stratum also produces new habitats for several groups of native species.

Although most information suggests that the impact of SPS on biodiversity in Patagonia is low, there is still poor information about the effect of livestock grazing inside plantations on wildlife. Grazing can cause a decrease in the understorey cover and diversity, and hence affect native plant and animal species. A major crucial factor that affects the dynamics and health of the systems is the stocking rate (Rois et al. 2006). In this sense, most information of the impact of sparse plantations in Patagonia is based on sites with a low number of livestock or without livestock. Further research into SPS should assess the effect of grazing densities on biodiversity.

### 5.5.2 Water Use of Ponderosa Pine in Silvopastoral Systems

Soil water availability is one of the main growth limiting factors in NW Patagonia (Paruelo et al. 1998). As precipitation is concentrated during the cool season of the year, the annual primary productivity produced during the warm season depends basically on the amount of water that the soil can store. A similar situation occurs in NW USA where ponderosa pine is found naturally. So, in general, this species is well adapted to this climatic condition, which partly explains its general high productivity in NW Patagonia.

As the amount of water used by a pine plantation depends on the stocking density, trees may show symptoms of drought stress earlier during the growth season in a high density plantation than in a SPS even when both plantations are situated under similar ecological conditions (Gyenge et al. 2003). Research in NW Patagonia found that transpiration rates were lower in SPS than normal plantations but remained relatively unchanged over the entire growing season (Gyenge et al. 2003; Licata et al. 2008). While ponderosa pine is able to take soil water from deeper than 1.8 m, half of the soil water used came from the upper 80 cm of soil (Licata et al. 2008). In SPS, the soil water patterns showed a similar depletion of reserves in the top 80 cm of soil compared to the pattern of an open grassland site throughout the season. However, the deeper soil layers become drier under SPS than in the grasslands during the second half of the summer (Gyenge 2005). Stable isotope analyses reveal that in spring and autumn, grasses in SPS use approximately 90 % of the water from the top 20 cm of soil, whereas in summer, when soil water content is very low, this proportion decreases to 75 % (Fernández et al. 2008). In contrast, the water used by pines in SPS basically come from soil layers deeper than 20 cm (Fernández et al. 2008). In contrast, trees growing at the higher plantation densities used water from different soil layers, depending on its abundance (Licata et al. 2008). During spring, about 90 % of the water transpired came from the upper 20 cm of soil, but this decreased to 20 % during autumn, which coincided with the water depletion of the upper soil layers. These results support the hypotheses that pines and grasses are somewhat complementary in their soil water usage, and that pines in agroforestry systems use less shallow water than dense pines plantations (Fernández et al. 2008). It appears ponderosa pine is capable of redistributing the water in the soil profile by moving water from the deeper moister soil layers to upper drier ones by hydraulic lift (Fernández 2003). However, available studies were not able to decide if understorey grasses were able to use this water or if it resulted in a positive effect on grass growth.

As with individual tree productivity, trees of a similar size growing in plantations at low densities use more water than those growing in stands of high density. For example, ponderosa pine of 30 cm DBH used about 120 l water day$^{-1}$ in SPS, compared to 25 l day$^{-1}$ in a high density plantation (Gyenge 2005).

As mentioned above, trees in SPS show relatively stable transpiration rates (taking into account the potential evapotranspiration) throughout the warm season (Licata et al. 2008). Sap flow measurements showed that trees growing at low density are capable of sustaining a maximum sapflow density of 0.55 ml cm$^{-2}$ min$^{-1}$ for the 6 h of maximum radiation (Gyenge et al. 2003; Gyenge 2005). Trees growing at high density showed maximum values of sapflow density of about 0.2 ml cm$^{-2}$ min$^{-1}$ for approximately 2 h per day (Gyenge 2005). These differences indicated a high hydraulic resistance in trees growing at high density. This stability in transpiration rate in SPS stands supported the stability in water use throughout the season. In contrast, evapotranspiration in grasslands is highly variable between years and within seasons, depending on the amount of precipitation during the season (Gyenge et al. 2002; Gyenge 2005).

As we can see, both wood production and water consumption are higher in individual ponderosa pines growing in SPS than in high density plantations. However, wood production increases more than water use, indicating higher water use efficiency (WUE, kg wood l$^{-1}$ water) in SPS than in conventional forestry (Gyenge 2005). In a regional study using satellite images, Rivero et al. (2006) demonstrated that ponderosa pine plantations have higher WUE than grasses and shrublands (7.7 and 3.9 kg ha$^{-1}$ year$^{-1}$ mm$^{-1}$, respectively). Other studies have demonstrated that larger trees in the stand had higher WUE than smaller ones, but at stand level, WUE decreases or increases with intra-specific competition levels at sites with less or more soil water availability (Fernández and Gyenge 2009). As a result of differences in water use and WUE of trees of different sizes (i.e., with different growth rates), comparing stands of similar age, WUE is higher in those with higher mean DBH such as SPS. In spite of the fact that WUE is linearly related with the mean DBH of a stand, water use of the stand depends on the number of trees per unit area and their leaf area index (LAI) in a nonlinear function. For example, a stand with 1300 trees $ha^{-1}$ and LAI of 6.4, transpired 3.8 mm $day^{-1}$, whereas on the same site SPS stands transpired 3 and 2.1 mm $day^{-1}$ when they had 500 (LAI=1.8) and 350 (LAI=1) trees $ha^{-1}$, respectively (Gyenge et al. 2011). Under similar ecological conditions, open grassland had evapotranspiration of 0.9 and 2.7 mm $day^{-1}$ for a dry or a wet season, respectively (Gyenge et al. 2011 and references therein).

As a whole, these studies indicate that SPS may be a way of reducing water use compared to traditional high density plantations and would be helpful if water resources have to be used for purposes other than primary production activities at the site. Similarly, if the objective is to take the maximum advantage of available water resources, wood production under SPS would increase WUE. This implies that a given amount of water should produce more wood in SPS than in conventional forestry.

### 5.5.3 Silvopastoral Systems and Climatic Change in NW Patagonia

The Intergovernmental Panel on Climatic Change defined vulnerability "as the extent to which a natural or social system is susceptible to sustaining damage from climate change" (McCarthy et al. 2001). In this regard, it is possible to change the vulnerability of a particular system by management procedures (McCarthy et al. 2001). With forest systems, yields can be reduced from drought stress caused by both soil water deficits and/or high atmospheric demands (McCarthy et al. 2001). In the context of global climate change, forecasts predict more stressful conditions for some areas of the world of resulting in higher average temperatures and more variable precipitation (IPCC 2007). This has been seen in NW Patagonia where climate change has recently been noticed in terms of a higher frequency of drought events over the last two decades. Dendrochronological studies performed in this region have also demonstrated that ten severe drought events have occurred in the last 20 years, whereas only six events occurred during the previous 80 years (Mundo et al. 2010). This climatic phenomenon threatens the sustainability of natural and managed forest systems in the region.

In the particular case of ponderosa pine plantations in NW Patagonia, Fernández et al. (2012) has found that during a particularly dry growing season, mean growth decreased by 30–38 % and 58–65 % compared to the previous 5-years mean growth, in SPS vs. closed stands, respectively, indicating higher sensitivity in the latter. The ecophysiological basis for these differences in sensitivity to drought with pines growing under different plantation densities was studied by Gyenge et al. (2012), Gyenge and Fernández (2014). They found higher variability in stem and annual leaf area growth in a high density than in an open stands. Those growth patterns were explained by differences in canopy conductance between trees growing under different competition levels, which in turn resulted from differences in individual water status (pre-dawn water potential) between treatments. Pines growing in high density treatments reached very low predawn water values (around −2 MPa), and they could not open their stomata at all for many days during the drought season (Gyenge and Fernández 2014). This would have decreased their C fixation, and probably decreased their carbohydrate reserves, affecting their initial growth capacity in the following growing season (resilience to drought). However, the observed patterns of higher sensitivity to drought in the high density stands were opposite to those reported for the same species growing in managed native forests in USA (McDowell et al. 2006), where pines growing in dense forests were the least sensitive to drought, apparently as they had lower leaf area and a general hydraulic architecture that was more adapted to drought than that of trees growing in open stands. It is important to note, however, that even when variability was lower, mean growth was also lower in those stands than in open stands (McDowell et al. 2006). Patagonian ponderosa pine plantations had much higher growth rates than those measured by McDowell et al. (2006), even when their stand density index reached the maximum value (Gyenge and Fernández 2014). The synergy between drought and other environmental stress factors in US native forests might have contributed to producing the observed opposite patterns.

At the seedling stage, in spite of the fact that ponderosa pine is a species well adapted to drought periods within the growing season, specific studies analyzing the effects of the time of drought occurrence on growth and physiology of the plants showed that a drought event during spring reduced stem growth and biomass accumulation during the occurrence of the drought and afterwards, even when water status and carbon fixation capacity were recovered (Fernández et al. 2014). In contrast, a summer drought event had a significant but much smaller impact on plant growth (Fernández et al. 2014). A drought event in spring and another one in summer, separated by a period of good water availability in between, led to reduced growth (as in the spring drought) but also compromised plant survival (Fernández et al. 2014). This could have important implications to plantation establishment depending on the pattern of drought events.

Available information suggests that the low plantation density necessary to maintain an SPS scheme may reduce the susceptibility to drought of the trees and therefore may be a management option for increasing forestry adaptability to climate change in Patagonian region.

## Bibliography


Andenmatten E, Letourneau F (2003) Predicción y proyección del rendimiento de pino ponderosa de Chubut y Río Negro, Argentina. Quebracho 10:14–25

Andenmatten E, Gonda H, Letourneau F (2002) Proyección de distribuciones diamétricas en plantaciones de pino ponderosa en la provincia de Neuquén. Final Report PIA 15/98, PFD–SAGPyA, 29 pp

Andenmatten E, Letourneau F, Getar E (2007) Simulador forestal para Pseudotsuga MenziesII (Mirb) Franco y Pinus ponderosa (Laws) en Patagonia Argentina. En: Gonda H, Davel M, Loguercio G, Picco O (eds) 1ra. Reunión sobre Forestación en la Patagonia. Ecoforestar 2007, Esquel

Animut G, Goetsch AL (2008) Co-grazing of sheep and goats: benefits and constrains. Small Rumin Res 77:127–145

Armas C, Rodríguez-Echeverría S, Pugnaire FL (2011) A field test of the stress-gradient hypothesis along an aridity gradient. J Veg Sci 22(5):818–827

Belsky AJ (1994) Influences of trees on savannah productivity: tests of shade, nutrients, and tree-grass competition. Ecology 75:922–932

Bertiller M, Defossé GE (1990) Grazing and plant growth interactions in a semiarid Festuca pallescens grassland (Patagonia). J Range Manag 43(4):300–303

Breshears DD, Rich PM, Barnes FJ, Campbell K (1997) Overstorey-imposed heterogeneity in solar radiation and soil moisture in a semiarid woodland. Ecol Appl 7:1201–1215

Brockerhoff E, Jactel H, Parrotta J, Quine C, Sayer J (2008) Plantation forests and biodiversity: oxymoron or opportunity? Biodivers Conserv 17:925–951

Bonino N, Bonvissuto G, Pelliza-Sbriller A, Somlo R (1986) Habitos alimenticios de los herbivoros en el area ecologica central de las sierras y mesetas occidentales de la Patagonia. Rev Arg Prod Anim 6(5/6):275–287

Caballé G (2013) Efecto interactivo de la defoliación del estrato herbáceo y la cobertura del estrato arbóreo sobre el crecimiento del estrato herbáceo en sistemas silvopastoriles. Doctoral thesis, University of Buenos Aires, 216 pp

Caballé G, Dezzotti A, Sbrancia R, Stecher G, Reising C, Bonvissuto G, Fernández ME, Gyenge J, Schlichter T (2009) Estudio de caso: Interacción entre el pastizal natural, la plantación de pino y el ganado caprino en el sistema silvopastoril experimental de Mallín Verde (Neuquén). I Silvopastoral Systems National Congress, Posadas, pp 1–7

Caballé G, Reising C, Cohen L (2010) Valor nutritivo y disponibilidad de materia seca de pino ponderosa en sistemas silvopastoriles. First Patagonian International Agroforestry Congress, Coyhaique

Caballé G, Borrelli L, Avila M, Castañeda S, Inostroza L, Muñoz O, Reising C (2011a) Interacción planta-animal: la cabra criolla y su preferencia por el pino ponderosa. II Silvopastoral Systems National Congress, Santiago del Estero

Caballé G, Fernández ME, Gyenge J, Aparicio A, Schlichter T (2011b) Modeling leaf maximum net photosynthetic rate of the dominant perennial grass of Patagonian silvopastoral systems. Agrofor Syst 83:13–24

Callaway RM, Walker LR (1997) Competition and facilitation: a synthetic approach to interactions in plant communities. Ecology 78:1958–1965

Callaway RM, Delucia EH, Schlesinger WH (1994) Biomass allocation of montane and desert ponderosa pine: an analog for response to climate change. Ecology 75:1474–1481

Corley JC, Sackmann P, Rusch V, Bettinelli J, Paritsis J (2006) Effects of pine silviculture on the ant assemblages (Hymenoptera: Formicidae) of the Patagonian steppe. For Ecol Manag 222:162–166

Curtis RO (1982) A simple index of stand density for Douglas-fir. For Sci 28:92–94

Danklmaier C (2004) Análisis de los factores socioculturales que influyen la performance de los programas de promoción forestal en Chubut, Argentina. Proyectos

Federales de Innovación Productiva PFIP 2004-1. Modelos de desarrollo forestal para la diversificación de los sistemas agrarios del Noroeste del Chubut. Diagnóstico Social, 37 pp

Davel M, Tejera L, Honorato M, Sepúlveda E (2001) Plantación de pino ponderosa. Control de malezas, tipos de plantas y fertilizantes. INTA Forestal 1: 4 pp

Fernández ME (2003) Influencia del componente arbóreo sobre aspectos fisiológicos determinantes de la productividad herbácea en sistemas silvopastoriles de la Patagonia Argentina. Doctoral thesis, Comahue National University, 240 pp

Fernández ME, Gyenge J (2009) Testing Binkley's hypothesis about the interaction of individual tree water use efficiency and growth efficiency with dominance patterns in open and close canopy stands. For Ecol Manag 257:1859–1865

Fernández ME, Gyenge JE, Dalla Salda G, Schlichter T (2002) Silvopastoral systems in NW Patagonia: I. growth and photosynthesis of Stipa speciosa under different levels of *Pinus ponderosa* cover. Agrofor Syst 55:27–35

Fernández ME, Gyenge JE, Schlichter TM (2004) Shade acclimation in the forage grass Festuca pallescens: biomass allocation and foliage orientation. Agrofor Syst 60(2):159–166

Fernández ME, Rusch V, Gyenge JE, Schlichter TM (2005) La heterogeneidad de la vegetación en plantaciones forestales en el N.O. de la Patagonia. In: León R, Oesterheld M, Aguiar M, Ghersa C, Paruelo J (eds) La Heterogeneidad de la Vegetación de los Agroecosistemas. F.A.U.B.A, Buenos Aires, pp 413–428

Fernández ME, Gyenge JE, Schlichter TM (2006a) Growth of the grass Festuca pallescens in silvopastoral systems in a semi-arid environment, Part 1: Positive balance between competition and facilitation. Agrofor Syst 66(3):259–269

Fernández ME, Gyenge JE, Schlichter TM (2006b) Growth of the grass Festuca pallescens in silvopastoral systems in a semi-arid environment, Part 2: Parameterization of models of stomatal conductance and leaf photosynthesis. Agrofor Syst 66(3):271–280

Fernández ME, Gyenge JE, Schlichter TM (2007) Balance of competitive and facilitative effects of exotic trees on a native Patagonian grass. Plant Ecol 188(1):67–76

Fernández ME, Gyenge J, Licata J, Schlichter T, Bond BJ (2008) Belowground interaction between trees and grasses in a temperate semiarid agroforestry systems. Agrofor Syst 74:185–197

Fernández ME, Gyenge J, Varela S, De Urquiza M (2012) Improving adaptability to climate change in forestry species: drought effects on growth and wood anatomy of ponderosa pines growing at different competition levels. For Syst 21(1):162–173

Fernández ME, Gyenge J, Varela S, De Urquiza M (2014) Effects of the time of drought occurrence within the growing season on growth and survival of *Pinus ponderosa* seedlings. Trees Struct Funct 28:745–756

Garnier E, Roy J (1988) Modular and demographic analysis of plant leaf area in sward and woodland populations of *Dactilys glomerata* and *Bromus erectus*. J Ecol 76:729–743

Gifford RM, Marshall C (1973) Photosynthesis and assimilate distribution in *Lolium multiflorum* Lam. following differential tiller defoliation. Aust J Biol Sci 26:517–526

Gonda H (2001) Manejo de Pino Ponderosa. CIEFAP-Patagonia Forestal, Año VII, N° 3, pp 7–10

González GL, Caballé G, Rossi CA, De Loof E, De Magistris A, Alonso DO (2013) Contenido de taninos en hojas de *Pinus ponderosa* en sistemas silvopastoriles en el norte de Neuquén. 4 to Congreso Forestal Argentino y Latinoamericano, Iguazú

Gyenge J (2005) Uso de agua y resistencia a la sequía de las principales especies forestales del noroeste patagónico, pino ponderosa y ciprés de la cordillera. Ph.D. thesis, Comahue National University, 364 pp

Gyenge JE, Fernández ME (2014) Patterns of resource use efficiency in relation to intra-specific competition, size of the trees and resource availability in ponderosa pine. For Ecol Manage 312:231–238. doi:10.1016/j.foreco.2013.09.052

Gyenge J, Fernández ME, Dalla Salda G, Schlichter T (2002) Silvopastoral systems in northwestern Patagonia II: water balance and water potential in a stand of *Pinus ponderosa* and native grassland. Agrofor Syst 55(1):27–35

Gyenge J, Fernández ME, Schlichter T (2003) Water relations of ponderosa pines in Patagonia Argentina: implications for local water resources and individual growth. Trees Struct Funct 17(5):417–423

Gyenge J, Fernández ME, Schlichter T (2009) Effect of pruning on branch production and water relations in widely spaced ponderosa pines. Agrofor Syst 77:223–235

Gyenge J, Fernández ME, Schlichter T (2010) Effect of stand density and pruning on growth of ponderosa pines in NW Patagonia, Argentina. Agrofor Syst 78(3):233–241

Gyenge J, Fernández ME, Licata J, Weigandt M, Bond B, Schlichter T (2011) Uso del agua y productividad de los bosques nativos e implantados en el NO de la Patagonia: aproximaciones desde la ecohidrología y la ecofisiología. Ecol Aust 21:271–284

Gyenge J, Fernández ME, Varela S (2012) Short and long-term responses to seasonal drought in ponderosa pines growing at different plantation densities in Patagonia, South America. Trees Struct Funct 26:1905–1917

Harper JL (1990) Population biology of plants. Academic, London

Hodgson J (1979) Nomenclature and definitions in grazing studies. Grass Forage Sci 34:11–18

Holmgren M, Scheffer M, Huston MA (1997) The interplay of facilitation and competition in plant communities. Ecology 78(7):1966–1975

Huang W, Xu Q (1999) Overyield of *Taxodium ascendens-intercrop* systems. For Ecol Manag 116:33–38

Huston JE, Shelton M, Ellis WC (1971) Nutritional requirements of the Angora goat. Tex Agr Exp Stat Bull 1105, 16 pp

IPCC (2007) Climate change 2007: the physical science basis. Summary for policymakers. Contribution of Working Group I to the Fourth Assessment Report. The Intergovernmental Panel on Climate Change. Cambridge University Press, Cambridge

Jobbagy E, Sala O (2000) Controls of grass and shrubs aboveground production in the Patagonian steppe. Ecol Appl 10:541–549

Joffre R, Rambal S (1988) Soil water improvement by trees in the rangelands of southern Spain. Oecol Plant 9:405–422

Kellas JD, Bird PR, Cumming KN, Kearney GA, Ashton AK (1995) Pasture production under a series of *Pinus radiata*-pasture agroforestry systems in South-west Victoria, Australia. Aust J Agric Res 46:1285–1297

Kho RM (2000) A general tree-environment-crop interaction equation for predictive understanding in agroforestry systems. Agric Ecosyst Environ 80:87–100

Kikvidze Z, Michalet R, Brooker RW, Cavieres LA, Lortie CJ, Pugnaire FI, Callaway FM (2012) Climatic drivers of plant-plant interactions and diversity in alpine communities. Alp Bot 121(1):63–70

Laclau P (1997) Los ecosistemas forestales y el hombre en el sur de Chile y Argentina. Boletín Técnico N° 37. Fundación Vida Silvestre Argentina, Buenos Aires

Laclau P, Andenmatten E (2005) Noroeste de la Patagonia: La calidad de sitio y el negocio forestal. IDIA XXI, Forestales. Ediciones INTA, Buenos Aires, pp 230–233

Laclau P, Pozo LM, Huerta G, Andenmatten E, Letourneau F (2002) Rentabilidad de las forestaciones con pino ponderosa (Pinus ponderosa (Dougl.) Laws) en el noroeste de la Patagonia, Argentina. Bosques 23(1):21–36

Lambers H, Chapin FS III, Pons TL (1998) Plant physiological ecology. Springer, New York, 540 pp

Landsberg JJ, Waring RH (1997) A generalized model of forest productivity using simplified concepts of radiation-use efficiency, carbon balance and partitioning. For Ecol Manag 95:209–228

Lantschner MV, Rusch V (2007) Impacto de diferentes disturbios antrópicos sobre las comunidades de aves de bosques y matorrales de *Nothofagus antarctica* en el NO Patagónico. Ecol Aust 17:99–112

Lantschner MV, Rusch V, Peyrou C (2008) Bird assemblages in pine plantations replacing native ecosystems in NW Patagonia. Biodivers Conserv 17:969–989

Lantschner MV, Rusch V, Hayes JP (2011) Influences of pine plantations on small mammal assemblages of the Patagonian forest-steppe ecotone. Mammalia 75:249–255

Lantschner MV, Rusch V, Hayes JP (2012) Habitat use by carnivores at different spatial scales in a plantation forest landscape in Patagonia, Argentina. For Ecol Manag 269:271–278

Lantschner MV, Rusch V, Hayes JP (2013) Do exotic pine plantations favor the spread of invasive herbivorous mammals in Patagonia? Aust Ecol 38:338–345

León RJC, Aguiar MR (1985) Deterioro por uso pastoril en estepa herbácea patagónica. Phytocoenologia 13:181–196

León RJC, Bran D, Collantes M, Paruelo JM, Soriano A (1998) Grandes unidades de vegetación de la Patagonia extra andina. Ecol Aust 8:125–144

Letourneau FJ, Adenmatten E (2007) Crecimiento de Pino ponderosa en el corto y mediano plazo: respuesta a la remoción de vegetación en el sitio de plantación. Actas Ecoforestar 2007:25–27

Letourneau FJ, Medina AA, Andia IR, Andenmatten E, De Agostini N, Mantilaro N (2014) Caracterización Xilotecnológica de la madera de una plantación adulta de Pinus ponderosa de la Patagonia Argentina. Rev Investig Agropecuarias 40(2): 6 pp

Letourneua FJ, Caballé G, Andenmatten E, De Agostini N (2010) Simulación de manejo silvícola en base a umbrales de cobertura en sistemas silvopastoriles compuestos por *Festuca pallescens* y *Pinus ponderosa*. First Patagonian International Agroforestry Congress, Coyhaique

Licata J, Gyenge JE, Fernandez ME, Schlichter TM, Bond BJ (2008) Increased water use by Ponderosa pine plantations in northwestern Patagonia, Argentina compared with native vegetation. For Ecol Manag 255:753–764

Lindenmayer DB, Fischer J (2006) Habitat fragmentation and landscape change: an ecological and conservation synthesis. Island Press, Washington, DC

Loguercio GA, Deccechis F (2006) Forestaciones en la Patagonia andina: potencial y desarrollo alcanzado. Patagonia Forestal, Año XII N° 1. Revista editada por la Secretaría de Extensión y Divulgación del Centro de Investigación y Extensión Forestal Andino Patagónico (CIEFAP)

Martínez-Meier A, Gyenge JE, Dalla Salda G, Fernández ME, Rozenberg P (2013) Implicancias del manejo forestal en la cantidad y calidad de madera producida en sistemas silvopastoriles de pino ponderosa. II Jornadas Forestales de Patagonia Sur, 2° Congreso Internacional Agroforestal Patagónico, Calafate, Santa Cruz, Argentina, pp 110

McCarthy JJ, Canziani OF, Leary NA, Dokken DJ, White KS (2001) Climate change II: impacts, adaptation and vulnerability. http://www.grida.no

McDowell N, Adams H, Bailey J, Hess M, Folb T (2006) Homeostatic maintenance of ponderosa pine gas exchange in response to stand density changes. Ecol Appl 16:1164–1182

Minson JD (1990) Forage in ruminant nutrition. Academia Press Limited, San Diego, p 483

Mundo I, El Mujtar V, Perdomo M, Gallo L, Villalba R, Barrera M (2010) Austrocedrus chilensis growth decline in relation to drought events in northern Patagonia, Argentina. Trees 24:561–570

Nicora EG (1978) Flora patagónica. Parte III: Gramíneas. Colección Científica del INTA, Buenos Aires, 563 pp

Nowak RS, Caldwell MM (1984) A test of compensatory photosynthesis in the field: implications for herbivory tolerance. Oecologia 61:311–318

Ong CK, Leakey RRB (1999) Why tree-crop interactions in agroforestry appear at odds with tree-grass interactions in tropical savannahs. Agrofor Syst 45:109–129

Paruelo JM, Bertiller MB, Schlichter T, Coronato FR (1993) Secuencia de deterioro en distintos ambientes patagónicos: Su caracterización mediante el modelo de estados y transiciones. Convenio Argentino-Alemán, Cooperación técnica INTA-GTZ. Lucha contra la Desertificación en la Patagonia a través de un sistema de monitoreo ecológico (LUDEPA-SME), 110 pp

Paruelo JM, Beltrán A, Jobbágy EG, Sala OE, Golluscio RA (1998) The climate of Patagonia: general patterns and controls on biotic processes. Ecol Aust 8:85–101

Paruelo JM, Golluscio RA, Guerschman JP, Cesa A, Jouve VV, Garbulsky MF (2004) Regional scale relationships between ecosystem structure and functioning: the case of the Patagonian steppes. Glob Ecol Biogeogr 13:385–395

Pelliza Sbriller A, Bonino NA, Bonvissuto G, Amaya JN (1984) Composición botánica de la dieta de herbívoros silvestres y domésticos en el área de Pilcaniyeu, Río Negro. Revista IDIA, Ediciones INTA, Buenos Aires, pp 429–432

Peri PL, Monelos HL, Bahamonde HA (2006) Evaluación de la continuidad del estrato arbóreo en bosques nativos de *Nothofagus antartica* (G. Forst) Oerst bajo uso silvopastoril con ganado ovino en Patagonia Sur, Argentina. Actas IV Congreso Latinoamericano de Agroforestería para la Pecuaria Sostenible, Varadero, 6 pp

Pfister JA, Adams D, Randall C, Wiedmeier D, Cates RG (1992) Adverse effects of pine needles on aspects of digestive performance in Cattle. J Range Manag 45(6):528–533

Poorter H, Nagel O (2000) The role of biomass allocation in the growth response of plants to different levels of light, CO2, nutrients and water: a quantitative review. Aust J Plant Physiol 27:595–607

Pugnaire FI, Armas C, Maestre FT (2011) Positive plant interactions in the Iberian Southeast: mechanisms, environmental gradients, and ecosystem function. J Arid Environ 75(12):1310–1320

Rivero D, Gyenge J, Fernández ME, Licata J, Schlichter T, Bond BJ (2006) Water use efficiency of plant community in NW Patagonia Argentina: effects of land use change and availability of water resources. Ecology in an Era of Globalization: challenges and opportunities for environmental scientist in the Americas. The Ecological Society of America (ESA), Merida

Rois M, Mosquera-Losada MR, Rigueiro-Rodriguez A (2006) Biodiversity indicators on silvopastoralism across Europe. European Forest Institute, EFI Technical Report 21, 61 pp

Rusch V, Sarasola M, Corley JC, Schlichter TM (2004) Sustentabilidad de las plantaciones de coníferas introducidas en la región Andino Patagónica: Biodiversidad e Invasión. Informe final PIA 01/00. Bariloche

Sackmann P, Villacide J, Corley JC (2008) La importancia del manejo de las plantaciones de pinos en la conservación de la diversidad de insectos epígeos. In: Villacide, Corley (eds) Serie Técnica: Manejo Integrado de Plagas Forestales. INTA Bariloche, Bariloche

SAGPyA (2001) Secretaría de Agricultura, Ganadería, Pesca y Alimentos. Proyecto Forestal de Desarrollo. Inventario nacional de Plantaciones Forestales, Argentina, 63 pp

Sand P (2004) 3PGPJS a user-friendly interface to 3-PG, the Landsberg and Waring model of forest productivity. Technical report 140. CRC for sustainable production forestry. CSIRO, 26 pp

Schlichter T (2002) Viabilidad ecológica y económica de las forestaciones con coníferas en sistemas silvopastoriles. Informe final PIA 14/98, Segunda Etapa, Proyecto de Desarrollo Forestal, SAGPyA

Schroth G, Da Fonseca GAB, Harvey CA, Gascon C, Vasconcelos HL, Izac AMN (2004) Agroforestry and biodiversity conservation in tropical landscapes. Island Press, Washington, DC

Somlo R (1997) Atlas dietario de herbívoros patagónicos. Prodesar INTA GTZ, 109 pp

Somlo R, Durañona C, Ortiz R (1985) Valor nutritivo de especies forrajeras patagónicas. Rev Argent Prod Anim 5:588–603

Soriano A (1983) Deserts and semi-deserts of Patagonia. In: West NE (ed) Temperate deserts and semi-deserts. Elsevier, Amsterdam, pp 423–459

Soriano O, Movia C (1986) Erosión y desertización en la Patagonia. Interciencia 11:77–83

Steffen W, Persson Å, Deutsch L, Zalasiewicz J, Williams M, Richardson K, Crumley C, Crutzen P, Folke C, Gordon L, Molina M, Ramanathan V, Rockström J, Scheffer M, Schellnhuber H, Svedin U (2011) The Anthropocene: from global change to planetary stewardship. Ambio 40:739–761

Veblen TT, Lorenz DC (1988) Recent vegetation changes along the forest/steppe ecotone in northern Patagonia. Ann Assoc Am Geogr 78:93–111

Villacide JM, Corley JC (2012) Ecology of the woodwasp Sirex noctilio: tackling the challenge of successful pest management. Int J Pest Manage 58:249–256

Waghorn GC, Jones WT, Shelton ID, McNabb WC (1990) Condensed tannins and the nutritive value of herbage. Proc N Z Grassl Assoc 51:171–176

Woledge J (1977) The effects of shading and cutting treatments on the photosynthetic rate of Ryegrass leaves. Ann Bot 41(6):1279–1286

# Silvopastoral Systems Under Native Forest in Patagonia Argentina

6

Pablo L. Peri, Nidia E. Hansen, Héctor A. Bahamonde, María V. Lencinas, Axel R. von Müller, Sebastián Ormaechea, Verónica Gargaglione, Rosina Soler, Luis E. Tejera, Carlos E. Lloyd, and Guillermo Martínez Pastur

**Abstract**

In Patagonia, silvopastoral systems in *Nothofagus antarctica* (ñire) forest has become an economically, ecologically and socially productive land-use system. Patagonian experience with silvopastoral systems in the past 15 years is reviewed in this chapter. The productivity and nutritive value (crude protein content and dry matter, DM, digestibility) of understorey grassland were dependent on the interaction of environmental (mainly soil water availability and light intensity) and system level management factors. Planned thinning in secondary forest stands provide wood production and also improve the undestorey DM production by increasing incoming radiation. Within a Management Plan, two thinning intensities, depending on stand water stress conditions, are proposed. In addition, the use of Reineke's Stand Density Index (SDI) is recommended to be used when deciding thinning intensities for different canopy covers. Livestock production is the main source of annual income from silvopastoral systems in N. antarctica forest where cattle and mixed livestock production (cattle + sheep) is the main activity. Animal performance at the whole-farm scale is described by comparing a traditional extensive grazing management with an adaptive silvopastoral management. The management factors that favor adoption of silvopastoral system are strategic separation of homogenous

P.L. Peri (✉)
Universidad Nacional de la Patagonia Austral (UNPA), Instituto Nacional de Tecnología Agropecuaria (INTA), CONICET, CC 332, 9400 Río Gallegos, Argentina
e-mail: peri.pablo@inta.gob.ar

N.E. Hansen • A.R. von Müller • L.E. Tejera
C.E. Lloyd
EEA INTA Esquel, Chubut, Argentina

H.A. Bahamonde • V. Gargaglione
EEA INTA Santa Cruz,
CC 332, (9400) Río Gallegos, Santa Cruz, Argentina

Universidad Nacional de la Patagonia Austral (UNPA), Río Gallegos, Argentina

M.V. Lencinas • R. Soler • G.M. Pastur
CONICET, Buenos Aires, Argentina

S. Ormaechea
EEA INTA Santa Cruz,
CC 332, (9400) Río Gallegos, Santa Cruz, Argentina



P.L. Peri et al. (eds.), *Silvopastoral Systems in Southern South America*, Advances in Agroforestry 11, DOI 10.1007/978-3-319-24109-8_6

areas (grass steppe, forest and riparian meadows), stocking rate adjustments based on grassland net primary productivity and the protection of regeneration from herbivores browsing by using individual tree guard. Also, data from litter decomposition, nutrients dynamic and carbon storage are informed. Finally, aspects related to criteria and indicator (C&I) to assess ñire forest's sustainability under silvopastoral use and biodiversity conservation issues are also presented.

**Keywords**
Animal production • Biodiversity conservation • Carbon storage • Nutrients • Nutritive value • Silviculture • Understorey grassland

## 6.1 Introduction

Patagonia region includes five provinces (Neuquén, Río Negro, Chubut, Santa Cruz and Tierra del Fuego) with an area of 197 million hectares and extends from latitudes 37 to 52° 30′ S (Fig. 6.1). There are four main ecosystems within the region: the steppe (representing ~93 % of total area) where extensive sheep production is the main activity, the Andes Mountains where most of the native forests grow (~1.8 %), the ecotone defined as forest-steppe boundary (~3.7 %), and the valleys (~1.5 %) where agricultural production is an important alternative (Peri 2009a).

Rainfall decreases from 4000 to 200 mm year$^{-1}$ from west to east across the Andes Mountains that act as an orographic barrier to moist winds coming from the west. This distinct precipitation gradient, together with local edaphic and topographic variations, substantially influences patterns of vegetation distribution. In the late 1800s, agricultural activities in the 'Pampas' region near Buenos Aires expanded rapidly, and as a result, livestock production (mainly sheep) was pushed into marginal areas such as Patagonia (Table 6.1). Erosion and degradation processes have occurred in several areas of Patagonia due to an overestimation of the carrying capacity of these rangelands, inadequate distribution of animals in very large and heterogeneous paddocks, and year-long continuous grazing (Golluscio et al. 1998). There are more than 6.5 million hectares affected by desertification (del Valle et al. 1995), where annual pasture production does not exceed 40 kg DM ha$^{-1}$.

The southern beech, ñire (*Nothofagus antarctica* (G. Forster) Oerst.), one of the main deciduous native species in this area, covers 751,640 ha (SAyDS 2005) and where much of the silvopastoral activity is concentrated. These forests occur naturally in different habitats such as poorly drained sites at low elevations, exposed windy areas with shallow soils, depressions under cold air influence, or in drier eastern sites near the Patagonian steppe (Veblen et al. 1996). Approximately 70 % of ñire native forest in Patagonia has been used as silvopastoral systems (Peri 2012a). Silvopastoral systems, that combine trees and grasslands or pastures under grazing in the same unit of land, became an economical, ecological and social productive alternative in Patagonia. These systems can provide diversification of farm income, either directly from the sale of timber and animals, and/or indirectly by the provision of stock shelter, enhancing animal welfare, and beneficial effects on soil conservation. There are ecological and economic interactions (positive and/or negative) between the woody, non-woody and animal components of the system. This sustains mainly the sheep and cattle production, and secondly provides a range of wood products including poles, firewood and timber for rural construction purposes (Peri 2005). Several silvopastoral production systems coexist in this region, with small farmers utilizing public lands in northern Patagonia, medium and large landowners ensconced in southern Patagonia and settler and native communities in National Parks arid Protected Areas (Somlo et al. 1997).

In the last few years, management plans or legislation to maximize the forest's sustainable benefits have been included (Peri et al. 2009a). The

**Fig. 6.1** Main ecosystems of the Argentine Patagonian region

proposal involved with the development of a Management Plan for native ñire forests under a silvopastoral use that included, for example, forestry inventories, silvicultural practices, adjustment of stocking rates based on pasture assessments and strategies to achieve forest regeneration. The management plan also included guidelines environmental quality conservation, such as adequate road density plan, biodiversity maintenance and conservation of water quality in streams.

This chapter reviews research results and main management practices related silvopastoral systems in ñire forest in the temperate region of Patagonia.

## 6.2 Understorey Component

The productivity and nutritive value of a pasture in a silvopastoral system is dependent on the interaction of environmental and management

**Table 6.1** The increment in livestock numbers in the Patagonian Region, Argentina, between 1895 and 2008

| Year | Sheep | Cattle | Goats | Horses |
|---|---|---|---|---|
| 1895 | 1,790 | 297.1 | 79.3 | 117.3 |
| 1988 | 13,255 | 756.4 | 1,252.2 | 236.1 |
| 2002 | 8,193 | 896.1 | 959.8 | 196.1 |
| 2008 | 8,335 | 917.9 | 924.8 | 151.9 |

Source: National Agricultural Census, National Institute of Statistics and Censuses, INDEC
Values are '000s of animals

factors, and in turn determines animal performance. For grasslands, under a defined light regime, the main determinants of growth are temperature, water, nitrogen and regrowth duration (Peri et al. 2006a). The main aspects of the incoming radiation, which are modified by trees and affect dry matter (DM) production of the understorey are the light intensity and light quality (Peri et al. 2007). The extent of the effects of the environmental and management factors on DM production and pasture nutritive value depend on seasonal changes and development of trees over time.

### 6.2.1 Understorey Dry Matter Production

In the last 15 years, several studies in Patagonia have evaluated the understorey dry matter (DM) production in ñire silvopastoral systems under a wide range of environmental conditions (Table 6.2). The DM production in these ecosystems ranges from 140 to 3760 kg DM ha$^{-1}$.

Furthermore, in ñire silvopastoral system, it has been demonstrated that the productivity of pasture is dependent on the interaction of soil water availability and light intensity reaching the understorey sward (Peri 2005; Peri et al. 2005b). For pasture growing in moderate water stress site conditions, there is a positive exponential relationship between DM production and the light reaching pasture, decreasing from 2800 kg DM ha$^{-1}$ in the open (100 % transmittance) to 500 kg DM ha$^{-1}$ under sever shade (5 % transmissivity) (Fig. 6.2a). In this condition trees in silvopastoral systems may reduce soil moisture by creating a rain shadow, direct interception of rainfall and root competition. The proportion of fine roots (≤2 mm diameter) of ñire trees are mostly concentrated in the 10–50 cm soil depth which is also where 95 % of pasture species roots are distributed (Peri et al. 2006b; Peri 2011). Thus, competition between tree and pasture roots for water will occur whenever soil moisture drops below field capacity. However, under severe water stress conditions pasture production increased from 5 to 47 % transmissivity and then reached a maximum value of 1400 kg DM ha$^{-1}$ around 50–60 % transmissivity (Fig. 6.2a). Dry matter production then declined with a reduction in tree number of the silvopastoral system that increase light intensity reaching the understorey layer. In these dry conditions, intermediate crown cover may conserve soil moisture through a reduction in evapotranspiration mainly by reducing wind speed within the stand. Wind speed (of great importance in Patagonia) is reduced in ñire forest up to 80 % compared with adjacent open areas (Bahamonde et al. 2009).

Other study evaluated the dry matter production of grasses in different radiation levels and its variation over time, in ñire forests under silvopastoral use growing in contrasting environmental (photosynthetic active radiation, air and soil temperature, relative air humidity and soil moisture) conditions in Southern Patagonia (Bahamonde et al. 2012a). The authors reported that the forage production was negatively correlated with site class (Fig. 6.2b). The effect of environmental variables over forage production depended on each site and year; however in most cases the aboveground biomass inside the forest was equal or higher than that of the adjacent areas without trees. The results suggest that silvopastoral use of ñire forests at intermediate crown covers may be desirable from the standpoint of forage production.

Understorey DM production response to thinning and pruning has been evaluated previously for main woodlands types in Chubut (Fertig et al. 2007, 2009). While in high ñire woodlands, DM production after 4 years of forest intervention increased 977 kg DM ha$^{-1}$ year$^{-1}$ compared with non-intervened forests, in shrubby woodland, the increase was 45 kg DM ha$^{-1}$ year$^{-1}$ (Table 6.3).

**Table 6.2** Mean annual dry matter (DM) production in ñire silvopastoral systems under different environmental conditions in main Patagonian provinces

| Province | Main site characteristics | kg DM ha$^{-1}$ | Source |
|---|---|---|---|
| Río Negro (41° 39′ SL) | Two locations and three stand densities (low density was 300–400 trees ha$^{-1}$) | 470–2173 | Somlo et al. (1997) |
| | Two soil moisture condition and two crown cover (30 and 60 %) | 1106–2575 | Sarasola et al. (2008a) |
| Chubut (42° 54′ SL) | Humid (tree dominant high 15 m) and dry (4 m) stands with contrasting crown cover | 186–2002 | Fertig et al. (2007, 2009) |
| Santa Cruz (50° 33′–51° 19′ SL) | Different radiation levels in contrasting environmental conditions | 140–3180 | Peri et al. (2005a) and Bahamonde et al. (2012a) |
| Tierra del Fuego (54° 10′–54° 21′ SL) | Different radiation levels in contrasting environmental conditions | 250–3760 | Peri et al. (2005a) and Bahamonde et al. (2012a) |

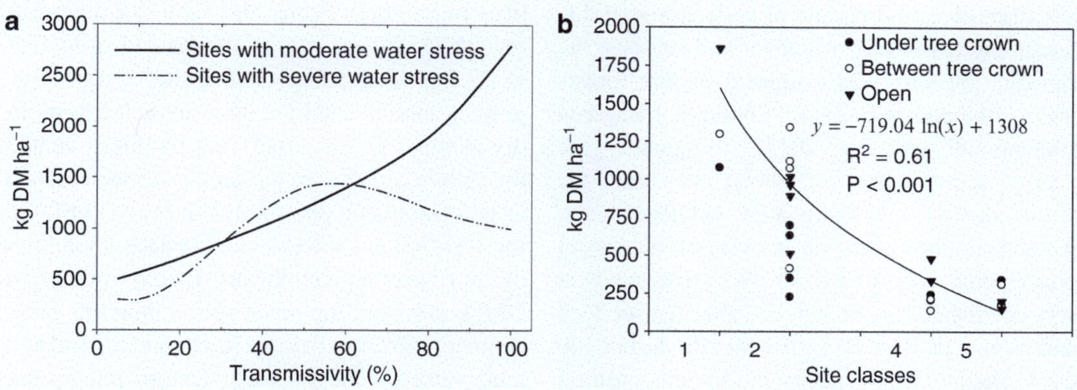


**Fig. 6.2** (**a**) Mean dry matter (DM) grass understorey production in ñire silvopastoral systems under different light intensities (determined by a gradient of crown cover) and adjacent open areas (100 % transmissivity). Severe water stress sites had a mean soil volumetric water content (VWC) in the top 250 mm less than 18 % during the main growing season (October–April) and moderate water stress sites had a soil VWC > 18 %. (**b**) Relation among grass DM production and site class of *N. antarctica* forests in Southern Patagonia

## 6.2.2 Grassland Assessment at Paddock Level

Determining stocking rates is one of the most important decisions in developing a grazing plan for silvopastoral systems. In Southern Patagonia (Santa Cruz and Tierra del Fuego provinces), a method has been developed for carrying capacity estimation in *N. antarctica* forests under silvopastoral use at paddock level (Peri 2009b). The Ñirantal Sur (Saint George) method is based on the Potential Aboveground Net Primary Production (PANPP) estimation for different forest conditions and date of use (spring or biomass peak, summer, autumn and winter), weighted by the area of each homogenous forest unit. This method is easy to use because it only needs three field measurements for PANPP estimation (Table 6.4): crown cover (CC), site class expressed as the mean height of dominant trees (SC) and wood debris cover (D). PANPP ranged from 85 (grassland grown in a SC III forest, CC of 5–30 % and D of 30–50 % in winter) to 2200 kg DM ha$^{-1}$ year$^{-1}$ (grassland grown in a SC I forest, CC of 5–30 % and D of 5–10 % in spring). Details about field measurements and sampling intensity and

**Table 6.3** Forage dry matter production (kg DM ha$^{-1}$ year$^{-1}$) after 3–4 years of ñire forest intervention (pruning and thinning) for different woodlands types in Chubut province

| Treatment | High woodland | Middle woodland | Shrubby woodland |
|---|---|---|---|
| Control | 791 ± 111 | 457 ± 103 | 101 ± 18 |
| | (n = 12) | (n = 25) | (n = 12) |
| Intervention | 1768 ± 499 | 769 ± 222 | 146 ± 32 |
| | (n = 15) | (n = 10) | (n = 12) |

errors are provided in Peri 2009c. An independent validation (n = 20) indicated that this method accounted for 83 % of the variation in PANPP.

Based on PANPP of this grassland assessment method (Table 6.4) and energy concentration values from digestibility data of understorey at different site classes, crown cover and seasons, the carrying capacity was estimated in ñire forests under silvopastoral use in Southern Patagonia (Bahamonde and Peri 2013). To quantify the energy requirements of sheep we used the "Equivalente Oveja Patagónico" (EOP) concept, which is defined as the mean energy requirement on an annual basis of a sheep 49 kg live weight at service, sheared in September, that feat or feed and wean a lamb of 20 kg live weight during 100 days. The carrying capacity in ñire silvopastoral systems varied according to site class, crown cover, woody debris and season of the year. For example, for CC between 30 and 60 %, in summer, the carrying capacity ranged between 27 and 9 EOP ha$^{-1}$ 30 day$^{-1}$ in forests growing at the best and poorest site class, respectively. In the same situations but in winter, the carrying capacity fluctuated between 16 and 6 ha$^{-1}$ 30 days$^{-1}$. This information is important as basis to determine the stocking rate in ñire forest under silvopastoral use with a nutritional criterion. However, for an adequate estimation of the stocking in these systems it is necessary to find a balance between the nutritional requirements of the animals and the conservation of soil and regeneration of the tree stratum. Therefore, this new grassland assessment method is showing promise and is being recommended for *N. antarctica* ecosystems as it also provides an objective technique that avoids overestimation of carrying capacity.

### 6.2.3 Forage Feeding Value

It is well recognised that feeding value of forage is defined as the animal production response to the total herbage consumed, which is a function of voluntary intake, digestibility and efficiency of use of absorbed nutrients. Digestibility is one of the major components of nutritive value because it defines quantitatively the nutrient availability per unit of feed intake.

Peri and Bahamonde (2012) evaluated the monthly variation in understorey organic matter digestibility (OMD) in ñire silvopastoral systems growing at four contrasting environments (dominant tree height from 5 to 12 m) with different crown covers and also in adjacent areas without trees in Southern Patagonia. Values of understorey OMD in these ecosystems ranged from 43.7 to 78.5 % depending on the site, time of the growing season and light intensity being lower in dry summer period. Analysing the mean annual digestibility in grasses measured parameters such as photosynthetic photon flux density (PPFD) in the 400–700 nm waveband and relative humidity did not have any significant (P > 0.05) effect on OMD, however, the mean soil volumetric water content (VWC) in the top 30 cm and air and soil temperature were the main factors influencing OMD variation (Table 6.5). Similarly, Peri et al. (2007) reported small differences between open and shaded treatments for OMD in cocksfoot pastures. In Chubut, understory (grasses and herbs) grazed by livestock in ñire silvopastoral systems during summer showed a mean dry matter digestibility of 57.8 ± 1.7 %, but this decreased during winter (mean value of 48.2 ± 1,9 %) when ñire forests are used as deferred forage (Fertig et al. 2009). Many results have shown that in vitro digestibility of grasses is reduced under low light intensity and this is associated with reduced digestibility of cell wall constituents (Hight et al. 1968). However, other studies have obtained less conclusive results with both increases and decreases in digestibility of herbage grown in a shaded environment (East and Felker 1993; Senanayake 1995).

Linear relationship ($y = -4.8x + 78.5$; $P < 0.001$; $R^2$: 0.84) between OMD% and site

**Table 6.4** Mean Potential Aboveground Net Primary Production (PANPP) (kg DM ha$^{-1}$ year$^{-1}$) (± standard deviation) of grassland under silvopastoral use growing at different conditions of *Nothofagus antarctica* forests (site class (*SC*), crown cover (*CC*), wood debris cover (*D*)) and date of use (spring or biomass peak, summer, autumn and winter) in Southern Patagonia

| | Spring (Nov.–Dec.) | | | Summer (Jan.–Feb.) | | | Autumn (Apr.–May) | | | Winter (June–Aug.) | | |
|---|---|---|---|---|---|---|---|---|---|---|---|---|
| | Debris 5–10 % | Debris 10–30 % | Debris 30–50 % | Debris 5–10 % | Debris 10–30 % | Debris 30–50 % | Debris 5–10 % | Debris 10–30 % | Debris 30–50 % | Debris 5–10 % | Debris 10–30 % | Debris 30–50 % |
| **SC I** | | | | | | | | | | | | |
| CC 5 a 30 % (n=110) | 2200±500 | 1850±425 | 1620±380 | 1630±370 | 1370±315 | 1195±280 | 1430±325 | 1200±275 | 1050±240 | 770±175 | 645±150 | 565±130 |
| CC 30 a 60 % (n=105) | 1200±350 | 1055±370 | 955±275 | 935±275 | 820±290 | 745±215 | 805±230 | 705±235 | 640±185 | 455±135 | 400±140 | 360±105 |
| CC >60 % (n=90) | 750±285 | 680±230 | 580±160 | 605±230 | 550±185 | 470±130 | 510±195 | 442±150 | 375±105 | 315±120 | 285±95 | 240±65 |
| **SC II** | | | | | | | | | | | | |
| CC 5 a 30 % (n=101) | 1100±370 | 990±310 | 870±235 | 780±260 | 705±220 | 620±165 | 615±205 | 555±170 | 485±130 | 308±105 | 275±85 | 245±65 |
| CC 30 a 60 % (n=108) | 745±285 | 655±245 | 565±190 | 545±205 | 480±180 | 415±140 | 430±165 | 380±140 | 325±110 | 245±95 | 215±80 | 185±60 |
| CC >60 % (n=92) | 340±130 | 310±140 | 270±115 | 260±100 | 235±105 | 205±90 | 205±80 | 185±85 | 160±70 | 135±50 | 125±55 | 110±45 |
| **SC III** | | | | | | | | | | | | |
| CC 5 a 30 % (n=102) | 390±125 | 345±130 | 280±90 | 265±85 | 235±80 | 190±60 | 195±60 | 175±60 | 145±45 | 115±35 | 105±40 | 85±30 |
| CC 30 a 60 % (n=107) | 485±210 | 430±185 | 375±110 | 370±160 | 325±140 | 285±85 | 265±110 | 240±100 | 200±60 | 170±75 | 150±65 | 130±40 |
| CC >60 % (n=85) | 280±160 | 250±130 | 220±85 | 220±125 | 200±100 | 175±65 | 190±105 | 170±90 | 150±50 | 130±75 | 115±60 | 100±35 |

Adapted from Peri (2009b, c)

Site Class I: mean height of dominant trees >12 m, Site Class II: mean height between 7 and 12 m, and Site Class III: mean height <7 m.

*n* number of dry matter cuts for each group of crown cover and Site Classes (n total 900 cuts)

**Table 6.5** Coefficient of determination ($R^2$) from simple linear regression between organic matter digestibility (OMD) and main environmental variables (*PPFD* photosynthetic photon flux density, *MAT* mean annual temperature, *MST* mean soil temperature, *RH* relative humidity, *VWC* mean soil volumetric water content at 0–30 cm depth) at different locations in Southern Patagonia

| Site | PPFD | MAT | MST | RH | VWC |
|---|---|---|---|---|---|
| Cancha Carrera[a] | 0.002 ns | −0.46*** | −0.48*** | 0.130* | 0.61*** |
| Tres Marías[b] | 0.001 ns | −0.30*** | −0.43*** | 0.040 ns | 0.39*** |
| Catalana[c] | 0.002 ns | −0.33*** | −0.30*** | 0.012 ns | 0.19** |
| Indiana[d] | 0.005 ns | −0.24** | −0.25** | 0.01 ns | 0.20* |

Linear regression significance, *P<0.05, **P<0.01
*ns* not significant
[a]Location: 51° 13′ 21″ SL, 72° 15′ 34″ WL (Santa Cruz province), Site Class IV: mean height of dominant trees 6–8 m, mean annual temperature (MAT): 6.1 °C, mean annual precipitation (MAP): 360 mm
[b]Location: 51° 19′ 05″ SL, 72° 10′ 47″ WL (Santa Cruz), Site Class V: <6 m, MAT: 5.8 °C, MAP: 320 mm
[c]Location: 54° 10′ 50″ SL, 67° 16′ 02″ WL (Tierra del Fuego province), Site Class II: 10–12 m, MAT: 4.3 °C, MAP: 430 mm
[d]Location: 54° 21′ 47″ SL, 67° 27′ 05″ WL (Tierra del Fuego), Site Class II: 10–12 m, MAT: 4.8 °C, MAP: 480 mm

classes (1–5) indicated highest values in the best forest sites. In conclusion, the interaction of the environmental variables that determined the productive capacity of the forest (clime, soil) also affected the OMD values of the understorey grasses.

On another hand, it has been reported that mean annual undestorey crude protein content (CP%) in four sites of ñire forest under silvopastoral use in Southern Patagonia ranged from 8.2 to 12.2 % (Peri et al. 2005a). In these ecosystems, mean values of CP percentage increased as light intensity declined from 9.9 % in open (100 % light transmission) to 11.2 % under severe shade (10 % transmissivity) showing an interaction (P<0.05) with site quality. The increase in CP% as PPFD declined may be attributed to either a decrease in photosynthates, with a consequent rise in the nitrogen concentration, or to an increase in soil organic matter mineralisation under trees that provided more nitrogen for grass uptake (Peri et al. 2007). Also, the CP% of grasses in ñire silvopastoral systems at different radiation levels and its variation over time, growing in contrasting environmental conditions in Southern Patagonia has been reported previously (Bahamonde et al. 2012a). In this study, there was a seasonal decrease in CP percentage when water stress was severe during the summer period (December–February) (Fig. 6.3). There was an interaction (p<0.05) in CP percentage between light intensities and site quality being higher in grasses growing under severe shade (~25 % transmissivity) and low site quality (Fig. 6.3). Fertig et al. (2009) reported that CP% decreased from 8.9 % in unmanaged ñire stand to 6.9 % in thinned silvopastoral stand (50 % crown cover) in the northwest of Chubut province due to mainly a dilution of mass-based leaf N in understorey higher DM growth rate.

### 6.2.4 Improved Pasture in Ñire Silvopastoral Systems

One approach to improve forage availability and quality in spite of local precipitation in ñire silvopastoral systems is to introduce high productive forage species. Therefore, the introduction of legumes is important because they can serve dual purpose; a source of nitrogen and high quality forage for animal grazing. Peri et al. (2005a) reported that the pasture production can be increased by 20–35 %, depending on light availability, in silvopastoral systems through the introduction of mixed improved pastures, in which clover made up 32 % of the annual dry matter. Furthermore, Peri et al. (2012a) evaluated the monthly production and quality of pasture (crude protein, %CP and in vitro digestibility) during

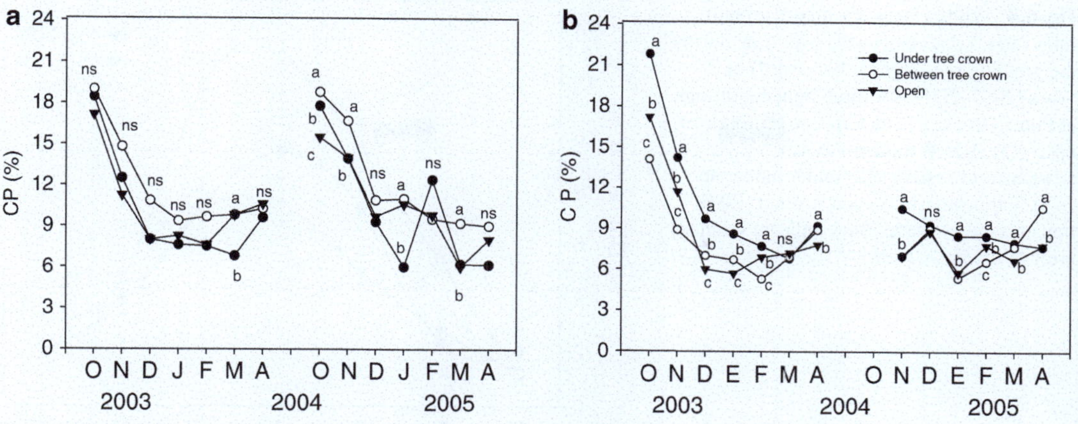


**Fig. 6.3** Understorey crude protein (CP%) over time for three shade conditions at two contrasting silvopastoral sites (**a**) Catalana station, 54° 10′ 50″ SL, 67° 16′ 02″ WL, Site Class II: mean height of dominant trees 10–12 m, MAT: 4.3 °C, MAP: 430 mm; (**b**) Tres Marías station, 51° 19′ 05″ SL, 72° 10′ 47″ WL, Site Class V: <6 m, MAT: 5.8 °C, MAP: 320 mm) in ñire forest (Southern Patagonia, Argentina): open (100 % transmissivity), between trees crown (~50 % transmissivity) and under trees (~25 % transmissivity). Different letters within a date indicate significant differences based on Tukey test at 0.05 probability

**Table 6.6** Total yield and botanical composition (kg MS ha$^{-1}$ año$^{-1}$) of improved pasture with White clover (*Trifollium repens*), and mean crude protein values (%PB) and in vitro digestibility (%DIVMO) during two growing seasons (2007–2008 and 2008–2009) in ñire silvopastoral systems (SS) with 20 and 70 % transmissivity and adjacent open sites (100 % transmissivity) in Southern Patagonia

| | Clover | Herbs | Grasses | Senescent | Total yield | PB | DIVMO |
|---|---|---|---|---|---|---|---|
| | (kg DM ha$^{-1}$ year$^{-1}$) | | | | | (%) | (%) |
| Open | 1380 a | 620 ab | 1020 a | 710 a | 3730 a | 20.9 a | 79.2 a |
| SS 70 % | 1360 a | 990 a | 890 a | 830 a | 4070 a | 20.3 a | 80.2 a |
| SS 20 % | 890 b | 480 b | 410 b | 390 b | 2170 b | 18.1 b | 79.1 a |

Values with different letters within a column are significantly different based on Tukey test at 0.05 probability

the 2007–2008 and 2008–2009 growing seasons (October–April) improved with *Trifolium repens* (white clover) growing under different light levels (20 and 70 % transmissivity) in a ñire silvopastoral system and in an adjacent pasture without trees (100 % transmissivity) at Punta Gruesa, Río Turbio, Santa Cruz province (51° 33′ 10″ SL, 72° 07′ 35″ WL). Total dry matter production varied according to light level being higher with 70 % transmissivity and in the open compared with the pasture grown under 20 % transmissivity related to better soil moisture conditions (Table 6.6). With rotational grazing clover survived well, contributing between 33 and 41 % of the total annual yield. In these ecosystems, maximum white clover growth rate occurred in January with values of 7.5, 9.4 and 12.9 kg DM ha$^{-1}$ day$^{-1}$ for 20, 70 and 100 % transmissivity, respectively (Fig. 6.4). While pasture CP% decreased at low light intensities, digestibility was influenced only by soil water content at the upper 20 cm depth. These results highlighted the adaptation of white clover to ñire silvopastoral systems and their ability to improve the quality of natural pastures.

In order to increase forage DM production in post-fire intermediate ñire woodlands in Chubut, strip clear-cutting (9 m wide separated by 20 m forest) were sown with consociated *Dactylis glomerata – Festuca sp.- Trifolium pratense* pastures (Hansen et al. 2013a). This increased the DM production up to 2178 kg DM ha$^{-1}$. However, the environmental impact of this practice has not been evaluated.

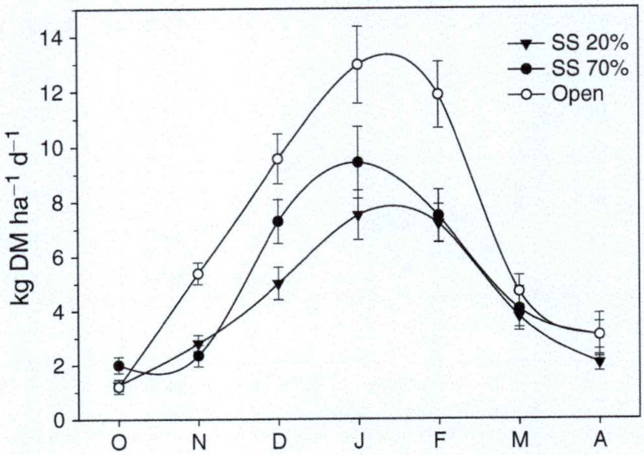


**Fig. 6.4** White clover dry matter growth rates (30±3 days regrowth) over time during the growing season (October–April) of 2 years (2007–2009) for three light conditions at Punta Gruesa (Santa Cruz, Argentina): open (○) (100 % transmissivity), silvopastoral system (SS) under moderate (~70 % transmissivity) and severe (~20 % transmissivity) shade. *Bars* indicate standard error of the mean (sem)

### 6.2.5 Simulation Model of DM Production and CP% of Grasses under Silvopastoral Use

The complexity of a silvopastoral system, in terms of factors that interact with each other, makes it difficult to interpret the effects of several factors acting simultaneously on the results of forage DM production or its crude protein (CP%) concentration, even more when these "results" change over time. To improve the understanding of these processes a tool that integrates existing information with systems approach is required, ie studying the relationships between variables connected to each other through causal relationships, including feedbacks between variables (Haraldsson 2000). This systemic approach also allows describing the dynamics (change over time) of the variables and/or relationships (Haraldsson and Sverdrup 2004). In this context, Bahamonde et al. (2014) generated a simulation model of DM production and CP% of forage in *N. antarctica* silvopastoral systems (Fig. 6.5), using the software STELLA, which simulates the systems dynamics through flows and accumulations. This model considered environmental variables (air and soil temperature, air humidity, soil moisture, site quality, crown cover) and it was focused at stand level of homogeneous even aged ñire forest under Southern Patagonia conditions. To test the model, an independent validation was carried out by comparing real values of dry matter and crude protein with estimated values from the model. The results had a significant lineal correlation for DM ($R=0.52$; $p<0,001$, average underestimation 22 %) and CP ($R=0.47$; $p<0,001$, average overestimation 24 %). Bahamonde et al. (2014) concluded that the results of the simulations were coherent with empirical data in regard to the environmental variables and management of the stand affecting DM and CP% of the understory, and suggested that the model is acceptable for use as a guiding tool in the management of understorey grasses in ñire forests.

## 6.3 Tree Component

Large-scale canopy disturbance in *N. antarctica* forests may occur mainly as a result of blowdown in small gaps scale, and to a lesser extent from snow and/or fire damages resulting from human activities. This contributes to abundant regeneration both from seed and/or root sprouts (e.g., 100,000 seedlings ha$^{-1}$ less than 1 m tall and 20 years of age) followed by self thinning due mainly to light competition resulting in a final stand density of 200–350 trees ha$^{-1}$ in a matured forest (more than 180 years of age). In the province of Chubut, it is possible to distinguish three types of ñire woodland with particular characteristics and uses (Quinteros et al. 2010). These types of ñire woodlands differ from the ones found in the province of Rio Negro (Reque et al.

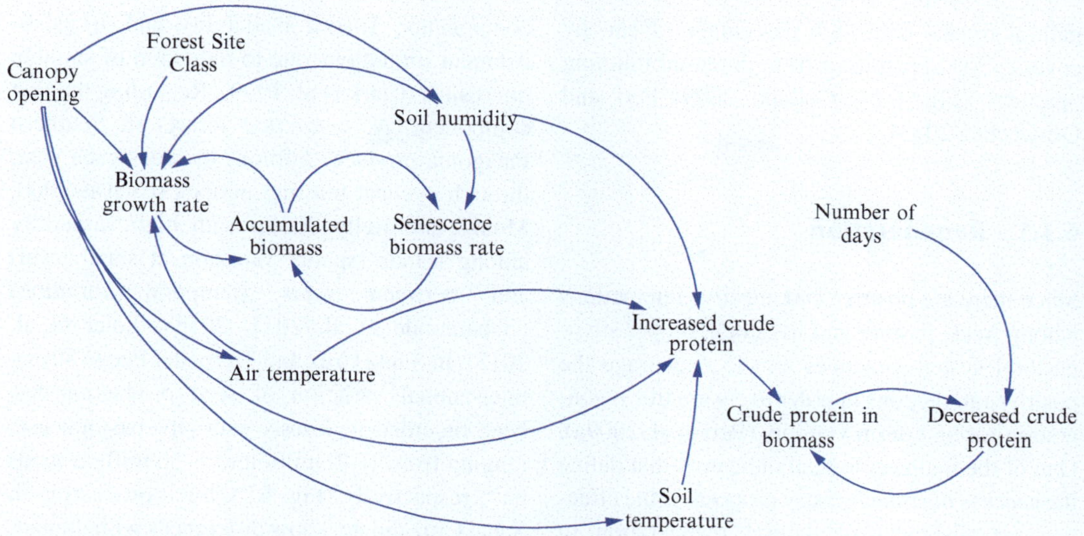


**Fig. 6.5** Loops diagram describing the relationship between different components in a simulation model predicting dry matter production and crude protein concentration in grasses in a silvopastoral system with *Nothofagus antarctica*

2007). For example, high woodland is characterized by tall trees (8–12 m height) with a mean tree density from 500 to 2000 trees ha$^{-1}$, large tree diameters (quadratic mean diameter, QMD, 19.3 ± 1.2 cm) and high canopy cover (>75 %). This type of woodland is mainly covered by a non-regular forest with gaps where tree regeneration relies on mast seeding years and mainly used for wood or firewood production. However, when gaps are created by harvesting, the patches became grasslands for livestock grazing. Middle ñire woodlands are the main forest type cover in Chubut is characterised by a regular structure, a tree density of 1500–3000 trees ha$^{-1}$, mean dominant height of 7.3 ± 0.8 m and 15–30 m$^2$ ha$^{-1}$ basal area. This forest type is mostly used as silvopastoral systems due to more opened canopy (50.1 ± 7.7 %) that allows a higher radiation reaching understory and grass production (Hansen et al. 2008) and timber from harvesting used mainly for firewood due to smaller tree diameters (QMD = 12.7 ± 0.8 cm) with a consequent lower economic income. In contrast, shrubby ñire woodlands on shallow soils with rock near surface are characterized by smaller trees (<4 m height, QMD <12 cm) and the tree density is more than 3000 trees per hectare. In these types of woodlands silvopastoral systems are not established due to limitations in forage production (light and soil limitations). In Santa Cruz most of the forest (>60 % of total area) correspond to irregular structure (uneven-aged stands), they are at mature growth developing phase (>100 years), growing in low site quality (dominant height <8 m), having and average basal area of 10–40 m$^2$ ha$^{-1}$ and <150 m$^3$ ha$^{-1}$ total volume (Peri and Ormaechea 2013). In Tierra del Fuego, around 55 % of ñire forest are characterised by a complete canopy cover (80–100 %) from mature (>100 years) or young growth phase (50–100 years) stands, demonstrating 7–9 m of mean dominant height and a basal area of 10–40 m$^2$ ha$^{-1}$ (Collado 2009).

For silvopastoral systems, planned thinning in secondary forest stands may reduce the time required to yield products of a desired quality, contribute to concentrated growth on selected trees and thereby increase wood production and also improve the undestorey DM production (and consequently increase animal production) by increasing incoming radiation (Peri et al. 2004). In this context, information from provincial inventories indicate that there are approximately 21,480 ha in Santa Cruz province (13.5 % of total forest area) and 108,430 ha in Tierra del Fuego province (59 % of total forest area) of natural forests with high percentage of canopy cover (>70 %) resulting from potential thinning practices (Fig. 6.6) (Collado 2009; Peri and Ormaechea 2013).

### 6.3.1 Regeneration

Since there are positive and negative interactions among trees, pasture and livestock, proper silvopastoral system practices aim to encourage the positive interactions in order to ensure tree regeneration for long-term viability (Peri et al. 2009a). One of the main ecological indicators that define the success of some forestry proposal is the effective establishment of natural regeneration to ensure the continuity of the tree layer (Martínez Pastur et al. 2009). Seedling stage constitutes the earliest and most critical stage of the tree life cycle. It also defines the capacity of forest ecosystems to sustain itself in time and space. Under natural conditions there are several limitations for forest regeneration, from flowering stage to seedlings establishment and survival (Jordano et al. 2008; Soler et al. 2013). These limitations could be intensified in those forests impacted by productive activities. Usually studies related to regeneration limitations emphasize two main components: (i) seed availability, and (ii) establishment limitations due to reduction of suitable microsites (Clark et al. 1999). Regarding the first component, *N. antarctica* forests in Southern Patagonia produces millions of seeds each year, through cyclical masting periods (Cuevas 2000; Monks and Kelly 2006), with high variability among stands (spatial variation at stand level) and between years (temporal variation) (Bahamonde et al. 2011, 2013a; Soler et al. 2013). In Santa Cruz and Tierra del Fuego strong inter-annual variation of seed production has been recorded in forests with silvopastoral use, ranging from 1–40 million to 1–56 million seeds $ha^{-1}$, respectively (Fig. 6.7). Such production in Santa Cruz did not show differences with unmanaged primary forests, where the mean production and inter-annual variations showed similar patterns than in managed sites. On the other hand, in Tierra del Fuego fruiting and seed production were more successful in unmanaged forests. A reduction of basal area in silvopastoral stands generated 40–50 % of empty seeds, which greatly reduces the natural potential for installation at early stage of the regeneration process. Empty seeds could be due to self-incompatibility. As noted in previous studies, *Nothofagus* species prevent inbreeding among populations with

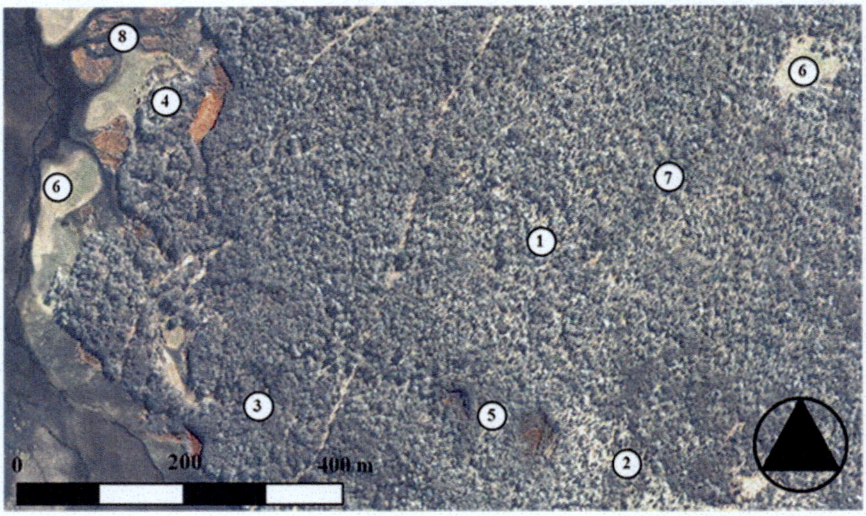


**Fig. 6.6** *Nothofagus antarctica* forests and their natural associate environments in Southern Patagonia: *1* old-growth closed forests, *2* old-growth open forests, *3* secondary forests, *4* forest advances over grasslands, *5* forest advances over peatlands, *6* grasslands, *7* riparian forests in a small stream, and *8* peatlands

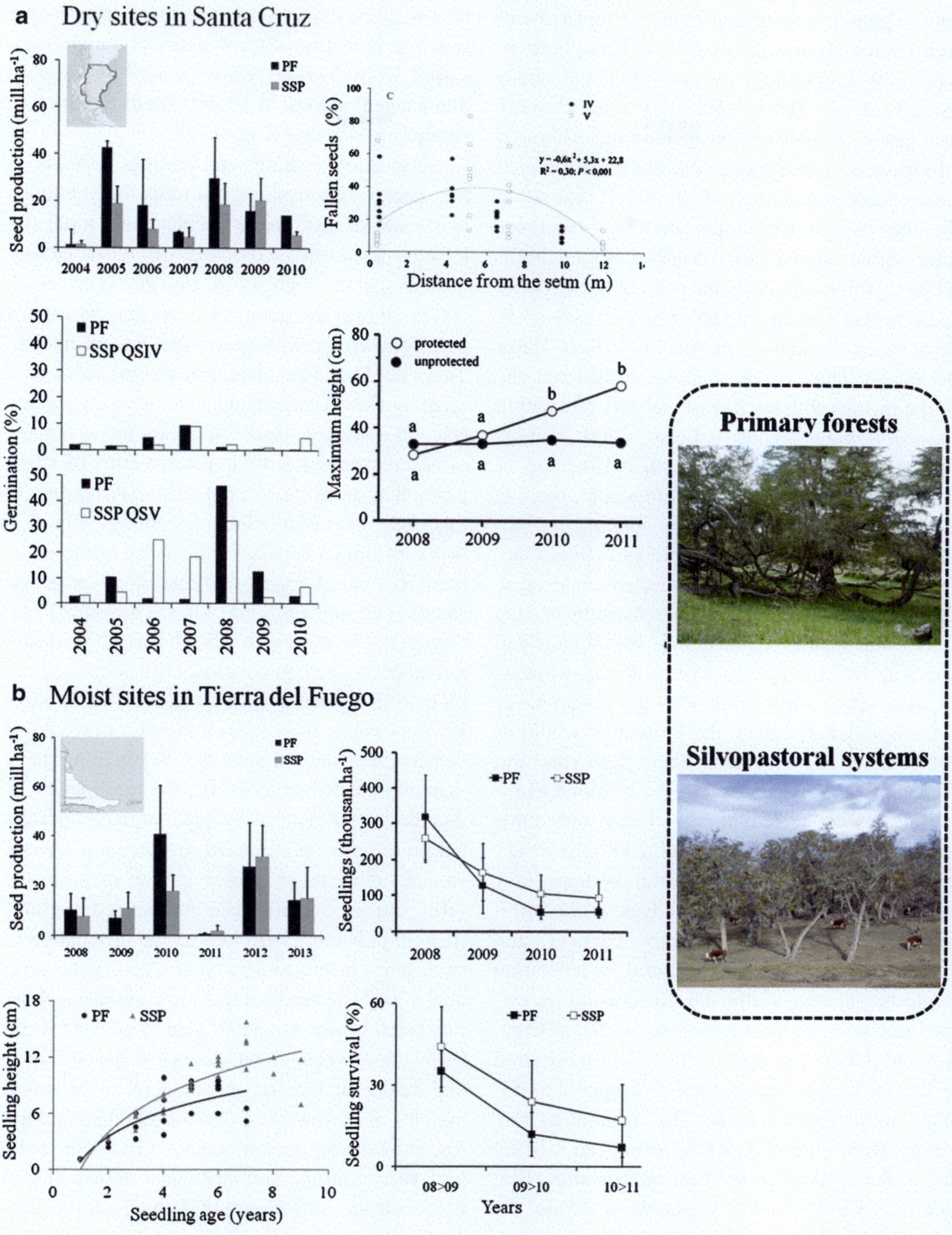


**Fig. 6.7** Seeding patterns and natural regeneration dynamic of *Nothofagus antarctica* in primary unmanaged forests (PF) and silvopastoral systems (SSP) in (**a**) Santa Cruz, and (**b**) Tierra del Fuego, Argentina

similar genetic structure through barriers to avoid fertilization (Donoso et al. 2006). These barriers would be highlighted by environmental stress (Steinke et al. 2008), induced by microclimatic changes generated by forest thinning. Although silvopastoral management did not increase such losses according to previous studies, it is possible that the proportion of empty seeds could increase after more severe interventions (e.g., high-intensity thinning). Also, the percentage of viable seeds is very low and ranged from only 8.5–29 % of full seeds in Santa Cruz and 11–30 % in Tierra del Fuego. Thus, *N. antarctica* is considered one of the species with the lowest viability rate within the *Nothofagus* genus (Cuevas 2000; Burgos et al. 2008; Martínez Pastur et al. 2013).

Regarding to the second component, there is evidence to indicate that canopy opening in silvopastoral systems modifies biotic and abiotic factors in *N. antarctica* forests (Bahamonde et al. 2009; Soler et al. 2013) and thereby influence the micro site quality for seedling installation and survival. Natural dynamics of *Nothofagus* forests is associated with small canopy disturbances (gaps), which stimulate the growth of seedlings established in the forest floor (Rebertus and Veblen 1993), by increasing the resource availability (e.g., solar radiation, soil moisture, nutrients dynamic) (Heinemann et al. 2000). However, canopy opening in silvopastoral systems may also generate more drastic changes on the original overstory (e.g., 50 % canopy openness), and therefore the success of natural regeneration could be negatively affected. Recent studies carried out in *N. antarctica* forests of Santa Cruz, showed that higher seedling installation occurred in forests with silvopastoral use compared to primary unmanaged forests. The installation rate ranges from 10 to 270,000 seedlings ha$^{-1}$, being the highest values in the best quality sites. But, commonly at the end of the growing season, the survival rate is low (<5 % of survival). On the other hand, in Tierra del Fuego, every year installed seedling survives and they range from 2 to 270,000 seedlings ha$^{-1}$ in primary unmanaged forests, and 1–175,000 seedlings ha$^{-1}$ in forests under silvopastoral use. In this site, the first year the survival rate was 15–38 %, which contributed to a well established seedling bank where total seedling densities, after 4 years of monitoring, varied from 54,000 seedlings ha$^{-1}$ in primary unmanaged forests to 93,000 seedlings ha$^{-1}$ in silvopastoral stands (Fig. 6.7).

Survival of seedlings and saplings depends on the species' ecophysiological traits. *N. antarctica* is considered less 'shade tolerant' from a physiological perspective compared to other closely related species such as *N. pumilio* (Peri et al. 2009b). It provides competitive advantages for this tree species to grow in open areas. Recent study in Tierra del Fuego has demonstrated the success of seedling establishment and survival at the sapling stage in managed stands indicating the advantage of *N. antarctica* for the implementation of silvopastoral systems (Soler et al. 2013). According to this study, the mean height of seedlings (Fig. 6.7) was very similar between unmanaged primary forests (10.5 cm at 6 years old) and in silvopastoral stands (11.5 cm at 8 years old). On the other hand, domestic livestock can also have a significant influence on tree regeneration. Using enclosed test plots in silvopastoral systems in Tierra del Fuego, it was possible to observe an increase in seedlings density three times higher (295,000 seedlings ha$^{-1}$) than unprotected sectors (102,000 seedlings ha$^{-1}$). Another experiment in Santa Cruz using individual protections on advanced regeneration demonstrated higher mean annual growth of protected (10.1 cm in height) than unprotected saplings (0.4 cm in height), mainly after the installation of enclosures in the second year. Also, Echevarria et al. (2012) determined that high animal stocking rate (>0.6 bovine units ha$^{-1}$ year$^{-1}$) affected negatively the survival, growth, crown shape (shrubby) and health of ñire regeneration in silvopastoral systems at northwest of Chubut. On the contrary, low to moderate stocking rate (<0.16 bovine units ha$^{-1}$ year$^{-1}$) during summer range allows ñire to recover from cattle browsing through a compensatory growth mechanism. Also, in Chubut, Tejera et al. (2005) reported a gradient of ñire seedlings establishment at different crown covers varying from 2 to 80 seedlings m$^{-2}$ in the adjacent open site at 10 m from trees and between crowns of a silvopastoral system, respectively. However, after the first growing season, grazing and environmental conditions determined the survival rate and there was more than 90 % seedling mortality rate. There is a general consensus in Patagonia on the need to control effectively the damage induced by livestock (e.g., browsing, trampling) on natural regeneration in silvopastoral plans (Tejera et al. 2005; Reque et al. 2007; Peri 2009d; Echevarria et al. 2012; Peri and Ormaechea 2013), in order to carry out a sustainable forest management for *N. antarctica* forest.

Another less known regeneration strategies of *N. antarctica*, is the great ability of agamic regeneration (e.g., stump regrowth and suckers from roots) under natural conditions (Steinke et al. 2008), which represents a reproductive advantage over the other *Nothofagus* species. Under some natural conditions in which *N. antarctica* lives (e.g., peatlands), regrowth is the only way of reproduction (Donoso et al. 2006). Vegetative reproduction is a key mechanism of persistence for those species facing natural anthropic disturbances that cause partial loss of above-ground biomass (e.g., thinning, browsing damage) (Vesk 2006). This vegetative reproduction has been identified in Chubut and Santa Cruz and it is particularly evident after disturbances like fire and tree logging. In Chubut, it is commonly found in intermediate and shrubby ñire woodlands. In Santa Cruz, 25 % of ñire forest with high regeneration cover (>25 %) originated from root suckers and it was referred to fire disturbance (Peri and Ormaechea 2013).

This ability for regrowth has been demonstrated by Bahamonde et al. (2013b) working on thinning experiment in *N. antarctica* forests of Santa Cruz. Here, the stumps sprouted 1 year after thinning by forming groups of large thin branches having a mean height of 11 cm year$^{-1}$ of growth and a maximum regrowth of 25 cm. Similarly, vegetative reproduction in Tierra del Fuego has been observed mostly in forests with silvopastoral use (25 % of the seedling bank) with 6 cm year$^{-1}$ of mean height growth and this being higher than the growth rate of seedlings >1 year old in the natural forest stand (Soler et al. 2012).

To date, the knowledge about the natural regeneration of *N. antarctica* showed evidence of ecological advantages of this forest species for implementation of silvopastoral systems in Patagonia. In such case, the climatic parameters for each region (e.g., precipitation range) should be integrated into the management plan. Further monitoring and long-term knowledge of regeneration processes would allow us to determine whether the quantity and quality of young trees surviving at managed forests are enough to guarantee the sustainable silvopastoral production in Southern Patagonia forests (Peri et al. 2006c).

### 6.3.2 Silviculture and Wood Production

Intermediate treatment schedules have been proposed for ñire forests, and several thinning trials were established in the Patagonian region. For example, in the central basin of the Foyel River (Río Negro province), Reque et al. (2007) have developed a typologic key the silvicultural stand characterization of the different forest types and local dendrometric equations. Based on the current development stage of the stands, the authors suggested that it is possible to apply forestry models based on density regulation and complete canopy cover. Within a Management Plan for natural ñire forests under silvopastoral use, Peri et al. (2009a) proposed two thinning intensities depending on water stress conditions. For stands growing under severe water stress condition (drier sites near the Patagonian steppe) where dominant trees reach a total final height between 4 and 8 m, a moderate thinning intensity leaving at least 50–60 % of canopy cover is recommended. Usually this remaining crown cover provides protection from desiccant strong winds improving the microclimate conditions for understorey production, tree regeneration and plant biodiversity conservation. In contrast, for stands growing in better site qualities (final height of dominant trees >8 m) with higher annual precipitation regime (>350 mm year$^{-1}$) and deeper soils (>0.5 m depth), more intensive thinning is promoted leaving a 30–40 % of canopy cover. The response of understorey dry matter production to the openness of the original overstorey canopy cover by thinning represented an increment that varied from 300 ± 150 kg DM $ha^{-1}$ for moderate thinning to 1400 ± 250 kg DM $ha^{-1}$ for intensive thinning in better sites (Peri 2009d). Other important consideration of silviculture practices in these forests under silvopastoral use is to guarantee the continuity/presence of tree strata in the long term. One of the major obstacles associated with successful establishment and growth of seedlings is browsing by herbivores (such as rabbits, hare and livestock) and competition with grasses/weeds for light. In several areas of ñire forest under grazing, trees regeneration was completely damaged and tree dynamics was interrupted due to the above indicated challenges (Peri and Ormaechea 2013). In this context, trees may be protected as individual specimens or as small groups by using individual tree guard or small fences (Peri et al. 2009c; Ormaechea and Peri 2010). We suggest to protect a final number of 250 seedlings $ha^{-1}$ for dry sites and 150 seedlings $ha^{-1}$ for better site conditions until regeneration trees reach >2 m height. The results indicated that tree guard can be effectively used to protect individual seedlings from cattle by enhancing height growth (10.0 cm $year^{-1}$ for protected trees vs. 1.8 cm $year^{-1}$ for unprotected). In Chubut, Hansen et al. (2008) reported that it is possible the conservation of established ñire saplings in silvopastoral systems by adjusting the stocking rate and monitoring. They found a forage production threshold of 1200 kg DM $ha^{-1}$ and below this the lateral shoot damage of ñire saplings increased up to 30 %. As an example, in Fig. 6.8, a silvicultural schedule for ñire forest under silvopastoral use is shown (Martínez Pastur et al. 2013). This includes the openness of overstorey canopy cover (30–60 % of the original cover) to enhance undestorey dry matter production (including improvement by introducing pastures such as *Dactylis glomerata* and *Trifolium repens*), thinning and pruning practices, and regeneration protection.

Tree growth in natural unmanaged forest is greatly influenced by the landscape, where higher site quality stands occupy the moister environments with deeper soils, mainly close to *N. pumilio* forests (Fig. 6.6). The lower site quality stands are found in heavy wind exposed or in the ecotone adjacent to the steppe in the drier areas. *N. antarctica* trees rarely live more than 200 years compared to other *Nothofagus* species that reach up to 450 years. The diameter growth is also influenced by the vertical structure, where dominance significantly affects this variable. The annual diameter growth of the dominant trees in the higher site qualities can reach to 0.37 cm $year^{-1}$, while in the lower site class it is 0.22 cm $year^{-1}$. In the same way, dominant trees can grow more (0.22–0.37 cm $year^{-1}$) compared to the suppressed trees (0.14–0.22 cm $year^{-1}$) along the site quality gradients.

Another alternative in the region for forest management planning has been the use of Reineke's Stand Density Index (SDI) to assist in the definition of thinning intensities for different canopy covers (Ivancich et al. 2009). From 266 ñire forest inventory plots in Santa Cruz and Tierra del Fuego, density and mean diameter were obtained, and a model of maximum relative density was fitted, achieving a $SDI_{25} = 1435$ trees $ha^{-1}$ ($\log_{10}N = \log_{10}897000 - 2*\log_{10}QMD$), where N is the number of trees per hectare and QMD is the quadratic mean diameter of the trees. Based on this stand density equation and a previously developed model that estimates stand crown cover (CC%) using stand basal area (BA) ($m^2\ ha^{-1}$) and an independent variable ($CC\% = 99.45 * 1\text{-}e^{-0.0264*BA}$; $R^2 = 0.95$) (Peri 2009b), a new density model was fitted for different canopy covers. This final model predicts the parameter used in an equation to estimate the SDI for a particular QMD at any given canopy cover. This model simplifies data collection during forest inventories, such as the stand density and basal area, which are needed to estimate the thinning intensity to achieve a desired crown cover. This model is independent of stand age and site quality, and could be applied at several forested landscapes in Southern Patagonia.

As part of PEBANPA network (Biodiversity and Ecological long-term plots in Southern Patagonia), permanent plots (1 ha) were established in 2008 in a young pure even age *N. antarctica* stand (41 ± 6 years old, 4055 ± 48 trees $ha^{-1}$, DBH = 5.2 ± 1.8 cm, BA = 29 ± 5.8 $m^2\ ha^{-1}$, $SI_{50} = 7.2$ m) located in Santa Cruz province

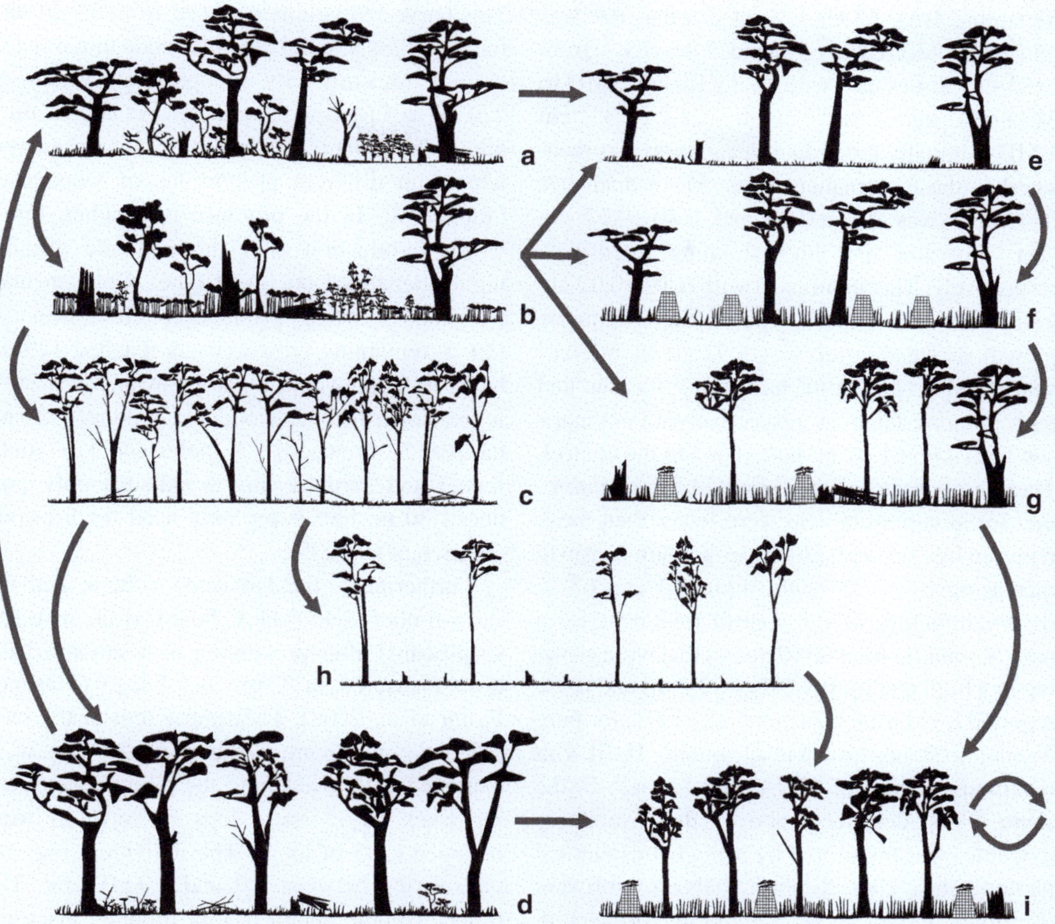


**Fig. 6.8** Silvicultural schedule for ñire forest under silvopastoral use (**a**) old-growth irregular stand, (**b**) old-growth stand with abundant regeneration (>20,000 trees ha$^{-1}$ of 20–40 years old), (**c**) regular stand at optimal growth phase (800–3000 trees ha$^{-1}$ of 60–90 years old), (**d**) mature regular stand (150–250 trees ha$^{-1}$ of >100 years), (**e**) managed stand based on original canopy cover, (**d**) old-growth regular stand, (**f**) managed stand based on original canopy cover with protection of regeneration by using individual tree guards, (**g**) mixed forest structure (original and secondary forest) with protection of regeneration, (**h**) secondary forest with thinning and pruning practices, and (**i**) managed secondary stand with protection of regeneration

(51°13′20″ SL, 72°15′24″ WL). Peri et al. (2012b, 2013a) indicated that thinning (final stocking of 1551 ± 35 trees ha$^{-1}$, crown cover 40 %) for silvopastoral use determined the mean tree DBH growth rate, which increased by approximately 40 % compared with the control treatment. The total over bark volume growth rate at stand level was 4.3 ± 0.65 m$^3$ ha$^{-1}$ year$^{-1}$ and 3.7 ± 0.43 m$^3$ ha$^{-1}$ year$^{-1}$ for the thinned stand and control, respectively. In Tierra del Fuego province, a trial was established in 2009 with two thinning intensities in an area of 5 ha of pure even age *N. antarctica* forest growing at a good site quality ($SI_{50}$ = 12.3 m) in Cape San Pablo Ranch (54° 15′ 45″ SL, 66° 49′ 44″ WL), and also leaving a stand without intervention as a control treatment (Ivancich et al. 2010, 2012). In each treatment five permanent plots were established. In the low intensity thinning treatment (18 m$^2$ ha$^{-1}$ remaining BA), 53 % (±15.9 %) of BA was harvested, decreasing stand density from 2793 ± 448 trees ha$^{-1}$ with 13,4 ± 1.0 cm QMD to 681 ± 48.3 trees ha$^{-1}$ with 18.3 ± 1.7 cm QMD. After thinning, crown cover was 66.5 % and 53.0 % for low and high intensity, respectively. In the high intensity thinning (12 m$^2$ ha$^{-1}$

remaining BA), 65 ± 9.1 % of original BA was removed, decreasing stand density from 2183 ± 834 trees ha$^{-1}$ with 14.3 ± 1.9 cm QMD to 345 ± 63 trees ha$^{-1}$ with 21.2 ± 1.3 cm QMD. Annually individual tree growth was measured in these permanent plots. Mean diametric increment was 0.34 ± 0.05 and 0.40 ± 0.02 cm year$^{-1}$ for low and high thinning intensities, respectively. This contrasted with 0.19 ± 0.02 cm year$^{-1}$ for the control. The mean volumetric growth at stand level was 3.92 ± 0.88 m$^3$ ha$^{-1}$ year$^{-1}$ and 3.12 ± 0.2 m$^3$ ha$^{-1}$ year$^{-1}$ for low and high thinning intensity respectively, in comparison with 4.86 ± 1.01 m$^3$ ha$^{-1}$ year$^{-1}$ in the control. These growth increments reported for *N. antarctica* for silvopastoral use were lower than those reported by Peri et al. (2002) with volume growth increasing by 83 % (light thinning) and 65 % (heavy thinning) of the control treatment for a pure *N. pumilio* forests (67 years) that were growing in a high quality site ($SI_{60}$ = 23.2 m) and those reported by Martínez Pastur et al. (2001) for pure *N. pumilio* stands grown in site quality II–III with maximum growth of 12.7 m$^3$ ha$^{-1}$ year$^{-1}$. In the same plots described above, the overstorey dynamics was monitored by using hemispherical photos and Gap Light Analyzer software (Martínez Pastur et al. 2011). In Río Negro, Sarasola et al. (2008b) reported that the mean tree DBH growth rate varied from 0.18 to 0.49 cm year$^{-1}$ for dense and thinned stands, respectively.

These results for *Nothofagus* forest in Southern Patagonia highlights the importance to establish permanent plots covering different options of thinning treatments (intensity) across a site quality gradient for full growth response evaluation of the *N. antarctica* stands. The importance of long-term plots is that this will allow to determine the economical feasibility of the intermediate treatments, to define base-lines and impacts of different silvicultural treatments, and provide monitoring methodologies and establish demonstrative areas for forest management. Also, these plots provide areas to train professionals in forest management practices.

The volume of harvested logs in mature *N. antarctica* forests under silvopastoral use in Santa Cruz provinces was depended on site class and thinning intensity (Peri et al. 2005a, b, c). This harvested volumes ranged from 33 m$^3$ ha$^{-1}$ for Site Class V stand after moderate thinning (30 % light transmissivity silvopastoral system) to 220 m$^3$ ha$^{-1}$ for Site Class I stand at intense thinning (60 % light transmissivity silvopastoral system) with different proportions of wood uses (Table 6.7). In the province of Chubut, ñire's woody production after thinning and pruning highly depended on forest type. While in high woodlands, wood production after thinning (50 % remaining crown cover) reached 129 m$^3$ ha$^{-1}$ (78 % firewood, 20 % poles and 2 % stick fence), intermediate woodland produced 58 m$^3$ ha$^{-1}$ (92 % firewood, 7 % poles and 1 % sticks fence) and shrubby ñire woodlands only produced 30 m$^3$ ha$^{-1}$ being only used for firewood (Hansen et al. 2005).

Furthermore, the harvested volume and the sawn-timber yield (plank, board, strut, strip and strip-board) after processing in a sawmill have been measured in Tierra del Fuego (Martínez Pastur et al. 2008). The timber (cut at the base bole and minimum diameter of 20 cm) were extracted with skidders to the piling zone (loading truck bay), where logs for sawmill were obtained (3–5 m long). The individual log volume varied between 0.9 and 2.9 ± 0.3 m$^3$. The timber volume obtained after thinning discriminated by log quality categories are presented in Table 6.8.

Because of high defects in the low quality logs, the mean conversion factor of the sawmill (ratio of sawn wood over log volume) was 18 %. However the sawmill conversion efficiency varied according to log quality: 34.4 ± 4.5 % for *B1*, 27.0 ± 3.6 % for *B2*, 10.5 ± 7.4 % for *C1*, 10.4 ± 3.6 for *C2* and 3.7 ± 2.4 % for quality *D*. The sawmill conversion efficiency found for ñire timber was lower than those reported for *N. pumilio* timber in same type of sawmill with a mean value of 41 % conversion factor (Martínez Pastur et al. 2000). Conversion efficiency not only affects sawmill profits, but is also important to estimate existing supplies of standing timber. Therefore, to improve the productive model of silvopastoral systems, it is necessary to take into account the integral harvest of the forest products, the adaptation of the sawmill industry to the wooden resource, diversification of sawn

**Table 6.7** Main structure values of *Nothofagus antarctica* stands of primary forest (PF) and silvopastoral systems (SS) growing at three site classes. The harvested volume after thinning sorted by sawmill timber (*Vs*), pole (*Vp*) y firewood (*Vf*) for 30 (SS 30 %) and 60 % light transmissivity (SS 60 %) is shown

| | | | | | | | Harvested volume | | |
|---|---|---|---|---|---|---|---|---|---|
| Treatment | AB (m² ha⁻¹) | DBH (cm) | Density (n ha⁻¹) | Cob (%) | HD (m) | VT (m³ ha⁻¹) | Vs (m³ ha⁻¹) | Vp (m³ ha⁻¹) | Vf (m³ ha⁻¹) |
| **Site class I** | | | | | | | | | |
| PF | 56 | 27.2 | 950 | 98 | 16,1 | 382 | – | – | – |
| SS 30 % | 34 | 31.0 | 345 | 65 | 16.8 | 238 | 23 | 42 | 85 |
| SS 60 % | 28 | 35.1 | 260 | 38 | 16.5 | 176 | 34 | 66 | 120 |
| **Site class III** | | | | | | | | | |
| PF | 47.0 | 24.8 | 895 | 92 | 8.8 | 276 | – | – | – |
| SS 30 % | 30.1 | 29.4 | 420 | 67 | 8.5 | 187 | – | 29 | 68 |
| SS 60 % | 25.5 | 31.4 | 295 | 34 | 8.6 | 115 | – | 52 | 102 |
| **Site class V** | | | | | | | | | |
| PF | 43.5 | 19.8 | 962 | 90 | 5.1 | 110 | – | – | – |
| SS 30 % | 28.2 | 22.7 | 442 | 65 | 4.9 | 79 | – | 6 | 27 |
| SS 60 % | 20.1 | 24.2 | 312 | 31 | 5.4 | 55 | – | 8 | 56 |

*BA* basal area, *DBH* diameter at breast height, 1.30 m, *Cob* canopy cover, *HD* mean height of dominant trees, *VT* total remanet standing volume

**Table 6.8** Mean harvested timber volume (± standard deviation) of *Nothofagus antarctica* forest in Tierra del Fuego after thinning for silvopastoral use

| | *H* | *AB* | *ABrem* | *VB1* | *VB2* | *VC1* | *VC2* | *VD* | *VT* |
|---|---|---|---|---|---|---|---|---|---|
| | m | m² ha⁻¹ | m² ha⁻¹ | m³ ha⁻¹ | m³ ha⁻¹ | m³ ha⁻¹ | m³ ha⁻¹ | m³ ha⁻¹ | m³ ha⁻¹ |
| Mean | 13.3 | 58.7 | 29.9 | 30.1 | 19.98 | 23.12 | 17.05 | 11.98 | 102.20 |
| SD | 1.3 | 11.46 | 7.61 | 22.95 | 12.88 | 16.45 | 11.24 | 9.70 | 29.04 |

*H* stand dominant height, *AB* original stand basal area, *ABrem* remnant basal area after thinning, *V* logs volume without bark, *B-C-D* logs quality classification, *1–2* log size, *VT* total harvested log volume. Log quality *B1* small end diameter >30 cm, white rot (fungi break down the lignin in wood, leaving the lighter-coloured cellulose behind) <10 %, brown rot (fungi break down hemicellulose and cellulose in wood, leaving brown discoloration, and cracks into roughly cubical pieces) <30 %, sweep <3 cm m⁻¹, breakage <50 cm, knots from alive branches <5 cm diameter and no wobbly log; quality *B2* = small end diameter <30 cm, without white rot, brown rot <10 %, stained wood <10 %, sweep <1 cm m⁻¹, and without breakage, knots or wobbly log; quality *C1* = small end diameter >30 cm, white rot <30 %, brown rot <50 %, sweep <5 cm m⁻¹, stained wood, breakage, knots and wobbly log accepted; quality *C2* = small end diameter <30 cm, white rot <10 %, brown rot <20 %, stained wood accepted, sweep <3 cm m⁻¹, breakage <50 cm, knots from alive branches <5 cm diameter, no wobbly log; quality *D* = small end diameter >30 cm and other variables conditions not included in previous categories

products and increases the added value of commercial products. Thus, the high percentage of small diameter logs produced from ñire forest has a great market value and therefore markets for these products need to be found. Economical and value-added uses for removed small diameter timber can help offset forest management cost in ñire silvopastoral systems and provide economic opportunities for stakeholders. In Tierra del Fuego province, practical use for harvested timber volumes from ñire silvopastoral systems should include high value products used for furniture, flooring and millwork (windows, doors, paneling cabinetry, mouldings and other custom woodwork).

## 6.4 Animal Component

Livestock production is the main annual income from silvopastoral systems in *N. antarctica* native forest under grazing management. Approximately

90 % of total ñire forest area in Santa Cruz under grazing with sheep and cattle (Peri and Ormaechea 2013). In the province of Chubut, the main livestock production under silvopastoral systems has been historically developed for cattle rearing where animal production is 14–16 kg ha$^{-1}$ year$^{-1}$ (Guitart Fité et al. 2004). Animal production occurs mainly in ñire forest, but alternates with other forest types like lenga (*Nothofagus pumilio*) and Ciprés de la Cordillera (*Austrocedrus chilensis*) (Quinteros et al. 2012). The current livestock management on landholdings in Patagonia in ñire forests has a data base for improving productivity and conservation of these ecosystems (Ormaechea et al. 2009). The total number of landholdings with ñire forest and the proportion of these on an area basis were estimated from provincial GIS cadastre and forest inventory (Peri and Ormaechea 2013). There are 102 landholdings (60 in Santa Cruz and 42 in Tierra del Fuego province) practicing silvopastoral systems. While in Santa Cruz, the majority of ranches (55 %) have less than 10 % ñire forest cover, in Tierra del Fuego most ranches (64 %) have between 10 and 50 % of the area covered with ñire forest. The extent of these properties in Santa Cruz province is 21,553 ha and in Tierra del Fuego is 18,050 ha, and the largest area of 160,000 ha. In Chubut, there are 219 landholdings with ñire forest with an average size of 1720 ha (maximum size of 129,700 ha and minimum size of 30 ha). The ñire forest cover in these landholdings ranges from 40 % in large properties to 80 % in small ones.

Through personal surveys conducted in all ranchers in ñire forests, the main livestock characteristics and management were established (Table 6.9). Cattle (mainly Hereford breed) and mixed livestock production (cattle+sheep) were the main activities having mean stocking rates of 0.60–0.86 sheep ha$^{-1}$ year$^{-1}$. In Chubut, the mean annual cattle stocking rate is 0.35 bovine units ha$^{-1}$ year$^{-1}$, but stocking rate fluctuates from 0.24 bovine units ha$^{-1}$ in winter season (May–

**Table 6.9** Main characteristics of livestock production in ranches of Chubut, Santa Cruz and Tierra del Fuego provinces with ñire forest

| | | Santa Cruz | Tierra del Fuego | Chubut |
|---|---|---|---|---|
| Types and percentage (%) of principal animal production | Mixed[a] | 35 | 59 | 60 |
| | Cattle | 39 | 22 | 36 |
| | Sheep | 26 | 19 | 4 |
| Percentage (%) of sheep breeds in ranch[b] | Corriedale | 58 | 100 | 8 |
| | Merino | 31 | 0 | 90 |
| | Others | 11 | 0 | 2 |
| Percentage (%) of cattle breeds in ranch | Hereford | 97 | 97 | 90 |
| | Hereford + Angus | 0 | 3 | 8 |
| | Hereford + other breeds | 3 | 0 | 2 |
| Mean ranch stocking rate (ewes ha$^{-1}$ year$^{-1}$)[c] | | 0.65 ± 0.15 | 0.60 ± 0.10 | 0.86 ± 0.47 |
| Mean lambing percentage (%) for sheep | | 74 ± 5.2 | 76 ± 3.1 | 72 ± 3.4 |
| Percentage (%) of ranch that apply artificial insemination | | 13 | 31 | 10 |
| Mean lamb weights at slaughter on Dec.–March (Kg animal$^{-1}$) | | 12.0 ± 0.71 | 11.6 ± 0.66 | – |
| Percentage (%) of each type of shearing in ranch | Post-lambing | 79 | 96 | 48 |
| | Pre-lambing | 21 | 4 | 52 |
| Mean wool production per animal (kg animal$^{-1}$) | | 4.8 ± 0.28 | 4.5 ± 0.26 | 4.2 ± 0.52 |

Mean ± standard error

[a]Mixed refer to a combined production of sheep and cattle in a same ranch, where neither species participates with less than 10 % of total ranch stocking rate

[b]On the basis of the principal ranch breed. In the case of crossbreed, they were recorded as "Others"

[c]To standardize the stocking rate to ewes per hectare, cattle was multiplied by a coefficient of 6.3 (Cocimano et al. 1977)

November) to 1.11 bovine units $ha^{-1}$ at peak of pastures growth (von Müller et al. 2014). In Chubut, currently mean annual cattle stocking rates in ñire silvopastoral systems is about 0.31 bovine units $ha^{-1}$ $year^{-1}$, but stocking rates fluctuates from 0.24 bovine units $ha^{-1}$ in winter season (May–October) to 0.35 bovine units $ha^{-1}$ in summer season (December to April). However, at peak of pastures growth monthly stocking rate can reach up to 1.11 bovine units $ha^{-1}$ (von Müller et al. 2014). While the main sheep breeds in Sothern Patagonia is Corriedale, in Chubut province it is Merino. For sheep, the lambing rate was 72–76 % and the application of artificial insemination varied from 13 % in Santa Cruz province to 31 % in Tierra del Fuego. In Chubut province, artificial insemination in cattle was approximately 10 % and the percentage of calves weaned was 66 ± 2.9 %. Mostly, the shearing in ranches was carried out post-lambing. Between the provinces there were similar values of lamb weights at slaughter (11.6–12.0 kg $animal^{-1}$) and wool production (4.2–4.8 kg $animal^{-1}$). In Chubut, the mean beef calve (6–8 months) weight at slaughter (April–May) was 165 ± 12 kg $animal^{-1}$.

More than 75 % of the landholdings use winter low altitude ranges (*invernadas*) and summer high altitude ranges (*veranadas*) (Table 6.10). In Chubut province, the *veranada-invernada* grazing management represents 93 % of landholdings (6 % rotational and 1 % year round). However, when considering silvopatoral systems, available management practice include monthly rotational grazing in enriched natural grasslands, where livestock fodder consumption efficiency increased up to 57 % compared with 40 % in the traditional summer-winter ranges (Fertig 2006). Only 6 % of the total landholdings have used a grassland assessment method for carrying capacity estimation in ranches with ñire forests being the main criteria to define stocking rate their personal experience, the range condition or the historic stocking rate used in their respective ranches. Thus, ranches mainly make grazing management decisions based on subjective criteria and previous experience. Also, only a low the percentage of ranches (<17 %) make divisions in paddocks to achieve homogenous grazing based on vegetation types. Usually, vast areas with few paddocks restrict the potential for controlling grazing. In the region, landholdings are using the paddocks within ñire forest mainly for animal breeding or maintenance or without specific objectives. Stakeholder perception highlighted the shelter benefits for animals provided by ñire forest in paddocks.

### 6.4.1 Animal Production

In silvopastoral systems the reduction of light induces changes in nutritive value (foliar nitrogen and herbage in vitro digestibility), structure sward characteristics (height, bulk density, botanical composition) and distribution of morphological components within the canopy that may have an important influence on daily herbage intake. In general, cattle fattening production under silvopastoral systems in the province of Chubut was possible in areas with understory enriched pastures and under intensified livestock management (Fertig 2006). This practice demands heifers and steers post-weaning nutritional restriction during winter, when calves daily body mass gain are between 0.25 and 0.30 kg $animal^{-1}$ $day^{-1}$ until reach a weight of up to 200–230 kg $animal^{-1}$ (Guitart Fité et al. 2004). This activity demands an intensive grazing management along summer growing season in order to obtain a weight gain in steers and heifers up to 320–340 kg $animal^{-1}$. Experimental fattening under ñire silvopastoral systems with understory pastures of *Dactylis glomerata-Holcus lanatus-Trifolium repens-T. pratense* and intensive grazing management with electrified wire fences determined a heifers and steers daily gain of up to 1.0 kg $animal^{-1}$ $day^{-1}$ and 200 kg of meat per hectare from November to May (Fertig 2006).

At experimental spatial scale, animal performance of sheep (Peri 2008) and cattle (Peri et al. 2006d) grazing on natural grassland improved with cocksfoot and white clover pastures under two contrasting canopy cover (CC = 40 and 60 %) and two post-grazing pasture masses (optimum and low) in a ñire silvopastoral system was assessed. The experiments were conducted at silvopastoral site (2 ha) located in Nibepo Aike station (50°33′17″SL, 72°50′33″WL) near Argentino lake, Santa Cruz province. The climate

**Table 6.10** Livestock management and stakeholders perceptions of ranches with ñire forest in Santa Cruz and Tierra del Fuego provinces

| | | Santa Cruz | Tierra del Fuego |
|---|---|---|---|
| Percentage (%) of grazing management | Summer – Winter ranges | 77 | 78 |
| | Rotational[a] | 16 | 16 |
| | Year-round | 7 | 6 |
| Percentage (%) of ranches that apply rangeland assessment[b] | | 6 | 6 |
| Main criterion to define stocking rate by stakeholders (%) | Historic stocking rate | 19 | 47 |
| | Personal experience | 26 | 25 |
| | Range condition | 19 | 16 |
| | Animal condition | 13 | 3 |
| | Rangeland assessment | 6 | 3 |
| | Annual precipitation | 3 | 3 |
| | No specific criterion | 13 | 3 |
| Percentage of ranches that makes divisions to achieve homogenous grazing according to vegetation types (%) | | 6 | 16 |
| Percentage (%) of aim use in paddocks with ñire forest | Without specific objective | 42 | 40 |
| | Breed or maintenance | 42 | 16 |
| | Calving | 10 | 22 |
| | Variable | 3 | 19 |
| | Fattening | 3 | 3 |
| Stakeholders perception (%) related to advantages of paddocks with ñire forest | Shelter | 39 | 38 |
| | Good grass (quality and quantity) | 13 | 12 |
| | Do not know/Do not answer | 48 | 50 |
| Stakeholders perception (%) related to limitations of using paddocks with ñire forest | Inaccessibility, animal lost or herding difficulties | 32 | 7 |
| | Lack of grass biomass | 6 | 3 |
| | Fleece contamination | 3 | 0 |
| | No specific criterion | 59 | 90 |
| Percentage (%) of ranches that apply forestry practices | | 6 | 3 |

[a]It refers to management systems that allow paddocks rest at least one season every 2 year
[b]It refers to rangeland assessments practices at least one time every 2 year or more frequent

is described as temperate and subhumid with a long-term average rainfall of 660 mm. All thinned material was removed. For cattle evaluation Polled Hereford heifer (14 months old, mean weight $264 \pm 32$ kg) and Corriedale ewes (4 years old, mean weigh $42 \pm 5$ kg) were used. Stocking rate during grazing period (30 days), during maximum biomass production (December) under trees and in open, averaged 21 ewes $ha^{-1}$ for cocksfoot and 19 lambs $ha^{-1}$ for lucerne. Stocking rate adjustments were based on similar pasture allowances (mean pasture allowance: 3.1 kg DM $hd^{-1}$ $day^{-1}$ for ewes and 7.9 kg kg DM $hd^{-1}$ $day^{-1}$ for heifer). Liveweight gain per animal (LWG, g $head^{-1}$ $day^{-1}$) and liveweight gain at area basis (kg $ha^{-1}$ $day^{-1}$) were recorded. There was no significant difference ($p > 0.05$) in LWG per animal between crown cover treatments for both for both sheep and cattle at optimum post-grazing forage mass (Table 6.11). When pasture bulk densities were calculated from pre-grazing mass and plant height, density slightly decreased in more shaded stand (CC = 60 %). However, pre-grazing sward characteristics (bulk density and CP% values) in both stands apparently did not influence potential intake from grazing animals. Instead in both crown cover stands there was no difference in LWG per head, the highest pre-grazing mass in the 30 % crown cover silvopastoral system allowed a higher stocking rate and consequently a greater LWG per hectare (3.8 vs. 2.4 kg $ha^{-1}$ $day^{-1}$ for sheep and 29.8 vs. 17.1 kg $ha^{-1}$ $day^{-1}$ for

**Table 6.11** Mean (±standard deviation) values of pre-grazing and two post-grazing pasture masses (optimum and low) masses, pre-grazing pasture height, bulk density, mean crude protein content (CP%) and organic matter digestibility (OMD%), and sheep and cattle liveweight gains (per hectare and per individual animal) during maximum biomass production (December 2005 for cattle and December 2007 for sheep) of pasture grown at two contrasting crown cover (CC=30 and 60 %) in ñire silvopastoral systems in Southern Patagonia

| Treatment | Pre-gazing mass (kg DM ha$^{-1}$) | Pasture height (cm) | Bulk density (mg DM cm$^{-3}$) | CP (%) | Post-gazing mass (kg DM ha$^{-1}$) | Individual LWG (kg day$^{-1}$) | LWG/ha (kg ha$^{-1}$ day$^{-1}$) |
|---|---|---|---|---|---|---|---|
| **Sheep** | | | | | | | |
| CC 30 % | 2850±341 | 24.2±5.1 | 0.971 | 11.2±0.9 | 543±134 | 0.098 a | 3.81 a |
| | | | | | 374±42 | 0.041 b | 1.02 d |
| CC 60 % | 2050±330 | 29.7±4.2 | 0.825 | 15.3±1.4 | 439±76 | 0.102 a | 2.43 b |
| | | | | | 290±74 | 0.048 b | 1.84 c |
| **Cattle** | | | | | | | |
| CC 30 % | 3060±341 | 22.8±6.0 | 0.965 | 10.6±1.1 | 843±134 | 1.39 a | 29.8 a |
| | | | | | 305±33 | 0.20 b | 10.3 c |
| CC 60 % | 1949±330 | 31.7±3.7 | 0.855 | 13.4±1.8 | 864±76 | 1.48 a | 17.1 b |
| | | | | | 298±52 | 0.55 b | 2.3 d |

cattle under optimum post-grazing mass). Thus, the superior animal performance on CC 30 % pastures compared with pastures under more shaded environment was attributed to the greater pasture production rate, pre-grazing mass and bulk density which allowed higher stocking rates and pasture intake. On another hand, there was a strong influence of pos-grazing herbage mass on LWG (per animal and per hectare). This decreased up to 50–80 % for animals grazing until a low post-grazing mass (Table 6.11). It has been demonstrated that low post grazing residuals decreased sward quality (decreases in green leaf content and OMD%) and DM grazing intake reducing consequently animal production (Thompson and Poppi 1990).

From these experiments (spatial scale of 1 ha each treatment) we highlight the importance that to achieve a production target in silvopastoral systems, a grazing manager requires a knowledge of the relationship of animal productivity to pasture allowance (pre-grazing herbage mass) and quality derived from different stand crown cover (thinning intensities) and residual pasture mass. However, there is a need to take into account the whole-farm level requirements when designing grazing prescriptions at a field scale, in order to improve animal production, ecosystem conservation objectives and the modes of adaptation of farm management strategies (Gibon 2005; Peri 2012a). The approach to study silvopastoral systems at the whole farm scale should be applied in developing an understanding of the diversity of livestock farmer decision-making, and should increasingly be used for assessing the economical and/or environmental impacts of change in grassland use in reference to grazing management practices. In this context, an study was carried out to compare a traditional extensive grazing management (TEGM) with an adaptive silvopastoral management (ASM) that included strategic separation in homogenous areas (grass steppe, forest and riparian meadows), stocking rate adjustment to grassland net primary production in order to increase the animal production at ranch level and the protection of regeneration from herbivores browsing (hare and livestock) by using individual tree guard (Ormaechea et al. 2010, 2011). The trial (2008–2009) was conducted under real production conditions in Cancha Carrera ranch (51°13′21″SL, 72°15′34″WL) in the forest-grass steppe region southwest of Santa Cruz province. Grassland assessment in all paddocks (300–5000 ha) was carried out in both management treatments before animals entrance. The ASM group grazed a paddock with only forest between June and September and a riparian meadow paddock (without animals during the spring) during January. Each management treatment had a group of 1000 ewes, where 300 animals were selected randomly to measure live weight gain and corporal condition in four moments of each evaluated year. During the pre-lamb shearing, we weighed 300 fleeces of each group and at the end of the annual season the lambing rate and the weight of 300 lambs was measured. The ASM paddocks were grazed with a stocking rate below its carrying capacity, whereas the TGM registered an overgrazed situation during winter. We found that there were no significant differences in individual animal performance between grazing managements (Table 6.12). Lambing rate and fleece weight for ASM sheep had slightly higher values compared with TGM, probably because the reasonable stocking rate used and the effective use of the forest and the riparian meadow grasses in moments of high quality compared with the rest of areas. In addition, the forest provided protection from cold winds and snowstorms. This advantage was probably the reason for winter weight loss reduction for the ASM group (Table 6.12). However, the main impact of the adaptive silvopastoral management provided an improvement on animal productivity at area basis (Table 6.12). Thus, lamb, meat and wool production in ASM per hectare increased by 30–40 % compared with TGM. Studies of animal distribution using global positioning system (GPS) technology provide important information for grazing plans since it allows long-term, uninterrupted monitoring of grazing animals at low cost. GPS data also allow easy pairing with numerous spatial parameters recorded in a geographic information system (GIS). In this context and for the study area, Ormaechea et al. (2012a) determined the preferred environments used by sheep using with GPS collars. Positions were stored as Gauß-Krüger coordinates of latitude and longitude with information on date, time, and distance from the previous point. Spatial position of the animals was monitored from February 2009 to January 2010 resulting in 50,400 position records for sheep. Grazing frequencies of sheep for each major vegetation type in paddocks were used to determine Ivlev's electivity index (IEI) (Ivlev 1961) and percentage of use by animals of each vegetation type related to paddock area. The IEI takes values between 1 (highly preferred) and –1 (completely avoided). IEI value will be zero (0) when the fraction of feeding time equals the fraction of area covered by the respective vegetation type (random grazing). The results showed a clear animal preference for the forest environment associated with better thermal conditions and good forage quantity and quality. Thus, in those paddocks with mixed vegetation types (ñire forest, riparian meadows and grass steppe), the forest environment had an IEI between 0.23 and 0.63 and sheep using 65–73 % of this vegetation type area. In contrast, sheep avoided the riparian meadow environment during waterlogged periods (spring) and frozen soils (winter) with IEI ranged between –0.02 and –0.24.

Similarly, Peri et al. (2012c) compared the cattle production in a Traditional Grazing Management (TGM) with an Integral Silvopastoral Intensive Management (ISIM) under real production conditions in San Pablo ranch (54°15′46″SL, 66°59′41″WL) in Tierra del Fuego province. While ISIM included strategic separation in homogenous areas (forest and riparian meadows), short rotation grazing and stocking rate adjustment to grassland net primary production, TGM was implemented in winter and summer extensive grazing paddocks (Table 6.13). We considered

**Table 6.12** Main site characteristics and animal performance (mean ± standard deviation) at ranch level of a traditional extensive grazing management (TEGM) and an adaptive silvopastoral management (ASM) over 2 years of measurements (2008–2009) in the forest-grass steppe region, southwest of Santa Cruz province, Patagonia

| | Extensive management | Adaptive management |
|---|---|---|
| Number of paddocks | 3 | 5 |
| Paddocks mean size (ha) | 2470 | 1140 |
| Forage allowance (kg DM ha$^{-1}$) | 301 ± 12.5 | 367 ± 20.9 |
| Stocking rate (ewes ha$^{-1}$ year$^{-1}$) | 0.3 ± 0.02 | 0.9 ± 0.20 |
| Winter weight loss (kg) | 12.4 ± 8.0 | 10.1 ± 8.9 |
| Lambing rate (%) | 84.9 ± 11.9 | 86.9 ± 4.9 |
| Lamb weight (kg) | 33.5 ± 1.9 | 27.6 ± 5.1 |
| Lamb production (lamb ha$^{-1}$) | 0.25 ± 0.02 | 0.41 ± 0.01 |
| Meat production (kg ha$^{-1}$) | 8.25 ± 0.2 | 11.35 ± 1.8 |
| Fiber diameter (μm) | 28.5 ± 0.4 | 28.5 ± 0.7 |
| Fleece weight (kg) | 4.0 ± 0.1 | 4.3 ± 0.4 |
| Wool production (kg ha$^{-1}$) | 1.10 ± 0.01 | 1.55 ± 0.07 |

that defoliation frequency is more important than defoliation intensity for proper grazing management. Each management treatment had a group of 160 Hereford heifers of 185 kg live weight where live weight gain was evaluated in different seasons. In general, while individual live weight gain was higher in TGM (except in autumn), cattle production on area basis was higher in the ISIM treatment (0.20 vs 0.91 kg $ha^{-1}$ $day^{-1}$) (Fig. 6.9). The parasitological results of nematode showed no infection problems.

In this study ranch area, Ormaechea et al. (2012b) analyzed the spatial use of cattle in different paddocks under intensive and extensive management using GPS collar. The results showed higher homogeneity use in smaller paddocks under intensive management achieving a better use of the forage resource during the growing season (Table 6.14, Fig. 6.10). Instead, there was no significant difference in daily walked distances by sheep between seasons and management grazing system (paddocks with different size vegetation types), animals in ISIM explored less area compared with TGM (Table 6.14). This determined that sheep grazing under ISIM used 50–90 % of paddocks area, while animals in the TGM used less than 25 % of paddocks area. We also observed that the cattle have been tamed under intensive management facilitating the handling of them.

**Table 6.13** Cattle grazing schedules for Traditional Grazing Management (TGM) and an Integral Silvopastoral Intensive Management (ISIM) at ranch level in Tierra del Fuego province, Patagonia

| Grazing period | Main vegetation type | N° paddocks | Paddock area (ha) | Pasture allowance (kg MS/ha) | Stocking rate (animal/ha) |
|---|---|---|---|---|---|
| **ISIM** | | | | | |
| Oct.–Dec. | Ñire forest (rotational grazing) | 9 | 20 | 800 | 1.3 (6.9)[a] |
| Jan.–March | Riparian meadow (rotational grazing) | 6 | 40 | 1580 | 0.9 (4.1)[a] |
| April–Sept. | Ñire forest and meadow | 1 | 800 | 710 | 0.2 |
| **Traditional** | | | | | |
| Oct–March (summer ranges) | Ñire forest and meadow | 1 | 470 | 1045 | 0.3 |
| April–Sept. (winter ranges) | Ñire forest and meadow | 1 | 1200 | 650 | 0.08 |

[a]Between brackets instantaneous stocking rate in subdivisions within paddocks

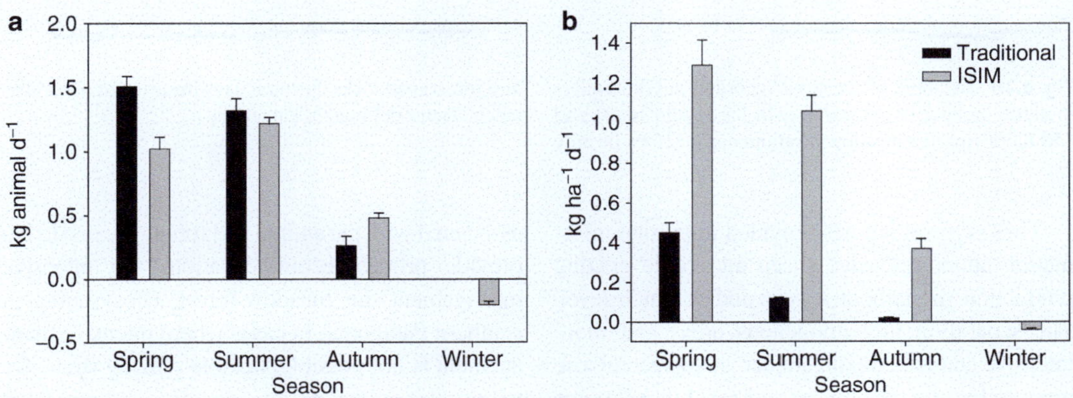


**Fig. 6.9** Cattle liveweight gains per individual animal (**a**) and per hectare (**b**) (mean ± standard deviation) at ranch level of Traditional Grazing Management (TGM) and an Integral Silvopastoral Intensive Management (ISIM) in Tierra del Fuego province, Patagonia

**Table 6.14** Mean and maximum (± standard deviation) daily walked distances and explored areas by cattle during two seasons under Traditional Grazing Management (TGM) and an Integral Silvopastoral Intensive Management (ISIM) in Tierra del Fuego province, Patagonia

| | | | Walked distance (km day$^{-1}$) | | Explored area (ha day$^{-1}$) | |
|---|---|---|---|---|---|---|
| Treatment | Vegetation type | Season | Mean | Maximum | Mean | Maximum |
| TGM | Mix Ñire forest and riparian meadow (1 paddock of 496 ha) | Spring | 4.5 (1.58) | 6.9 (1.90) | 78.9 (48.43) | 202.7 (170.08) |
| | | Summer | 5.3 (1.47) | 8.7 (0.05) | 118.7 (21.00) | 272.1 (99.14) |
| ISIM | Ñire forest (3 paddocks of 7 ha) | Spring | 4.8 (1.60) | 7.8 (2.57) | 6.3 (0.26) | 9.5 (0.47) |
| | Riparian meadow (2 paddocks of 42 ha) | Summer | 4.0 (1.80) | 5.9 (2.52) | 15.3 (6.61) | 21.4 (8.27) |

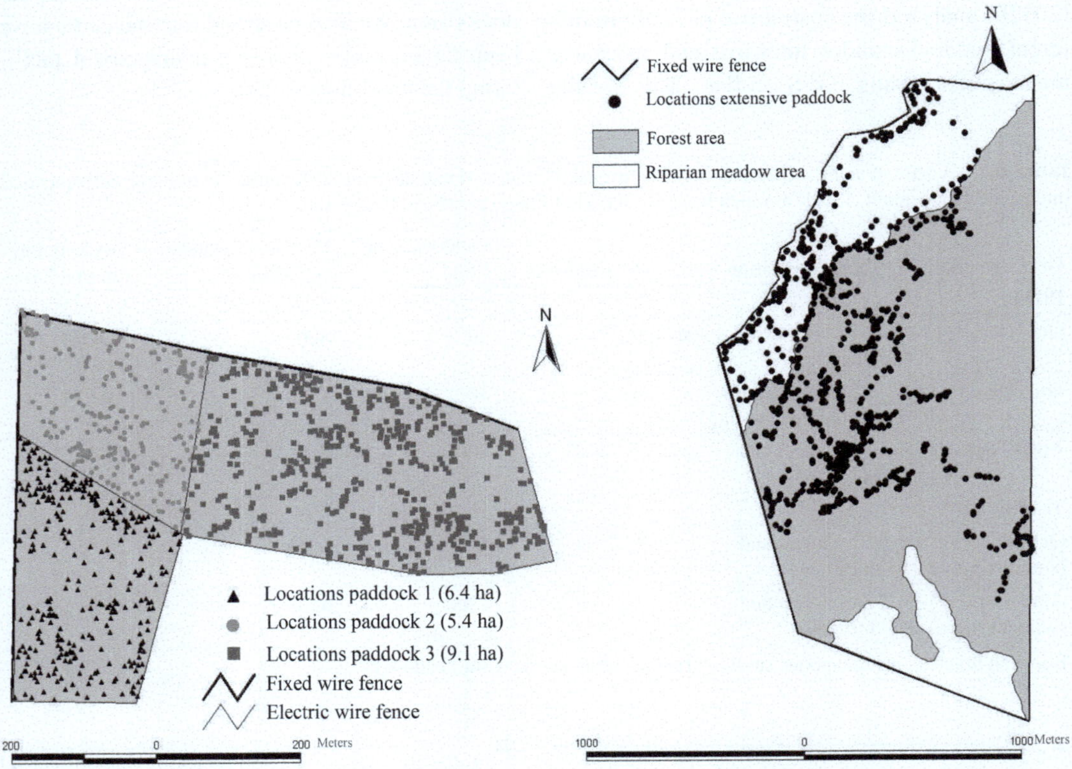


**Fig. 6.10** Example of locations recorded in GPS collars in cattle grazing in extensive (348 ha of ñire forest and 150 ha of riparian meadow) and intensive (21 ha of only ñire forest) paddocks during 8 days in spring at San Pablo ranch, Tierra del Fuego, Patagonia

These type of studies provided important information about extensive sheep and cattle grazing production throughout a year and at ranch level. The separation in homogenous areas and more intensive animal grazing under silvopastoral use proposed at the whole year and ranch level seems to have promising results for ranchers in Southern Patagonia. However, further studies about variables associated with the animal and forage are needed to provide proper recommendations for intensive management in meadow-forest ecosystems in southern Patagonia. In cases where intensive management is not possible, flexible grazing strategies to optimize animal production and avoid grassland ecosystem deterioration is an option for ranchers (Figs. 6.11, 6.12, 6.13, 6.14, 6.15, and 6.16).

**Fig. 6.11** Extensive paddock with *Nothofagus antarctica* forests under silvopastoral use, Tres Marías ranch, Santa Cruz province, Argentina

**Fig. 6.12** Buffers zones for riparian habitats protection within a Management Plan for native ñire forests under a silvopastoral use, Tierra del Fuego province, Argentina

**Fig. 6.13** Thinned stand of *Nothofagus antarctica* in a silvopastoral system at San Pablo ranch, Tierra del Fuego province, Argentina

**Fig. 6.14** Shrubs removal in a *Nothofagus antarctica* silvopastoral system to improve forage production, El Foyel, Río Negro province, Argentina

**Fig. 6.15** Cattle grazing in a *Nothofagus antarctica* paddock under silvopastoral use at Ea. Morro Chico, Santa Cruz province, Argentina

**Fig. 6.16** Stakeholders training in silvopastoral management of ñire woodlands in north-western Chubut, Argentina

## 6.5 Carbon Storage

There is an increasing interest in research related to improve the understanding of carbon (C) sequestration mainly under Article 3.4 of the Kyoto Protocol of the United Nations Framework Convention on Climate Change where countries can count this sequestration as a contribution to reduce greenhouse gas emission (IPCC 2001). Data on C storage in forests is essential for understanding the importance of rapidly increasing level of $CO_2$ in the atmosphere and its potential effect on global climate change. Also, estimates of native forest C storage under different management practices are required for estimating regional and national greenhouse gas balance. In Southern Patagonia, mean maximum annual temperature is predicted to increase by 2–3 °C in 2080 between 46 and 52° 30′ SL (Kreps et al. 2012). Such an increase would have significant effects on Patagonian ecosystems. In this context, secondary indigenous forests are considered efficient C sink ecosystems. Previous research has highlighted the importance of stand age, site quality and crown classes on the magnitude of C pools in both forest above- and below-biomass and forest floor pools (Peri et al. 2005c, 2008a, 2010). It is important to emphasise that roots in these forest ecosystems can contribute up to two times more biomass and C storage than above-ground components in young growth phases. There are few studies on above- and below-ground C pools in Patagonian *Nothofagus* forests under different disturbance and management regimes. In this context, it is also believed that forest ecosystem C pools and fluxes are strongly affected or influenced by forest management.

### 6.5.1 Carbon Storage in Ñire Silvopastoral System

Peri et al. (2010) and Peri (2011) showed that C storage in tree components (leaves, stems, branches, roots) and forest floor change as a result of different forest structure determined by the proportion of crown classes, development stages (age) and the site quality where trees grow. Aboveground and belowground C sequestration for different components of trees and pasture (green and dead leaves, pseudostem and coarse and fine roots), and the C storage in litter floor and at different soil horizons (from 0 to 0.6 m depth) in a *N. antarctica* silvopastoral systems grown in Southern Patagonia at three site classes are shown in Table 6.15. Mean stand density was 175 ± 20 trees $ha^{-1}$ (80 % dominant trees and 20 % codominant trees) at matured stage (190 ± 15 years). In these ecosystems, the C concentration was higher in the rot tree component

**Table 6.15** Mean carbon storage (Mg C $ha^{-1}$) of different components of *Nothofagus antarctica* silvopastoral systems growing at three sites classes (SC) in Southern Patagonia

| System component | SC V | SC IV | SC III |
|---|---|---|---|
| *Soil* | | | |
| Litter | 4.4 | 7.4 | 8.2 |
| Organic horizon (0–03 m) | 9.4 | 13.6 | 16.0 |
| Mineral horizon (0.03–0.1 m) | 12.2 | 17.6 | 18.4 |
| Mineral horizon (0.1–0.3 m) | 44.3 | 81.3 | 85.9 |
| Mineral horizon (0.3–0.6 m) | 55.9 | 98.6 | 104.3 |
| *Total soil* | *126.2* | *218.5* | *232.8* |
| *Trees* | | | |
| Leaves | 0.2 | 0.5 | 0.7 |
| Small branches | 0.6 | 1.0 | 2.4 |
| Sapwood | 4.1 | 7.3 | 10.4 |
| Heartwood | 5.0 | 8.2 | 7.3 |
| Bark | 1.7 | 3.7 | 4.1 |
| Rotten wood | 0.4 | 0.5 | 1.3 |
| Fine roots | 0.6 | 0.1 | 2.0 |
| Coarse roots | 7.7 | 8.7 | 9.3 |
| *Total trees* | *20.3* | *30.1* | *37.5* |
| *Pasture* | | | |
| Green leaves | 0.2 | 0.5 | 0.9 |
| Dead leaves | 0.05 | 0.1 | 0.2 |
| Pseudostem | 0.2 | 0.4 | 0.7 |
| Roots | 1.3 | 2.5 | 2.8 |
| *Total pasture* | *1.7* | *3.5* | *4.6* |
| ***Total silvopastoral system*** | ***148.2*** | ***252.1*** | ***274.9*** |

Site Class III: stands where the mean total height of dominant mature tree (H) reach 10.2 m, Site Class IV: H = 7.8 m, Site Class V: H = 5.3 m

(55 %) and lower in the dead leaves of pasture (40 %). The C concentration decreased from 51 % in floor litter to 0.5 % at 0.6 m mineral soil depth in the low site quality stand. At the silvopastoral stand level, the total C stored ranged from 148 to 252 Mg C $ha^{-1}$ distributed approximately 85 % in soil, 12 % in trees and 3 % in pasture (Table 6.15). Belowground biomass represented an important C storage pool in the ecosystem with mean values of 7.7–9.3 and 1.7–4.6 Mg C $ha^{-1}$ for trees and pasture roots components, respectively. The importance of root as C sink in these forests (Fig. 6.17) has been reported as an adaptation to improve water and nutrient uptake in dry environments and to provide better support in windy sites with shallow soils, compared to other *Nothofagus* species (Peri et al. 2010). While total C accumulation in trees growing in the best site quality (37.5 Mg C $ha^{-1}$) followed the order sapwood > coarse roots > heartwood > bark > small branches > fine roots > rotten wood > leaves, for trees growing in a low site quality (20.3 Mg C $ha^{-1}$) the total C accumulation followed the order coarse roots > heartwood > sapwood > bark > small branches/fine roots > rotten wood > leaves (Table 6.15). This is consistent with Dube et al. (2011) who reported that the carbon storage of *Pinus ponderosa* silvopastoral system was 224 Mg C $ha^{-1}$ in the in the Chilean Patagonia being more efficient in C sequestration than tree plantations or natural prairie.

Carbon storage estimation of aerial and subterraneous (roots) components of *N. antarctica* forests and soil (litter, organic and mineral horizons, 0.6 m depth) at a regional level was carried out under a provincial inventory (355 plots) framework (Peri et al. 2013b). Taking into account the total forest area and the mean C sink values of each stand, we estimated that in Santa Cruz province *N. antarctica* forests accumulate approximately 45 million tons of C, where 20 % corresponded to tree components (aerial and roots) and 80 % in soils. The mean annual C accumulation rate in most of the forest (73 %) was less than 1 Mg $ha^{-1}$ $year^{-1}$ due to the mature growth

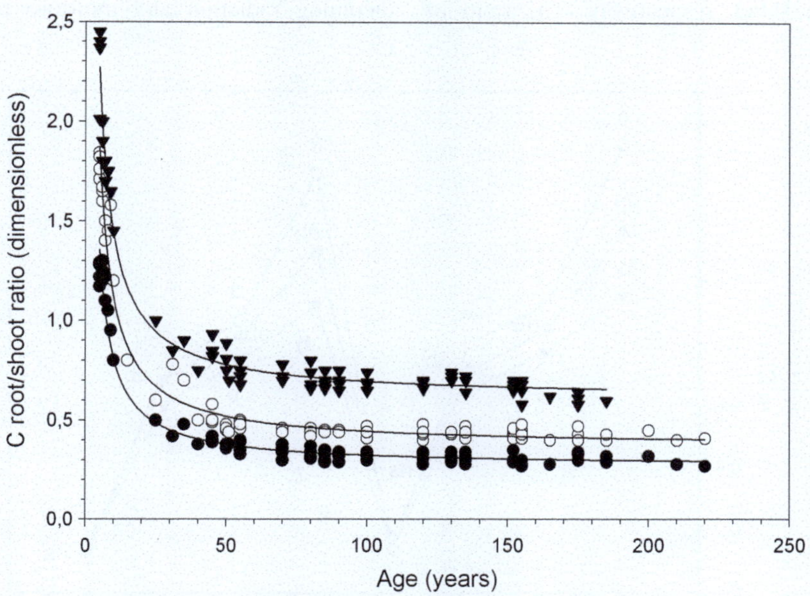


**Fig. 6.17** Carbon root/shoot ratio for *Nothofagus antarctica* trees against age. (●) Site Class III, (○) Site Class IV, and (▼) Site Class V. Site Class III: stands where the mean total height of dominant mature tree (H) reach 10.2 m, Site Class IV: H = 7.8 m, Site Class V: H = 5.3 m

phase state of these forests growing in low site qualities. Carbon sequestration rates ranged from 0.12 to 0.21 Mg $ha^{-1}$ $year^{-1}$. These values are lower than those reported by Kraxner et al. (2003) for typical temperate forests with long-term C storage of 2.5 Mg $ha^{-1}$ $year^{-1}$. Instead, the C sequestration rate in ñire forest is low, and they can accumulate C for long periods (>120 years).

### 6.5.2 Carbon Storage and Soil Respiration for Different Land Uses

The differences in total C storage (above- and belowground biomass components and soil at 0.6 m depth) and soil respiration related to land-management practices were carried out in three study areas of 10 ha each in a *N. antarctica* forest under silvopastoral use (centre of study area located at 51° 13′ 21″ S, 72° 15′ 34″ W), primary forest (51° 13′ 09″ S, 72° 16′ 13″ W) and adjacent open grassland site (51° 12′ 54″ S, 72° 08′ 29″ W). Total C pool was 161.8, 205.6 and 83.8 t C $ha^{-1}$ in the primary forest, silvopastoral system and sole grassland, respectively. The ratio of above to belowground C pools was approximately 1:8, 1:10 and 1:24 for the primary forest, silvopastoral system and adjacent open grassland, respectively. Measurements of $CO_2$ resulting from soil respiration (both from roots and micro-organisms) were taken from each site in main seasons: spring (November), summer (January–February), autumn (April) and winter (July) during 2011–2013 using the soda lime method (Edwards 1982). Soil respiration in silvopastoral system and primary forest was higher ($p<0.01$) than grassland in the adjacent open site (Fig. 6.18). Soil respiration ranged from 0.11 g $CO_2$ $h^{-1}$ $m^{-2}$ in winter to a maximum of 1.43 g $CO_2$ $h^{-1}$ $m^{-2}$ in spring in the silvopastoral system grassland. There was an interaction ($p<0.001$) between land uses and time expressed by seasonal fluctuations in soil respiration, with no differences in soil respiration among sites during winter (July–August). Soil carbon concentration (depth 0.2 m), litter cover and depth and bare soil cover were the main factor explaining 83 % of soil respiration. Also, the trend of higher soil respiration in silvopastoral system compared with primary forest may be due to an improvement of microclimate (soil temperature, incoming radiation and moisture regime) that

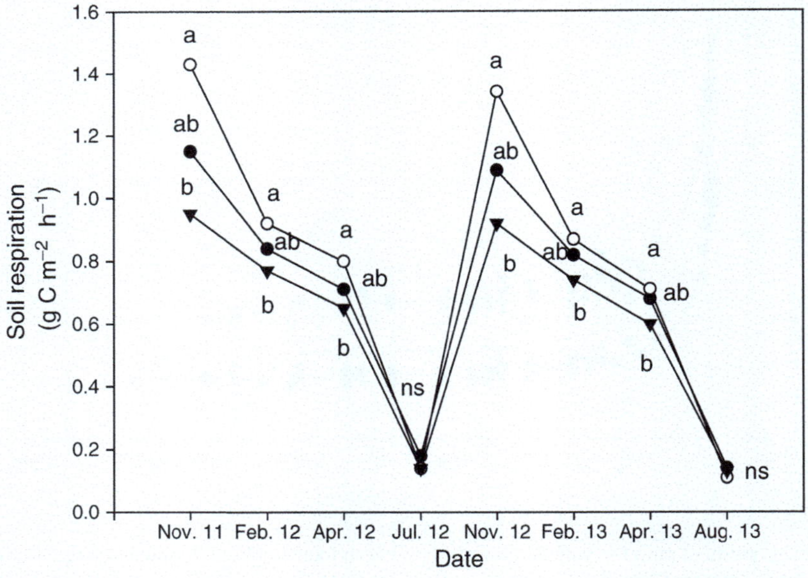


**Fig. 6.18** Seasonal variation of soil respiration for primary forest understorey (●), silvopastoral system (○) and open grassland (▼) in Santa Cruz province (Southern Patagonia, Argentina) Within dates, different lower-case letters indicate significant ($p<0.05$) differences among land uses

enhanced organic matter decomposition. This has been confirmed by Bahamonde et al. (2012b) who determined that total transmitted radiation, soil and air temperatures were the main environmental factors explaining 61 % and 40 % of the variation of litter (grasses and tree leaves) decomposition in *N. antarctica* forest in Patagonia under silvopastoral use.

These studies improved the understanding on the potential of C sequestration of *N. antarctica forests* under silvopatoral management and highlights the importance of these forests as efficient carbon sink ecosystems in Patagonia.

## 6.6 Nutrient Dynamics

A key aspect that determines the sustainability of a forest is maintaining nutrient cycles over time. These cycles may be modified by management practices adopted in different forestry systems (Chapin et al. 2002). However, the effect of silvicultural practices on nutrient cycling may differ depending on the species composition, structure of the forest, age and the environmental conditions (Begon et al. 2006).

### 6.6.1 Litter Decomposition

It is well known that the organic matter decomposition (OMD) is controlled by the chemical composition of the litter, soil organisms and the environmental conditions (Swift et al. 1979). Similarly, special attention should also be paid to N cycling, because N is considered the most limiting element in temperate forests (Fisher and Binkley 2000). In this context, it is desirable to know the influence of silvicultural practices, carried out to improve the forage production in silvopastoral systems, on litter decomposition and nutrients dynamics. In Southern Patagonian ecosystems, the litterfall decomposition and nutrient dynamics in *N. antarctica* forests under silvopastoral use and growing at two site conditions have been reported (Bahamonde et al. 2012b). In both site quality stands, the organic matter decomposition in ñire leaves after 480 days increased with the canopy openness from 20 to 30 % under and between crowns, respectively. Similarly, increased senescent grass decomposition was found in the open without trees when compared with decomposition rates under crown, and the rates were 27 and 38 % under crown and in the adjacent area without trees, respectively. These results suggested that a higher intensity of thinning in these forests (for silvopastoral use) also may increase the decomposition of herbaceous stratum. In practical terms, these results could provide information on the number of years the decomposing material remains in the soil as almost original material. For example, the ñire leaves decomposing under higher tree coverage (under crown) could take between 23 and 27 years to decompose 99 % of organic matter, while in intermediate crown coverage (between trees) that time could be reduced to 15 and 17 years, respectively (Table 6.16). The same pattern was found when the rates of decomposition were transformed in years in the decomposing grasses (Table 6.17).

When the environmental factors driving decomposing processes in silvopastoral systems of ñire forests were analyzed, Bahamonde et al. (2012b) found that the main factor contributing towards the decomposition was the total transmitted radiation for both decomposing materials (ñire and grasses leaves) (Table 6.18).

Bahamonde et al. (2012b) highlighted the importance of radiation level, which is positively influenced by the tree removal that in turn affects decomposition rates. Austin and Vivanco (2006) also have reported that solar radiation may have a direct action on decomposition via photo degradation processes or indirectly by facilitating the action of microorganisms on litter decomposition. Related to this issue, Peri et al. (2008b) found that the removal of trees through thinning practices in forest used for silvopastoral ñire caused a decrease between 35 and 50 % in litterfall to the forest floor, which also would influence lower potential return nutrients. However, as the litter in these disturbed forests will be subjected to a higher rate of decomposition, it could result in a more rapid cycling of nutrients, and consequently decrease the accumulation of litter along with other fine and coarse debris (Frangi et al. 1997).

**Table 6.16** Decay constants ($k$) and total time of decomposition of 99 % ($t_{99}$) organic matter for *N. antarctica* tree leaves growing at two site classes (SC) in Southern Patagonia

| Crown cover | Model[a] | Significance[b] | $R^2$ | $k^c$ (year$^{-1}$) | $t_{99}^c$ (years) |
|---|---|---|---|---|---|
| **SCIV** | | | | | |
| Under crown | $X_1/X_0=92.0e^{-0.112\,t}$ | ** | 0.88 | 0.20b | 23.0a |
| Between crown | $X_1/X_0=90.0\,e^{-0.21\,t}$ | *** | 0.93 | 0.31a | 14.9b |
| **SCV** | | | | | |
| Under crown | $X_1/X_0=89.4\,e^{-0.06\,t}$ | ** | 0.89 | 0.17b | 27.1a |
| Between crown | $X_1/X_0=90.9\,e^{-0.17\,t}$ | *** | 0.96 | 0.27a | 17.1b |

$R^2$ = coefficient of determination
** ($P<0.01$); *** ($P<0.001$)
[a]Model, decay model according to Olson (1963)
[b]Significance, decay model significance
[c]Among crown cover, means followed by the same letter do not differ significantly ($P>0.05$). Analyses were performed separately for each Site Class (SC)

**Table 6.17** Decay constants ($k$) and total time of decomposition of 99 % ($t_{99}$) organic matter for grass leaves growing at two site classes (SC) in Southern Patagonia

| Situation | Model[a] | Significance[b] | $R^2$ | $k^c$ (year$^{-1}$) | $t_{99}^c$ (years) |
|---|---|---|---|---|---|
| **SCIV** | | | | | |
| Under crown | $X_1/X_0=89.1\,e^{-0.14\,t}$ | *** | 0.94 | 0.26b | 17.7a |
| Between crowns | $X_1/X_0=87.8\,e^{-0.19\,t}$ | *** | 0.97 | 0.31b | 14.9a |
| Without trees | $X_1/X_0=84.6\,e^{-0.22\,t}$ | *** | 0.94 | 0.39a | 11.8b |
| **SCV** | | | | | |
| Under crown | $X_1/X_0=89.2\,e^{-0.15\,t}$ | *** | 0.98 | 0.25b | 18.4a |
| Between crowns | $X_1/X_0=89.2\,e^{-0.21\,t}$ | *** | 0.97 | 0.32ab | 14.4ab |
| Without trees | $X_1/X_0=86.1\,e^{-0.24\,t}$ | *** | 0.94 | 0.39a | 11.8b |

$R^2$ = coefficient of determination
** ($P<0.01$); *** ($P<0.001$)
[a]Model, decay model according to Olson (1963)
[b]Significance, decay model significance
[c]Among crown cover, means followed by the same letter do not differ significantly ($P>0.05$). Analyses were performed separately for each Site Class (SC)

**Table 6.18** Simple linear regression between decay constants (k) and main environmental variables

| | Grass leaves | | *N. antarctica* leaves | |
|---|---|---|---|---|
| Environmental variable | Significance[a] | $R^2$ | Significance[a] | $R^2$ |
| Total transmitted radiation | ** | 0.61 | ** | 0.49 |
| Average air temperature | ** | 0.40 | ** | 0.41 |
| Average soil temperature | ** | 0.40 | ** | 0.40 |
| Volumetric soil moisture | ns | | ns | |
| Air relative humidity | ** | 0.30 | * | 0.27 |

$R^2$ = coefficient of determination
*ns* no significant
* ($P<0.05$); ** ($P<0.01$)
[a]Significance, linear regression significance

Also, Bahamonde et al. (2012b) evaluated the mineralization-immobilization dynamics in decomposing leaves of ñire and grasses. The authors reported that N, P and Ca were mineralized during the first stage (60 days) and then got immobilized, while K was immobilized throughout the study period (480 days). This pattern – mineralization and then immobilization of nutrients – of nutrient dynamics was also seen under different crown coverage in the studied sites.

### 6.6.2 Soil N Mineralization

In two *N. antarctica* forests under silvopastoral use in Southern Patagonia a net nitrogen soil mineralization study was conducted (Bahamonde et al. 2013c). The authors reported that the predominant form inorganic N in the ñire soil forests was the $NH_4^+$, representing on average 75 % of the total etractable N. The net N mineralization varied significantly between crown cover and dates, depending on the studied stand. Similarly, when the values of annual N net mineralization per unit area were compared the results varied according to the site: in Site Class IV annual net mineralization was lower in the location without trees, while in Site Class V there were no differences among crown cover (Table 6.19).

When environmental factors were correlated to the N mineralization, the simple linear regression analysis showed that the volumetric soil moisture was the only parameter that explained variation in N mineralization at the studied sites ($P<0.05$). Also, the potential net nitrogen soil mineralization from these forests was evaluated under laboratory conditions using soil samples (0.3 m depth) taken at three different dates (Bahamonde et al. 2013c). As it was expected, net nitrogen mineralization measured under laboratory conditions was higher than *in situ* for both site class stands during two dates, which reflected the environmental (mainly temperature) limitations in the field. The authors attributed these results to other factors limiting the process, e.g. microbial biomass of soil on that date.

**Table 6.19** Total annual net nitrogen mineralization ($NH_4+NO_3$) (kg N ha$^{-1}$ year$^{-1}$) measured in *N. antarctica* forest soils (0–20 cm) at three crown cover and two site classes (SC)

| | Crown cover | | |
|---|---|---|---|
| | Under crown | Between crowns | Without trees |
| SCIV | 54.4 a | 72.4 a | 11.0 b |
| SCV | 51.6 a | 59.5 a | 59.1 a |

Different letters in the same Site Class indicate significant differences ($P<0.05$) between radiation levels

### 6.6.3 Nutrient Amounts in Silvopastoral Systems

Information on biomass and nutrient accumulation in different tree components are essential to evaluate the importance and impact of management practices (silviculture, silvopastoral systems, harvesting) on site productivity, mineral fertility, bio-element recycling and long-term effects on the mineral balance (Santa Regina 2000). The ability of *N. antarctica* to grow in different site qualities and environmental conditions (drier sites near the Patagonian steppe where dominant trees reach <4 m height and sites with MAP >600 mm where trees reach >10 m) is also associated with nutrient redistribution in trees biomass (Peri et al. 2006b, 2008a; Gargaglione et al. 2010, 2013). While biomass and nutrients partitioning to roots increased when trees grow in low quality sites, in better sites trees distribute more resources to aerial components, like stems and leaves. Site quality also influenced biomass and nutrients amounts of *N. antarctica*, where old dominant trees in better sites accounted for 2983, 2578, 839, 393, 365, 357 g tree$^{-1}$ of Ca, N, K, P, S and Mg, respectively, meanwhile in lower site qualities old dominant trees only accounted for 1039, 626, 341, 208, 129 and 133 g tree$^{-1}$ in the same order (Gargaglione et al. 2013). The lowest nutrient contents in worse sites was due to both less biomass accumulation and nutrient concentration in all trees component (Peri et al. 2006b, 2008a). Other factor that influence nutrient amounts in *N. antarctica* tree is the crown class

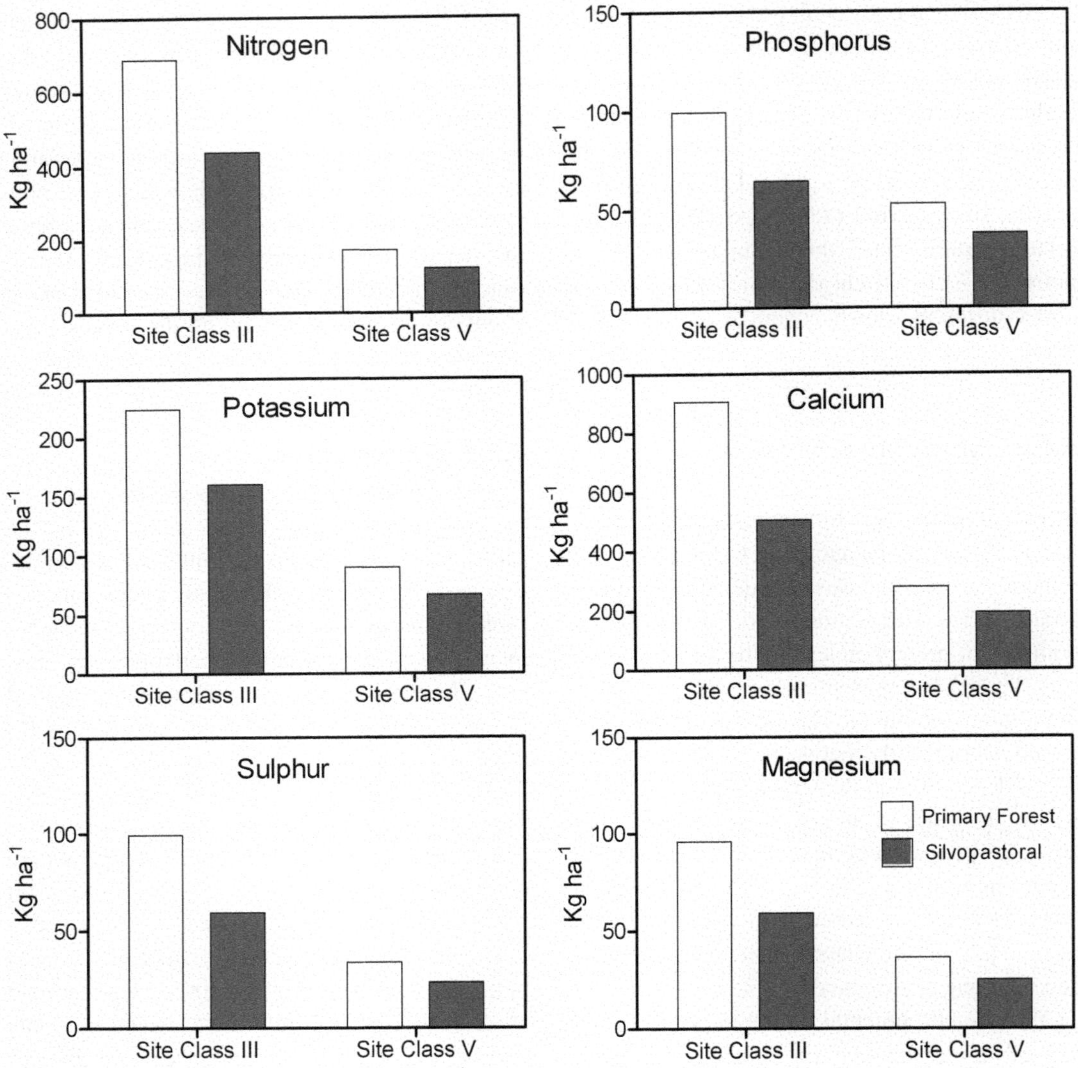


**Fig. 6.19** Mean nutrient amounts (kg ha$^{-1}$) in *Nothofagus antarctica* forests under two uses (primary forest and silvopastoral system) and two contrasting site qualities (Site Class, SC III where old dominant trees height reach 9 m and SC V with tree height of 5.3 m) in Southern Patagonia, Argentina. *D* percentage of dominant trees at stand level, *C* codominant, *I* intermediate, *S* suppressed trees

condition. As this species is not shade tolerant, the position of the trees in the canopy is very important in relation to nutrient accumulation.

Thinning practice in silvopastoral systems reduced the nutrient amounts retained in the biomass of the ecosystem mainly by removing trees from the stand and by changing the proportions of crown classes (dominant, codominant, intermediate and suppressed trees). Figure 6.19 shows nutrient allocation for a primary forest and under silvopastoral use at two different quality sites. Nutrients accumulation at stand level varied from 907, 688, 225, 100, 99 and 96 kg ha$^{-1}$ of Ca, N, K, P, S and Mg, respectively in a primary forest growing at site class III to 192, 121, 67, 38, 24 and 25 kg ha$^{-1}$ for the same nutrients for a silvopastoral system growing on a marginal site class. The nutrient accumulation in understorey grass biomass (2100 kg DM ha$^{-1}$ in SC III and 326 kg DM ha$^{-1}$ in SC V) did not compensate for the nutrient lost from trees removal during thinning management practice. In both sites, it is important to note that majority of the nutrients were being accumulated in leaves and small branches. Because these components usually show a rapid turnover in the ecosystem (structures with low lignin contents), thinning practices should aim to leave fine material (mainly leaves, small branches and bark) on the site to ameliorate nutrient removal and to avoid a decline of long-term yields. Furthermore, a good proportion of total nutrients are also located in roots, and in this case when trees are removed, the root system remains in the soil and this proportion of nutrients may be available for underneath grasses when roots decompose over a long period of time. This effect could be more important in lower quality sites, where trees allocate more resources to belowground components.

### 6.6.4 Nitrogen Facilitation in Silvopastoral Systems

Another important point to consider in silvopastoral systems is the interactions that exist between grasses and trees. There is evidence that positive (facilitation) and negative (competition) interactions occur between woody and non-woody (herbaceous understory) components. For example, in savannas it has been reported that trees and grasses compete for light, nutrients and water, but trees can increase soil fertility, microbial activity or improve soil structure, and also soil water availability by reducing water loss from reduced evapotranspiration in shaded conditions (Archer et al. 2001; Simmons et al. 2008). Positive mechanisms may act simultaneously with competitive mechanisms, and the overall effect of one plant species on another depends on which mechanisms are the most important in a given environment (Callaway et al. 2002). In our particular case of *N. antarctica* silvopastoral systems, there have been reports that grasses growing under tree canopy used the applied N fertilizer more efficiently (Gargaglione et al. 2014). Thus, a study with isotope $^{15}$N labeled fertilizer showed that grasses growing under tree canopy in silvopastoral systems accounted with higher N concentrations in aerial components (Fig. 6.20a) and also around 60 % of the total N was derived from the applied fertilizer (Ndff) throughout the growing season. In contrast, grasses growing on adjacent open sites without trees only had a 20–30 % of their N coming from the applied labeled N (Fig. 6.20b).

Likewise, in silvopastoral system some proportion of the applied N was also absorbed by *N. antarctica* trees, although the amounts were insignificant to those absorbed by grasses, with a maximum around of 6 % of N derived from the fertilizer applied, located mainly in leaves and small branches (Fig. 6.21). Gargaglione et al. (2014) postulated that possibly grasses absorbed more than

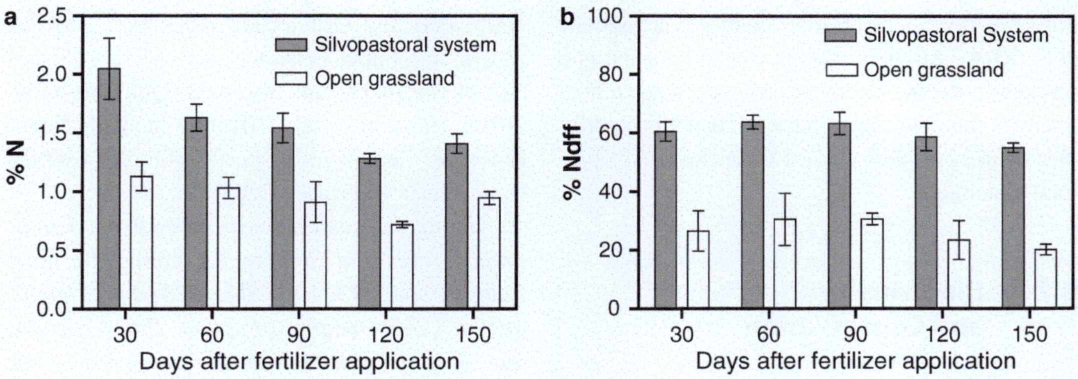


**Fig. 6.20** Nitrogen concentration (**a**) and percentage of nitrogen derived from fertilizer (Ndff, **b**) for aerial components of grasses growing in a *Nothofagus antarctica* silvopastoral system (*grey bars*) and in an open grassland (*white bars*) along the growing season in south Patagonia (Data derived from Gargaglione et al. (2014) from a study with labeled $^{15}$N fertilizer applied on spring (early November))

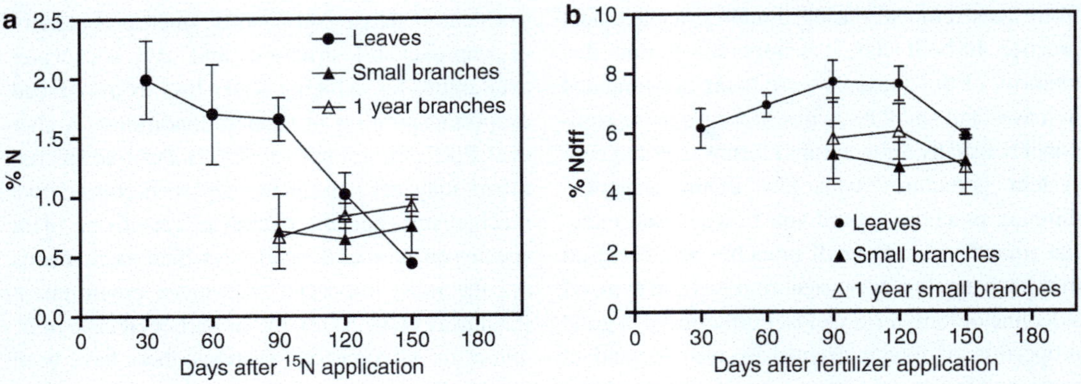


**Fig. 6.21** Nitrogen concentration (**a**) and Nitrogen derived from fertilizer (Ndff, **b**) in leaves and small branches of *Nothofagus antarctica* growing in a silvopastoral system in south Patagonia (Data derived from Gargaglione et al. (2014) from a study with labeled $^{15}$N fertilizer applied on spring (early November))

trees due to the higher growth rates compared with trees, and also because grasses accounted with a higher root density at 0–20 cm soil depth and started to grow earlier in the season. In conclusion, the silvopastoral systems recovered 65 % more of the applied N than the open grassland. These results indicate that a better and more efficient use of the applied N in silvopastoral system, where grasses absorbed twice as much the applied N than grasses in the open site. In this sense, trees may not strongly compete for N with grasses in this silvopastoral system or *N. antarctica* trees in the system may indirectly "facilitate" N absorption by underneath grasses improving environmental conditions like lower water stress (by protection of strong winds) or by reducing competition between soil microorganisms and grasses for inorganic N, since litterfall improves decomposing litter quality (Gargaglione et al. 2014). All these aspects should be taken into account in the management practices, since a high intensity thinning may reduce the benefits contributed by trees towards N use efficiencies in silvopastoral systems.

## 6.7 Biodiversity and Conservation

A sustainable management will be one that reconciles the productive needs and requirements of the local people with conservation of native forests. The biodiversity of ñire forest is considered the richest environment compared with other *Nothofagus* forests in Northern Patagonia (e.g., in vascular plant diversity, according to Speziale and Ezcurra 2008) or a medium rich environment in Southern Patagonia (Lencinas et al. 2011).

The species assemblages in ñire forests vary along their latitudinal distribution range and with habitat conditions, as humidity, with different characteristic species from North to South, and from xeric to humid environments. In Tierra del Fuego Province, 28 species of vascular plants were cited in the understory (Lencinas et al. 2008a), 251 species of insects (Lencinas et al. 2008b) and 18 species of birds (Lencinas et al. 2005). In Santa Cruz Province 225 species of vascular plants was observed, 112–165 species of insects and 10–21 species of birds (Gallo et al. 2005; Peri and Ormaechea 2013; Peri and Paz 2013). In Chubut Province there were 105 species of vascular plants observed (Quinteros et al. 2010; Hansen et al. 2013b). In Río Negro Province, there were 34 species of birds observed (Lantschner and Rusch 2007).

Among the main plant biodiversity in *N. antarctica* forests in Sothern Patagonia, the more frequent species are the shrubs *Berberis microphylla*, *Chiliotrichum diffussum*, *Ribes cucullatum*, *R.magellanicum*, *Schinus patagonicus*; the dwarf shrubs *Empetrum rubrum*, *Azorella monantha*, *Bacharis magellanica*; the grasses and graminoids *Carex andina*, *Deschampsia flexuosa*, *Holcus lanatus*, *Bromus sp.*, *Poa spp*,

*Dactylis glomerata*; the herbs *Potentilla chiloensis*, *Anemone multifida*, *Viola maculata Galium aparine*, *Osmorhiza chilensis*, *O. depauperata*, *Cotula scariosa*, *Acaena ovalifolia*, *A. magellanica*, *A. pinnatifida*, *Geranium sessiliflorum*, *Taraxacum officinale*, *Trifolium repens*, *Rumex acetosella* and *Vicia magellanica*. Among birds, the main species are *Aphrastura spinicauda* Gmelin, *Pygarrhichas albogularis* King. and *Turdus falcklandii* King. In relation to mammals, *Nothofagus* forests are considered an endemic zone, with 36 endemic species and 12 endemic genera (Patterson 1993), including several species of rodents, foxes and bats, but there are a lack of information on their specific assemblage in forests along the complete range of *N. antarctica* distribution. However, in a latitudinal gradient study, a greater mammal diversity was detected between 40 and 41°50′ SL (Murúa 1996). Likewise, in *N. antarctica* forests, usually there are several exotic and/or invasive species found, accidentally or intentionally introduced, including plants, insects and mammals. For example, *Holcus lanatus*, *Poa pratensis* and *Dactylis glomerata* provides forage for animal grazing, *Taraxacum officinale* is a widely distributed and naturalized weed, *Hieracium pilosella*, *H. praealtum* and *Vespula germanica* are aggressive invaders, *Castor canadensis* is an ecosystem engineer and threat, and *Mustela vison* is a competitive predator.

*N. antarctica* forests are usually immersed in a landscape spatial matrix that alternate with rangelands, peat-lands and other *Nothofagus* forests, with dominance over open or forested environments depending on the geographic zone. Biodiversity is partially shared among these environments, e.g. 79–93 % of their vascular plant species are usually common to all environments of the landscape matrix, with low values of exclusive vascular plant species in *N. antarctica* forests (7 % of vascular plant richness), being their composition quite similar to *N. pumilio* forests and grasslands (Lencinas et al. 2008a). Similar proportions were observed for insect and bird diversity, with 36–95 % of common species and 5 % of exclusive ones for insects (Lencinas et al. 2008b) and 67–78 % of common species and 22 % of exclusive species for birds (Lencinas et al. 2005). The existence of species that only occurs in *N. antarctica* forests denote the importance of this unique environment, which is usually undervalued due to the lack of natural provincial and national reserves that preserves habitat wilderness (Rusch et al. 2004).

*N. antarctica* forests are included in one of the last remaining pristine wilderness areas on the planet (Mittermeier et al. 2003). However, there are several human-related activities that incrementally modified their original structure and natural dynamic, e.g. livestock, harvesting and beaver activities, which influence the entire forest system, modifying biodiversity levels at understory, soil and canopy levels. The historic use of *N. antarctica* forests in Tierra del Fuego has modified its biodiversity status and has produced a gradient of low to high conservation value (Martínez Pastur et al. 2014), as demonstrated by vascular plants (Fig. 6.22a). Moreover, there are regional differences in biodiversity according to the zone where *N. antarctica* forests are growing. For example, differences in their understory vascular plant diversity were observed among north, south and east zones in Tierra del Fuego, as shown in a non-metrical multidimensional scaling (NMS) graphic (Fig. 6.22b).

On the other hand, variations in forest structure through modifications in overstorey canopies (more closed or open canopies) generate differences in richness and relative abundance of different organisms (Fig. 6.23a–c), which could be minima as for understory vascular plant, insect and bird richness and bird density, or very important as in vascular plant cover or insect abundance. Forest management activities also produces biodiversity modifications, as it was observed in vascular plants (Fig. 6.23d), with less richness and cover in secondary forests compared to closed and opened canopy stands, and higher richness in burned forests mainly due to the colonization of species from grassland and peatlands.

Similarly in Chubut, native understory species are more frequent in dense ñire woodland independently of its forest type. However, when considering vegetation changes along a disturbance

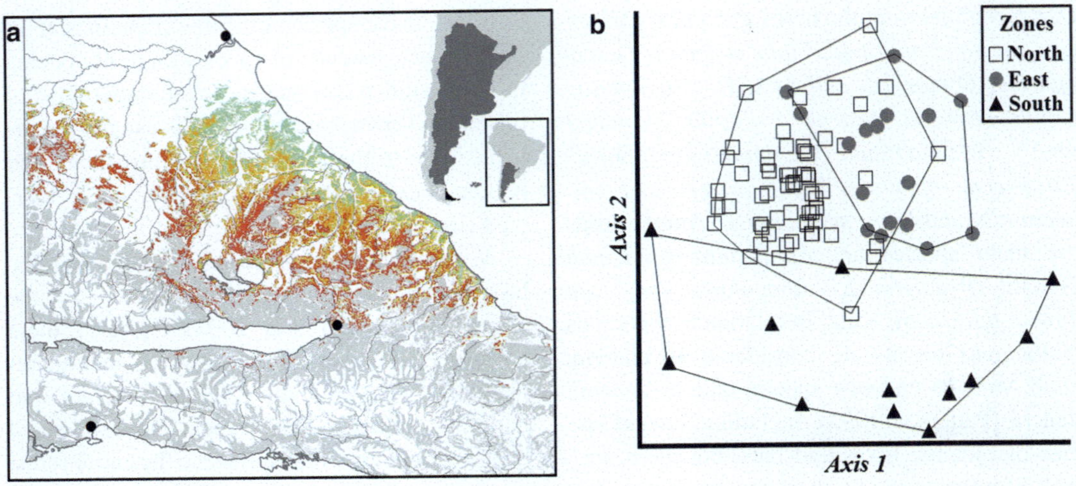


**Fig. 6.22** Understory vascular plant variations in *Nothofagus antarctica* forests of Tierra del Fuego: (**a**) Conservation value map (*red* color for high conservation value-high biodiversity forests, *orange* color for medium conservation value, and *green* for low conservation value-low biodiversity forests). (**b**) Differences in diversity at regional landscape level scale, highlighting north, south and east zones through multidimensional non metrical scaling analysis (NMS) (Color figure online)

gradient, in a more opened stand (canopy cover ~50 %) the understorey cover is mainly dominated by adventitious grasses and herbs. Thus, forest interventions that partially opened the canopy facilitate the introduction of adventitious species and together with native species resulted in an increase of plants richness and diversity supported by intermediate disturbance theory (Rusch et al. 2005). More intensive disturbance of ñire woodlands usually is accompanied by a higher frequency of degradation of key-species like *Acaena splendens*, *Rumex acetosella*, *Carduus thoermeri* and *Hypochaeris radicata* (Bran et al. 1998; Quinteros et al. 2010). Likewise, after ñire woodland thinning and pruning it was possible to increase grasses and herbs coverage for 3–4 years after interventions (Fertig et al. 2009), but after 10 years increased the shrubs cover (*Schinus patagonicus* in intermediate ñire woodlands and *Berberis microphylla* in high ñire woodland) and consequently decreased the forage production of the silvopastoral system.

Livestock use in *N. antarctica* forests usually diminishes the original biodiversity, both richness and relative abundance, mainly due to loss of sensitive species to animal pressure and selective grazing on more palatable species. But this effect could be masked by incoming species, which could generate similar values of richness. In Fig. 6.23e, f an example for impact of cattle use over vascular plants and coleopterans in *N. antarctica* forests of Tierra del Fuego is shown. Also, high livestock charge produced greater reduction in vascular plant richness (Fig. 6.23g), but not in cover, while in grasslands high and low livestock charges could not significantly influence vascular plant diversity (Lencinas et al. 2014).

The use of bioindicators may act as early warning indicators of environmental changes (environmental indicator), monitor a specific ecosystem stress (ecological indicator) or indicate levels of taxonomic diversity at a site (biodiversity indicator) (McGeoch 1998).

The increase in biodiversity research in *N. antartica* forests, including natural distribution and assemblage of inconspicuous species (e.g., coleopterons and arachnids), and their changes under different impacts (e.g., thinning, beaver and silvopastoral use), allowed to make progresses in the searching of convenient bioindica-

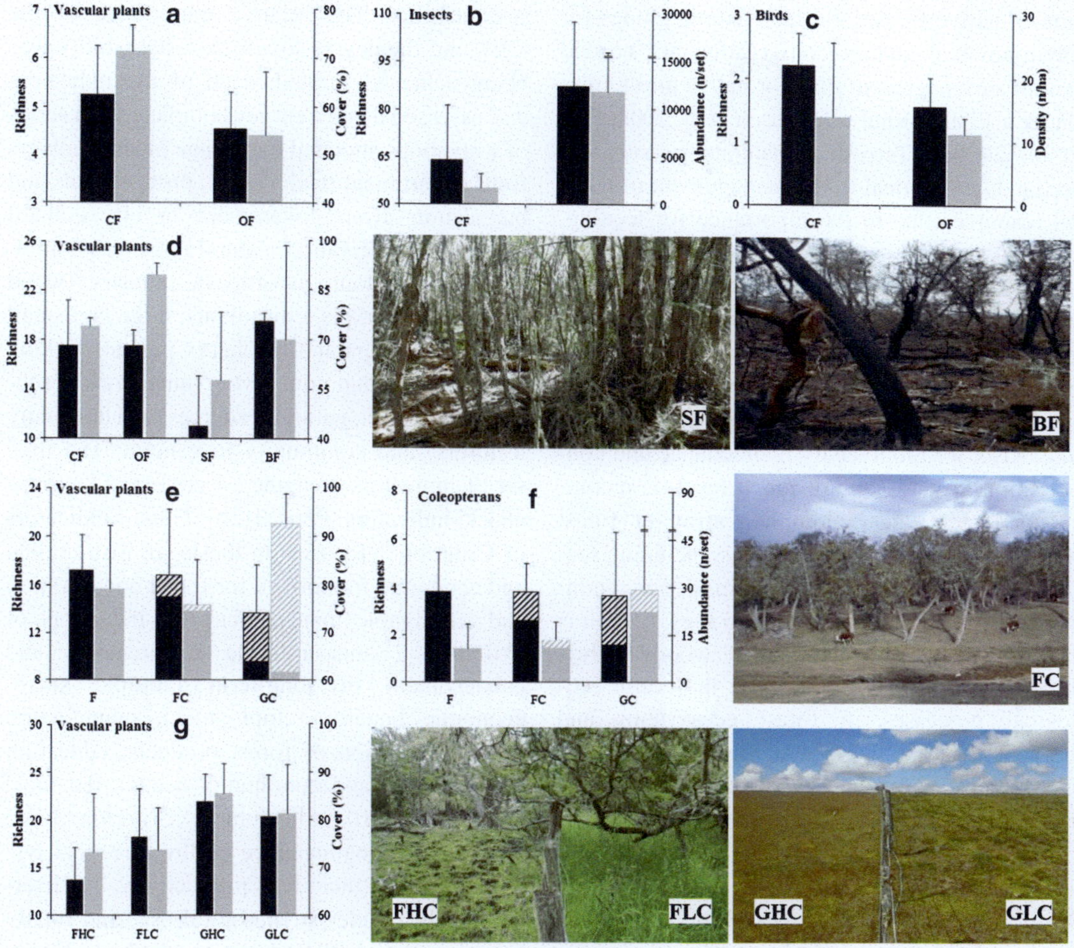


**Fig. 6.23** Biodiversity variations in *Nothofagus antarctica* forests: (**a–c**) Vascular plants (richness and cover), insects (richness and abundance) and birds (richness and density) for different forest structures (*CF* closed forests, *OF* opened forests). (**d**) Vascular plants (richness and cover) for forest management history (*CF* without thinning or closed forests, *OF* opened forests by thinning, *SF* secondary forests, *BF* burned forests). (**e**, **f**) Impact of livestock use over vascular plants (richness and cover) and coleopterans (richness and abundance), where slash bar patterns show incoming species (*F* forests without cattle, *FC* forests with cattle, *GC* grasslands with cattle). (**g**) Impact of different livestock charge over vascular plants (richness and cover) (*FHC* forests with high cattle charge, *FLC* forests with low cattle charge, *GHC* grasslands with high cattle charge, *GLC* grasslands with low cattle charge)

tors, which could reflect the state of the environment acting as early warning indicators of environmental changes (environmental indicator), monitor a specific ecosystem stress (ecological indicator) or indicate levels of taxonomic diversity at a site (biodiversity indicator) (McGeoch 1998). Bioindicators may also be used for conservation prioritisation using spatial comparisons of site value, monitoring of ecosystem recovery, or response to management. This information is required to set up explicit recommendations to preserve or to minimize impacts on natural ecosystems and over very sensitive or endangered species. In this context, *Threchisibus antarcticus* (Coleoptera) (Lencinas and Martínez Pastur 2010) and *Neochelanops michaelseni* Simon (Pseudoscorpionida) are considered as promising species, which are currently under study in the PEBANPA network.

Furthermore, a sustainable management will be one that reconciles the productive needs and requirements of the local people with conservation of native forests. In this context, guides for environmental quality conservation are needed within a Management Plan for native ñire forests under a silvopastoral use (Rusch et al. 2004; Peri et al. 2009a). Terrestrial habitats surrounding wetlands are critical to the management of natural resources due to its importance for feeding, nesting and conservation of water quality in streams, providing also substantial benefits as habitat or dispersal corridor. For this, buffers zones for wetland and riparian habitats protection are proposed by preserving a core terrestrial habitat from 15 to 60 m from the edge of the aquatic site. Also, for biodiversity maintenance and conservation at landscape or ranch level, is recommended to create a complete array of forest successional stages and structures, including sectors with old-growth forest conditions and retaining standing dead trees and fallen logs. Also, it is recommended to leave coarse woody debris because plays a substantial role in several ecological processes in forest ecosystems and because a large number of organisms (fungi, insects) are dependent on decaying wood for nutrients or habitat (Franklin et al. 1987).

## 6.8 Criteria and Indicator to Assess Ñire forest's Sustainability under Silvopastoral Use

The Santa Cruz government has a statutory responsibility to ensure that native forests are sustainably used under silvopastoral systems. Provincial regulations also require that a Regional Forest Plan and long-term forest policies are developed. To provide strategies to maximize the forest's sustainable benefits for society, Peri (2012b) developed indicators for sustainable management of *N. antarctica* forests. Multi-criteria methods were used to integrate different perspectives regarding environmental, social and economic aspects of the forest's management (Mendoza and Prabhu 2000). Starting with a range of internationally accepted criteria and indicator (C&I) schemes, a local set of C&I scheme was developed to assess the ñire forest's sustainability. The relative importance of the C&I, and the degree to which sustainability was being achieved through each of the indicators was assessed using a survey completed at a series of workshops attended by a range of stakeholders from government (technicians, professionals and bureaucrats from government or city council planning, conservation, natural resources administration and tourism offices), industry (wood processing workers, consultants, large and small company owners and ranchers), people working for environmental non-govt organisations, teachers (high school, natural resources, and university lecturers) and community in general. The final set of indicators contained 4 criteria, 11 groups and 55 indicators (Peri 2012b). Thus, 9 indicators of Criterion 1 focuses on the legal, institutional and economic framework for forest conservation and sustainable management, 17 indicators of Criterion 2 measure the maintenance and enhancement of long-term multiple socio-economic benefits (employment, manufacture and consumption of forest products, education and research, and community needs), the focal point of the 10 indicators of Criterion 3 are related to the maintenance of ñire forest ecosystem integrity (including productivity, biodiversity, health) under silvopastoral use, and finally Criterion 4 has 19 indicators which deal with planning, monitoring and assessment of the ñire forest resource use. The relative importance of the criterion and group level is shown in Table 6.20. The relative importance of the four criteria was quite similar. At the group level, the institutional framework was assessed as substantially more important (37 %) than the other two groups in Criterion 1. Employment (28 %) group of indicators was the most important ones in Criterion 2. Both two groups of Criteria 3 and 4 were evenly weighted.

An overall sustainability score for the whole set of C&I was calculated using the weighted results from the assessment of the relative importance of C&I and their individual sustainability scores. The results of the study show that based on the opinion of stakeholders, current management is achieving 1.72 on a range from 1.0 to 5.0, where 1.0 means that the aspect mea-

**Table 6.20** Relative importance of criteria and groups for sustainable use of ñire forest under silvopastoral systems in Santa Cruz province, Southern Patagonia, Argentina

| Level in the hierarchy | Criterion and groups text | Criterion level weights | Group level weights |
|---|---|---|---|
| **Criterion 1** | Legal, institutional and economic framework for ñire forest conservation and sustainable management | 31 % | |
| Group A | Legal framework | | 37 % |
| Group B | Institutional framework | | 33 % |
| Group C | Economic framework | | 30 % |
| **Criterion 2** | Maintenance and enhancement of long-term multiple socio-economic benefits to meet the needs of society | 26 % | |
| Group A | Employment | | 28 % |
| Group B | Silvopastoral forest products consumption and production | | 26 % |
| Group C | Education and research | | 25 % |
| Group D | Community needs | | 21 % |
| **Criterion 3** | Maintenance of ñire forest ecosystem integrity (including productivity, biodiversity, health) | 22 % | |
| Group A | Biodiversity | | 53 % |
| Group B | Protection | | 47 % |
| **Criterion 4** | Planning, monitoring and assessment of the ñire forest resource use | 21 % | |
| Group A | Long-term regional plan | | 51 % |
| Group B | Short to medium-term silvopastoral plans | | 49 % |

sured by the indicator is close to a critical threshold value and 5.0 is far from this threshold value. It provides a reference point for future change of ñire forest use in Santa Cruz province. The sustainability scores and weightings at group and criterion level can be used to focus efforts on improving the poorest performing groups of indicators. Criterion 4 had the lowest mean score (1.23) and Criterion 3, biodiversity and protection, showed the best performance (2.34). Groups of indicators with the lowest scores (less than 1.3) were legal and economic frameworks within Criterion 1, and long-term regional and short-term management plans (Criterion 4). In general, the forest ecosystem integrity and conservation indicators were evaluated as performing reasonably well.

The results also show that the key areas to work on are in policy and planning, resource administration of the forest. Finally, the most important factors affecting sustainable management of ñire forest under silvopastoral use were selected, by choosing indicators with a mean individual sustainability score less than or equal to 1.2, and at the same time considering the performance of groups of indicators (Table 6.21). Using this approach, one of the main areas of priority was policy and planning that revolve around the development of management plans and monitoring activities of the timber industry and pastoralists.

Multi-criteria analysis was used to incorporate a diverse range of variables related to silvopastoral forest management, involving the development of a list of indicators for sustainability assessment that were applicable to Santa Cruz. A multidisciplinary group of local experts assigned scores for each indicator based on its contribution to sustainability.

Similarly, Rusch et al. (2009a) developed C&I for silvopastoral system based on *N. antarctica* forests in south western Chubut. The authors

**Table 6.21** Top five lowest scoring indicators of silvopastoral forest use in Santa Cruz province

| Number | Indicator text | Group | Score |
|---|---|---|---|
| 1.2 | Extent to which the legal framework supports the conservation and sustainable management of forests under silvopastoral use | Legal framework | 1.06 |
| 2.8 | Investments in forest management under silvopastoral use are carried out (thinning, environment planning, livestock plan, etc). Access to financial support (application for credit) for stakeholders to improve the productive system including investments in timber technology | Forest products consumption and production in silvopastoral systems | 1.08 |
| 4.2 | Silvopastoral forest management planning and its implementation takes into account economic variables (costs, investments, revenue), as well as the results of the environmental and social impact assessments | Long-term regional plan | 1.08 |
| 4.7 | There is a regional land use Plan which reflects the different forested land uses promoting its integral use, and gives attention to ecosystem services such as carbon fixation and biodiversity | Long-term regional plan | 1.09 |
| 4.9 | Periodic monitoring is conducted to assess the effectiveness of the measures employed to maintain or enhance the applicable conservation attributes | Long-term regional plan | 1.10 |
| 4.14 | The silvopastoral forest management has a monitoring system for correct implementation. Data collection includes productive, economic and social variables | Short-term management plan | 1.11 |

found that at the management unit level, riparian areas were preserved and there were no signals of disequilibrium in the food web, except for the invasion of a rodent. However, there was no plan developed for wildlife connectivity, forest recover (regeneration), water resources and carbon conservation. Rusch et al. (2009b) analysing the maintenance of the productive capacity (especially in terms of wood and grass) and the maintenance or improvement of stakeholders' welfare, detected problems due to over-exploitation for firewood, poor juridical security for land tenure, loggers had low decision making power and their work and livelihood conditions were also not optimal. Such poor or low working conditions could be attributed to small farm sizes and low economic conditions.

## 6.9 Extension and Policy

Previously it was highlighted that one of the main areas to support the conservation and sustainable management of forests under silvopastoral use is the development of legal framework, policy and planning. In this context, in 2007, the Argentinean government enacted National Law 26.331 for the Environmental Protection of Native Forests and the objectives are to promote the conservation of native forests through land planning, sustainable management and tightening the regulations associated with land-use change. Law 26.331 requires all the provinces to develop a Land Use Planning Process (LUPP) with respect to native forests in a participatory fashion. Native forests have been classified according to three conservation categories (colours in maps) determined based on a number of technical and social criteria: Red (high conservation value forests for ancestral uses, gathering of non-timber forest products, scientific research, "respectful" tourism, conservation plans, ecological restoration), Yellow (medium conservation value forests for sustainable productive activities and tourism under the guidelines of management and conservation plans) and Green (low conservation value forest where land-use change is allowed). The enforcement authority of this law is the Secretary of Environment and Sustainable Development and the Federal Council on the Environment – COFEMA (Argentina's federal governmental institution) coordinates the development of the environmental policies between provinces. In the last few years, more than 50 % of the budget has been destined to silvopastoral system plans in the Yellow category that includes forestry inventories, silvicultural practices, adjustment of stocking rate and fencing for strategic separation in homogenous areas (grass steppe, forest and riparian meadows) covering planning areas of ñire forest from 22 to 13,000 ha. However, most ranchers in Patagonia have been slow to adopt an integral silvopastoral system management plan, possibly because of lack of convincing evidence of positive economic returns or long term benefits from ecosystem service values. Based on information related to ranch natural resources, constraints to production, and expected prices of inputs and outputs, a farm plan can be established. In this context, an integral economic analysis is required, quantifying timber products, annual animal production, soil conservation benefits, as well as the landscape values associated with ñire forests. Estimates of the cost of silvopastoral system implementation and maintenance must also be included in addition to the area designated for forest conservation. Financial analysis to calculate farm profit involves calculating present net value (PNV) and internal rate of return (IRR) for different discount rates, using computer models. This type of economic models is essential in extension programs, and for policy makers concerned with sustainable grazing areas in Patagonia in order to convince the ranchers.

Silvopastoral system development requires multi-agency, interdisciplinary, participatory and user-focused learning environments to encourage innovation. In this context, it is critical to involve landowners who practice silvopastoral systems to share their experiences related to adequate management and land use. Similarly, policy related to silvopastoral use of ñire native forest is best developed through a pluralistic learning process, focused on creating a better informed and more enabling environment for public and private decision-making. It is important also to increase awareness of the interdependence among stakeholders and key actors to address issues in common in silvopastoral practice. For example, mutual concerns by forestry, livestock and specialists about biodiversity conservation and water quality at landscape level may be addressed by the development of riparian buffer strips in several ranches.

In the region, extension advice is provided by the National Agricultural Institute (Instituto Nacional de Tecnologia Agropecuaria, INTA) and Provincial Estate institutions both of which have sub-offices in the countryside and small agricultural towns. Moreover, under the Cambio Rural Program (INTA) there are active consultants responsible for development of groups of farmers with a common agricultural activity (e.g. sheep, cattle, agriculture). However, in many situations the advisory function is fragmented among several agencies. In any extension programme, the perceptions of farmers in relation to the practice of silvopastoral systems and ñire forest on their farms are important. In addition to farmers and rural science researchers (from INTA, the National Universities of Patagonia, the National Scientific and Technical Research Council-CONICET), social scientists and policy makers can help towards improving farming practices through extension programmes. The use of research sites for demonstration developed in the Patagonian region by the university and agency-based research is an important additional resource for the learning process. They have an important role to play towards increasing the rate of new technology adoption by farmers. However, the conventional notion that 'research precedes extension' must change to a more systems-oriented view, where knowledge generation, exchange and application are multi-directional processes. Joint venture schemes with industry, government, or other financial sources can be fruitful, particularly when farmers have financial, marketing or knowledge issues. In conclusion, the challenges associated with the adoption of sustainable silvopastoral systems in ñire forest in Patagonia also provides the opportunity for an ecologically based approach to land management that can contribute to production diversity and long-term economic sustainability and profitability for farmers and society in this region.

## References


Archer S, Boutton TW, Hibbard KA (2001) Trees in grasslands: biogeochemical consequences of woody plant expansion. In: Schulze ED, Schimel D, Prentice IC, Heimann M, Harrison S, Holland E, Lloyd J (eds) Global biogeochemical cycles in the climate system. Academic, New York

Austin AT, Vivanco L (2006) Plant litter decomposition in a semi-arid ecosystem controlled by photodegradation. Nature 442:555–558

Bahamonde HA, Peri PL (2013) Receptividad ganadera en bosques de *Nothofagus antarctica* (ñire) bajo uso silvopstoril en Patagonia Sur basado en los requerimientos energéticos de los animales. In: Proceedings 4^to^ Congreso Forestal Argentino y Latinoamericano. 11 pp., Iguazú, Misiones

Bahamonde HA, Peri PL, Martínez Pastur G, Lencinas MV (2009) Variaciones microclimáticas en bosques primarios y bajo uso silvopastoril de *Nothofagus antarctica* en dos Clases de Sitio en Patagonia Sur. In: Proceedings of the 1st National Congress of Silvopastoral Systems, Misiones, Argentina. INTA Editions, pp 289–296

Bahamonde HA, Peri PL, Monelos LH, Martínez Pastur G (2011) Ecological aspects of regeneration from seeds of *Nothofagus antarctica* native forest in Southern Patagonia, Argentina. Bosque 32:20–29

Bahamonde HA, Peri PL, Alvarez R, Barneix A (2012a) Producción y calidad de gramíneas en un gradiente de calidades de sitio y coberturas en bosques de *Nothofagus antarctica* (G. Forster) Oerst. en Patagonia. Ecología Aust 22:62–73

Bahamonde HA, Peri PL, Alvarez R, Barneix A, Moretto A, Martínez Pastur G (2012b) Litter decomposition and nutrients dynamics in *Nothofagus antarctica* forests under silvopastoral use in Southern Patagonia. Agrofor Syst 84:345–360

Bahamonde HA, Peri PL, Monelos LH, Martínez Pastur G (2013a) Seed regeneration in native forests of *Nothofagus antarctica* under silvopastoral use in Southern Patagonia. Bosque 34:89–101

Bahamonde HA, Peri PL, Monelos L, Gargaglione V (2013b) Crecimiento en rebrotes vegetativos de *Nothofagus antarctica* (ñire, ñirre) desde el tocón después de un raleo. In: Peri PL (ed) 2nd Congress on Patagonian Agroforestry (International), El Calafate, Argentina. INTA Editions, p 93

Bahamonde HA, Peri PL, Alvarez R, Barneix A, Moretto A, Martínez Pastur G (2013c) Silvopastoral use of *Nothofagus antarctica* in Southern Patagonian forests, influence over net nitrogen soil mineralization. Agrofor Syst 87:259–271

Bahamonde HA, Peri PL, Mayo JP (2014) Modelo de simulación de producción de materia seca y concentración de proteína bruta de gramíneas creciendo en bosques de *Nothofagus antarctica* (G. Forster) Oerst. bajo uso silvopastoril. Ecología Aust 24:111–117

Begon M, Townsed CR, Harper JL (2006) Ecology: from individuals to ecosystems. Blackwell Publishing, Malden, Oxford

Bran D, Rusak S, Ayesa J, Moraga H (1998) Los matorrales de *Nothofagus antarctica* (ñire) en el contacto Bosque-Estepa. Estructura y regeneración después de una tala. Comunicación técnica N° 47, INTA, EEA Bariloche, Río Negro

Burgos A, Grez AA, Bustamante RO (2008) Seed production, pre-dispersal seed predation and germination of *Nothofagus glauca* (Nothofagaceae) in a temperate fragmented forest in Chile. For Ecol Manag 255:1226–1233

Callaway RM, Brookert RW, Choler P, Kikvidezs Z, Lortiek C, Michalet R, Paolini L, Pugnaire FI, Newingham B, Aschehoug ET, Armas C, Kikodze D, Cook BJ (2002) Positive interactions among alpine plants increase with stress. Nature 417:844–848

Chapin FS, Matson PA, Mooney HA (2002) Principles of terrestrial ecosystem ecology. Springer, New York

Clark JS, Beckage B, Camill P, Cleveland B, Hillerislambers J, Lichter J, McLachlan J, Mohan J, Wyckoff P (1999) Interpreting recruitment limitation in forests. Am J Bot 86:1–16

Cocimano M, Lange A, Menvielle E (1977) Equivalencias ganaderas para vacunos de carne y ovinos. Comisión de producción de carne, Departamento de estudios. AACREA, Buenos Aires, 24 pp

Collado L (2009) Clasificación de los ñirantales de Tierra del Fuego. In: Peri PL (ed) Relevamiento de los bosques nativos de ñire (*Nothofagus antarctica*) de Tierra del Fuego (Argentina) como herramienta para el manejo sustentable. Editorial Instituto Nacional de

Tecnología Agropecuaria (INTA), Buenos Aires, pp 10–27

Cuevas J (2000) Tree recruitment at the *Nothofagus pumilio* alpine timberline in Tierra del Fuego, Chile. Ecology 88:840–855

del Valle HF, Eiden G, Mensching H, Goergen J (1995) Evaluación del estado actual de la desertificación en áreas representativas de la Patagonia. Informe Final Fase I. Proyecto de Cooperación Técnica entre la República Argentina y la República Federal Alemana. Cooperación técnica argentino-alemana, proyecto INTA-GTZ, Lucha contra la Desertificación en la Patagonia a través de un sistema de Monitoreo Ecológico

Donoso C, Steinke L, Premoli A (2006) *Nothofagus antarctica* (G. Forster) Oerst. In: Donoso Zegers C (ed) Las especies arbóreas de los bosques templados de Chile y Argentina. Autoecología. Marisa Cuneo Editions, Valdivia, pp 401–410

Dube F, Thevathasan NV, Zagal E, Gordon AM, Stolpe NB, Espinosa M (2011) Carbon sequestration potential of silvopastoral and other land use systems in the Chilean Patagonia. In: Kumar BM, Nair PKR (eds) Carbon sequestration in agroforestry systems. Springer, Dordrecht, pp 101–127

East RM, Felker P (1993) Forage production and quality of 4 perennial grasses grown under and outside canopies of mature *Prosopis glandulosa* Torr. var. *glandulosa* (mesquite). Agrofor Syst 22:91–110

Echevarria D, von Müller A, Hansen N, Bava J, Escalona M (2012) Efectos del herbívoro sobre la regeneración de ñire bajo diferentes situaciones de manejo. In: Proceedings of the 2nd National Congress of Silvopastoral Systems, Santiago del Estero, Argentina. INTA Editions, pp 311–316

Edwards NT (1982) The use of soda-lime for measuring respiration rates in terrestrial systems. Pedobiologia 23:312–330

Fertig M (2006) Producción de carne bajo distintos sistemas de pastoreo en ñirantales del Noroeste del Chubut. Cartilla Técnica INTA EEA Esquel. Ganadería, pp 93–96, Chubut, Argentina

Fertig M, Hansen N, Tejera L (2007) Producción forrajera en bosques de *Nothofagus antarctica* (ñire). Actas IV Congreso Nacional sobre Manejo de Pastizales Naturales, San Luis, Argentina, p 19

Fertig M, Hansen N, Tejera L (2009) Productividad y calidad forrajera en raleos en bosques de ñire *Nothofagus antarctica*. In: Proceedings of the 1st National Congress of Silvopastoral Systems, Misiones, Argentina. INTA Editions, pp 358–363

Fisher RF, Binkley D (2000) Ecology and management of forest soils. Wiley, New York

Frangi JL, Richter LL, Barrera MM, Aloggia M (1997) Decomposition of *Nothofagus* fallen woody debris in forests of Tierra del Fuego, Argentina. Can J For Res 27:1095–1102

Franklin JF, Shugart HH, Harmon ME (1987) Tree death as an ecological process. The causes, consequences, and variability of tree mortality. Bioscience 37:550–556

Gallo E, Lencinas MV, Peri PL (2005) Desarrollo de sistemas silvopastoriles en bosques de *Nothofagus antarctica*: Biodiversidad en ñirantales. In: Alternativas de Manejo Sustentable para el Manejo Forestal Integral de los bosques de Patagonia. Informe Final del Proyecto de Investigación Aplicada a los Recursos Forestales Nativos (PIARFON), Dirección de Bosques de la Secretaría de Ambiente y Desarrollo Sustentable de Nación (SAyDS), Buenos Aires, Argentina, pp 645–669

Gargaglione V, Peri PL, Rubio G (2010) Allometric relations for biomass partitioning of *Nothofagus antarctica* trees of different crown classes over a site quality gradient. For Ecol Manag 259:1118–1126

Gargaglione V, Peri PL, Rubio G (2013) Partición diferencial de nutrientes en árboles de *Nothofagus antarctica* creciendo en un gradiente de calidades de sitio en Patagonia Sur. Bosque 34:291–302

Gargaglione V, Peri PL, Rubio G (2014) Tree–grass interactions for N in *Nothofagus antarctica* silvopastoral systems: evidence of facilitation from trees to underneath grasses. Agrofor Syst 88:779–790

Gibon A (2005) Managing grassland for production, the environment and the landscape. Challenges at the farm and the landscape level. Livest Prod Sci 96:11–31

Golluscio RA, Deregibus V, Paruelo JM (1998) Sustainability and range management in the Patagonian steppe. Ecología Aust 8:265–284

Guitart Fité E, Fertig M, Bottaro H, Buratovich O, Villa M (2004) Programa para el desarrollo local y competitividad de pequeñas empresas. BID-ATN/ME7295-AR. Componente N° 1: Fortalecimiento y diversificación productiva

Hansen N, Fertig M, Tejera L (2005) Desarrollo de sistemas silvopasoriles en bosques de *Nothofagus antarctica*. In: Alternativas de Manejo Sustentable para el Manejo Forestal Integral de los bosques de Patagonia. Informe Final del Proyecto de Investigación Aplicada a los Recursos Forestales Nativos (PIARFON), Dirección de Bosques de la Secretaría de Ambiente y Desarrollo Sustentable de Nación (SAyDS), Buenos Aires, Argentina, pp 671–680

Hansen N, Fertig M, Escalona M, Tejera L, Opazo W (2008) Ramoneo en regeneración de ñire y disponibilidad forrajera. In: Proceedings Segunda Reunión sobre *Nothofagus* en la Patagonia – EcoNothofagus 2008, Esquel, Chubut, Argentina, pp 137–142

Hansen N, Rios C, Dromaz M, De María G, Rôo G, von Müller A, Lloyd C, Raso M, Buduba C, Fetherson S (2013a) Ejecución de prácticas de manejo silvopastoril en bosque de ñire (*Nothofagus antarctica*). In: Peri PL (ed) 2nd Congress on Patagonian Agroforestry (International), El Calafate, Argentina. INTA Editions, p 92

Hansen N, Codesal P, Quinteros P, Gallardo C (2013b) Especies de sotobosque en ambientes de ñire en Chubut. INTA, Buenos Aires

Haraldsson HV (2000) Introduction to systems and causal loop diagrams. System dynamic course, Lumes. Lund University, Lund

Haraldsson HV, Sverdrup HU (2004) Finding simplicity in complexity in biogeochemical modelling. In: Wainwright J, Mulligan M (eds) Environmental modelling. Wiley, Chichester, pp 211–223

Heinemann K, Kitzberger T, Veblen T (2000) Influences of gap microheterogeneity on the regeneration of *Nothofagus pumilio* in a xeric old-growth forest of northwestern Patagonia, Argentina. Can J For Res 30:25–31

Hight GK, Sinclair DP, Lancaster RJ (1968) Some effects of shading and of nitrogen fertiliser on the chemical composition of freeze-dried and oven-dried herbage, and on the nutritive value of oven-dried herbage fed to sheep. N Z J Agric Res 11:286–302

IPCC (2001) Climate change 2001: the scientific basis. In: Houghton JT, Ding Y, Griggs DJ, Noguer M, Van Der Linden PJ, Dai X, Maskell K, Johnson CA (eds) Contribution of working group I to the third assessment report of the intergovernmental panel on climate change. Cambridge University Press, Cambridge

Ivancich H, Soler Esteban R, Martínez Pastur G, Peri PL, Bahamonde H (2009) Índice de densidad de rodal aplicado al manejo silvopastoril en bosques de ñire (*Nothofagus antarctica*) en Patagonia sur. In: Proceedings of the 1st National Congress of Silvopastoral Systems, Misiones, Argentina. INTA Editions, pp 245–250

Ivancich H, Martínez Pastur G, Peri PL, Soler Esteban R, Lencinas MV (2010) Primeros resultados de raleos en bosques de *Nothofagus antarctica* para el manejo silvopastoril en Tierra del Fuego (Argentina). 1st Congress on Patagonian Agroforestry (International), Coyhaique, Chile, p 10

Ivancich H, Martínez Pastur G,Lencinas MV, Soler Esteban R, Peri PL (2012) Respuesta del canopeo y crecimiento de *Nothofagus antarctica* bajo manejo silvopastoril. In: Proceedings of the 2nd National Congress of Silvopastoral Systems, Santiago del Estero, Argentina. INTA Editions, pp 270–275

Ivlev VS (1961) Experimental ecology of the feeding of fishes. Yale University Press, New Haven

Jordano P, Pulido F, Arroyo J, García-Castaño JL, García-Fayos P (2008) Procesos de limitación demográfica. In: Valladares F (ed) Ecología del bosque mediterráneo en un mundo cambiante, 2nd edn. EGRAF, Madrid, pp 231–250

Kraxner F, Nilsson S, Obersteiner M (2003) Negative emissions from BioEnergy use, carbon capture and sequestration (BECS)—the case of biomass production by sustainable forest management from semi-natural temperate forests. Biomass Bioenergy 24:285–296

Kreps G, Martínez Pastur G, Peri PL (2012) Cambio Climático en Patagonia Sur: Escenarios futuros en el manejo de los recursos naturales. Ediciones INTA, Buenos Aires

Lantschner MV, Rusch V (2007) Impacto de diferentes disturbios antrópicos sobre las comunidades de aves de bosques y matorrales de *Nothofagus antarctica* en el NO Patagónico. Ecología Aust 17:99–112

Lencinas MV, Martínez Pastur G (2010) Variable retention as strategy to protect insect diversity in managed South Patagonian forests. Symposium Arthropod conservation in managed landscapes, 24th International Congress for Conservation Biology: Conservation for a Changing Planet. Edmonton, Alberta, Canada

Lencinas MV, Martínez Pastur G, Medina M, Busso C (2005) Richness and density of birds in timber *Nothofagus pumilio* forests and their unproductive associated environments. Biodivers Conserv 14:2299–2320

Lencinas MV, Martínez Pastur G, Rivero P, Busso C (2008a) Conservation value of timber quality versus associated non-timber quality stands for understory diversity in *Nothofagus* forests. Biodivers Conserv 17:2579–2597

Lencinas MV, Martínez Pastur G, Anderson CB, Busso C (2008b) The value of timber quality forests for insect conservation on Tierra del Fuego Island compared to associated non-timber quality stands. J Insect Conserv 12:461–475

Lencinas MV, Martínez Pastur G, Soler R (2011) Distribución de plantas a nivel de paisaje: semejanzas y diferencias entre bosques y praderas. III Jornadas Argentinas de Ecología de Paisajes. Bariloche (Argentina)

Lencinas MV, Soler R, Martínez Pastur G, Kreps G, Peri PL (2014) Estudio comparativo de intensidades de uso ganadero por bovinos y ovinos en sotobosque de *Nothofagus antarctica* (Forster f.) Oersted y pastizales aledaños. In: Proceedings XVI Jornadas Técnicas Forestales y Ambientales- Facultad de Ciencias Forestales. Eldorado Misiones, Argentina, pp 497–507

Martínez Pastur G, Cellini JM, Peri PL, Vukasovic R, Fernández C (2000) Timber production of *Nothofagus pumilio* forest by a shelterwood system in Tierra del Fuego (Argentina). For Ecol Manag 134:153–162

Martínez Pastur G, Cellini JM, Lencinas MV, Vukasovic R, Vicente R, Bertolami F, Giunchi J (2001) Modificación del crecimiento y de la calidad de fustes en un raleo fuerte de un rodal en fase de crecimiento óptimo inicial de *Nothofagus pumilio* (Poepp. et Endl.) Krasser. Ecología Aust 11:95–104

Martínez Pastur G, Cellini JM, Lencinas MV, Peri PL (2008) Potencialidad de la cosecha y rendimiento industrial de bosques de *Nothofagus antarctica* en Tierra del Fuego (Argentina). In: Proceedings IV Congreso Chileno de Ciencias Forestales, Talca, Chile, 10 pp

Martínez Pastur G, Lencinas MV, Cellini JM, Peri PL, Soler Esteban R (2009) Timber management with variable retention in *Nothofagus pumilio* forests of Southern Patagonia. For Ecol Manag 258:436–443

Martínez Pastur G, Peri PL, Cellini JM, Lencinas MV, Barrera M, Ivancich H (2011) Canopy structure analysis for estimating forest regeneration dynamics and growth in *Nothofagus pumilio* forests. Ann For Sci 68:587–594

Martínez Pastur G, Peri PL, Lencinas MV, Cellini JM, Barrera MD, Soler Esteban R, Ivancich H, Mestre L, Moretto AS, Anderson C, Pulido F (2013) La producción forestal y la conservación de la biodiversidad en los bosques de *Nothofagus* en Tierra del Fuego y Patagonia Sur. In: Donoso P, Promis A (eds) Silvicultura en Bosques Nativos: Avances en la investigación en Chile. Argentina y Nueva Zelanda. Universidad Austral de Chile, Valdivia, pp 15–179

Martínez Pastur G, Schindler S, Lencinas MV, Peri PL (2014) Forest conservation strategies in Southern Patagonia: differences between theory and practice. IUFRO World Congress, Salt Lake City

Mcgeoch MA (1998) The selection, testing and application of terrestrial insects as bioindicators. Biol Rev 73:181–201

Mendoza G, Prabhu R (2000) Multiple criteria decision making approaches to assessing forest sustainability using criteria and indicators: a case study. For Ecol Manag 131:107–126

Mittermeier RA, Mittermeier CG, Brooks TM, Pilgrim JD, Konstant WR, Da Fonseca GAB, Kormos C (2003) Wilderness and biodiversity conservation. Proc Natl Acad Sci 100:10309–10313

Monks A, Kelly D (2006) Testing the resource-matching hypothesis in the mast seeding tree *Nothofagus truncata* (Fagaceae). Aust Ecol 31:366–375

Murúa R (1996) Comunidades de mamíferos del bosque templado de Chile. In: Armesto JJ, Villagrán C, Arroyo MK (eds) Ecología de los bosques nativos de Chile, 2nd edn. Editorial Universitaria, Santiago, pp 113–133

Olson JS (1963) Energy storage and the balance of producers and decomposers in ecological systems. Ecology 44:331–332

Ormaechea S, Peri PL (2010) Protección de renovales de ñire al ramoneo en el marco de un uso silvopastoril. In: Proceedings Jornadas Forestales de Patagonia Sur: El bosque como bien social, fuente de trabajo y bienestar. Ushuaia, Tierra del Fuego, Argentina, p 94

Ormaechea S, Peri PL, Molina R, Mayo JP (2009) Situación y manejo actual del sector ganadero en establecimientos con bosque de ñire (*Nothofagus antarctica*) de Patagonia sur. In: Proceedings of the 1st National Congress of Silvopastoral Systems, Misiones, Argentina. INTA Editions, pp 385–393

Ormaechea S, Peri PL, Cipriotti P, Aanchorena J (2010) Producción y calidad de lana bajo diferentes manejos ganaderos en un establecimiento con bosque de ñire en Santa Cruz. 1st Congress on Patagonian Agroforestry (International), Coyhaique, Chile, p 323

Ormaechea S, Peri PL, Anchorena J, Cipriotti P (2011) A comparison of two types of sheep grazing management in a forest-grass steppe ecotone in southern Patagonia. In: Proceedings IX International Rangeland Congress, Rosario, Santa Fé, Argentina, p 661

Ormaechea S, Peri PL, Cipriotti P, Anchorena JA, Huertas L (2012a) Uso de collares GPS para evaluar la preferencia de ovinos por diferentes ambientes en Patagonia Sur. In: Proceedings of the 2nd National Congress of Silvopastoral Systems, Santiago del Estero, Argentina. INTA Editions, pp 456–461

Ormaechea S, Peri PL, Ceccaldi E (2012b) Uso espacial de vacunos bajo dos tipos de manejo ganadero en establecimiento con bosque de ñire. In: Proceedings of the 2nd National Congress of Silvopastoral Systems, Santiago del Estero, Argentina. INTA Editions, pp 94–99

Patterson BD (1993) Geographical patterns of endemic among recent mammals from South America. VI International Theriological Congress Abstracts Book, Sydney, Australia, p 236

Peri PL (2005) Sistemas silvopastoriles en ñirantales. IDIA XXI For 8:255–259

Peri PL (2008) Respuesta de ovinos a pastizales creciendo en diferentes cobertura de copas en sistemas silvopastoriles de ñire (*Nothofagus antarctica*) en Patagonia Sur, Argentina. Zootec Trop 26:363–366

Peri PL (2009a) Sistemas silvopastoriles en Patagonia: revisión del conocimiento actual. In: Proceedings of the 1st National Congress of Silvopastoral Systems, Misiones, Argentina. INTA Editions, pp 10–26

Peri PL (2009b) Evaluación de pastizales en bosques de *Nothofagus antarctica* – Método Ñirantal Sur. In: Proceedings of the 1st National Congress of Silvopastoral Systems, Misiones, Argentina. INTA Editions, pp 335–342

Peri PL (2009c) Método Ñirantal Sur -San Jorge: una herramienta para evaluar los pastizales naturales en bosques de ñire. Technical report EEA INTA Santa Cruz, Sección 9. Producción Animal. Edición EEA Santa Cruz, pp 33–38

Peri PL (2009d) Implicancias prácticas del uso del ñirantal como sistema silvopastoril y su conservación. In: Peri PL (ed) Relevamiento de los bosques nativos de ñire (*Nothofagus antarctica*) de Tierra del Fuego (Argentina) como herramienta para el manejo sustentable. Editorial Instituto Nacional de Tecnología Agropecuaria (INTA), Buenos Aires, Argentina, pp 28–48

Peri PL (2011) Carbon storage in cold temperate ecosystems in Southern Patagonia, Argentina. In: Atazadeh I (ed) Biomass and remote sensing of biomass. InTech Publisher, Croacia, pp 213–226

Peri PL (2012a) Implementación, manejo y producción en SSP: enfoque de escalas en la aplicación del conocimiento aplicado. In: Proceedings of the 2nd National Congress of Silvopastoral Systems, Santiago del Estero, Argentina. INTA Editions, pp 8–21

Peri PL (2012b) Criterios e Indicadores para el manejo sustentable de sistemas silvopastoriles con ñire (*Nothofagus antarctica*) en Santa Cruz. In: Proceedings of the 2nd National Congress of Silvopastoral Systems,

Santiago del Estero, Argentina. INTA Editions, pp 399–410

Peri PL, Bahamonde H (2012) Digestibilidad de gramíneas creciendo en bosques de ñire (*Nothofagus antarctica*) bajo uso silvopastoril. In: Proceedings of the 2nd National Congress of Silvopastoral Systems, Santiago del Estero, Argentina. INTA Editions, pp 264–269

Peri PL, Ormaechea S (2013) Relevamiento de los bosques nativos de ñire (*Nothofagus antarctica*) en Santa Cruz: base para su conservación y manejo. Ediciones INTA, Buenos Aires

Peri PL, Paz F (2013) Reserva Provincial Punta Gruesa: Una Mirada al bosque nativo y los pastizales naturales. Ediciones Secretaria de Ambiente y Desarrollo Sustentable. INTA, Buenos Aires

Peri PL, Martínez Pastur G, Vukasovic R, Díaz B, Lencinas MV, Cellini JM (2002) Thinning schedules to reduce risk of windthrow in *Nothofagus pumilio* forests of Patagonia, Argentina. Bosque 23:19–28

Peri PL, Sturzenbaum MV, Monelos L (2004) Sistemas silvopastoriles en Santa Cruz. In: Alternativas de Manejo Sustentable para el Manejo Forestal Integral de los bosques de Patagonia. Informe Final del Proyecto de Investigación Aplicada a los Recursos Forestales Nativos (PIARFON), Dirección de Bosques de la Secretaría de Ambiente y Desarrollo Sustentable de Nación (SAyDS), Buenos Aires, Argentina, pp 617–644

Peri PL, Sturzenbaum MV, Monelos L, Livraghi E, Christiansen R, Moretto A, Mayo JP (2005a) Productividad de sistemas silvopastoriles en bosques nativos de ñire (*Nothofagus antarctica*) de Patagonia Austral. In: Proceedings III Congreso Forestal Argentino y Latinoamericano, Corrientes, Argentina, 10 pp

Peri PL, Martínez Pastur G, Monelos L, Allogia M, Livraghi E, Christiansen R, Sturzenbaum MV (2005b) Sistemas silvopastoriles en bosques nativos de ñire: una estrategia para el desarrollo sustentable en la Patagonia Sur. In: Zárate R, Artesi L (eds) Dinámicas Mundiales, Integración Regional y Patrimonio en Espacios Periféricos. Editorial Universidad Nacional de la Patagonia Austral, Río Gallegos, pp 251–259

Peri PL, Viola M, Martínez Pastur G (2005c) Estimación del contenido de carbono en bosques de ñire (*Nothofagus antarctica*) en Patagonia Sur. In: Proceedings III Congreso Forestal Argentino y Latinoamericano, Corrientes, Argentina, 9 pp

Peri PL, Moot DJ, McNeil DL (2006a) Validation of a canopy photosynthesis model for cocksfoot pastures grown under different light regimes. Agrofor Syst 67:259–272

Peri PL, Gargaglione V, Martínez Pastur G (2006b) Dynamics of above- and below-ground biomass and nutrient accumulation in an age sequence of *Nothofagus antarctica* forest of Southern Patagonia. For Ecol Manag 233:85–99

Peri PL, Monelos L, Bahamonde H (2006c) Evaluación de la continuidad del estrato arbóreo en bosques nativos de *Nothofagus antarctica* bajo uso silvopastoril con ganado ovino en Patagonia Sur, Argentina. In: Proceedings of the 4th Latin American Congress of Sustainable Agroforestry, Varadero, Cuba, 6 pp

Peri PL, Sturzenbaum MV, Rivera EH, Milicevic F (2006d) Respuesta de bovinos en sistemas silvopastoriles de ñire (*Nothofagus antarctica*) en Patagonia Sur, Argentina. In: Proceedings of the 4th Latin American Congress of Sustainable Agroforestry, Varadero, Cuba, 7 pp

Peri PL, Lucas RJ, Moot DJ (2007) Dry matter production, morphology and nutritive value of *Dactylis glomerata* growing under different light regimes. Agrofor Syst 70:63–79

Peri PL, Gargaglione V, Martínez Pastur G (2008a) Above- and belowground nutrients storage and biomass accumulation in marginal *Nothofagus antarctica* forests in Southern Patagonia. For Ecol Manag 255:2502–2511

Peri PL, Bahamonde H, Monelos L, Martínez Pastur G (2008b) Producción de hojarasca en bosques primarios y bajo manejo silvopastoril de *Nothofagus antarctica* en la provincia de Santa Cruz, Argentina. In: Proceedings Segunda Reunión sobre *Nothofagus* en la Patagonia – EcoNothofagus 2008, Esquel, Chubut, Argentina, pp 149–155

Peri PL, Hansen N, Rusch V, Tejera L, Monelos L, Fertig M, Bahamonde H, Sarasola M (2009a) Pautas de manejo de sistemas silvopastoriles en bosques nativos de *Nothofagus antarctica* (ñire) ñire en Patagonia. In: Proceedings of the 1st National Congress of Silvopastoral Systems, Misiones, Argentina. INTA Editions, pp 151–164

Peri PL, Martínez Pastur G, Lencinas MV (2009b) Photosynthetic and stomatal conductance responses to different light intensities and water status of two main *Nothofagus* species of south Patagonian forest. J For Sci 55:101–111

Peri PL, Ormaechea S, Huertas L (2009c) Protección de renovales de ñire. Carpeta Técnica EEA INTA Santa Cruz, Sección 4. Producción Forestal. Edición EEA Santa Cruz, pp 15–16

Peri PL, Gargaglione V, Martínez Pastur G, Lencinas MV (2010) Carbon accumulation along a stand development sequence of *Nothofagus antarctica* forests across a gradient in site quality in Southern Patagonia. For Ecol Manag 260:229–237

Peri PL, Mayo JP, Christiansen R (2012a) Producción y calidad del pastizal mejorado con trébol blanco en sistemas silvopastoriles de ñire en Patagonia. In: Proceedings of the 2nd National Congress of Silvopastoral Systems, Santiago del Estero, Argentina. INTA Editions, pp 70–75

Peri PL, Monelos L, Martínez Pastur G, Ivancich H (2012b) Propuestas de raleo para un bosque de *Nothofagus antarctica* con uso silvopastoril en Santa Cruz. In: Proceedings of the 2nd National Congress of

Silvopastoral Systems, Santiago del Estero, Argentina. INTA Editions, p 181

Peri PL, Ormaechea S, Ceccaldi E, Bahamonde H, Gargaglione V (2012c) Una cuestión de escala: Producción bovina a nivel de establecimiento con bosque de ñire en Tierra del Fuego. In: Proceedings of the 2nd National Congress of Silvopastoral Systems, Santiago del Estero, Argentina. INTA Editions, pp 46–51

Peri PL, Monelos L, Martínez Pastur G, Ivancich H (2013a) Raleo en bosque de *Nothofagus antarctica* para uso silvopastoril en Santa Cruz. In: Peri PL (ed) 2nd Congress on Patagonian Agroforestry (International), El Calafate, Argentina. INTA Editions, p 96

Peri PL, Ormaechea S, Martínez Pastur G, Lencinas MV (2013b) Inventario provincial del contenido de carbono en bosques nativos de ñire en Santa Cruz. In: Proceedings 4th Congreso Forestal Argentino y Latinoamericano, 10 pp, Iguazú, Misiones, Argentina

Quinteros P, Hansen N, Kutschker A (2010) Composición y diversidad del sotobosque de ñire (*Nothofagus antarctica*) en función de la estructura del bosque. Ecología Aust 20:225–234

Quinteros CP, Lopez Bernal P, Gobbi ME, Bava JO (2012) Distance to flood meadows as a predictor of use of *Nothofagus pumilio* forest by livestock and resulting impact, in Patagonia, Argentina. Agrofor Syst 84:261–272

Rebertus AJ, Veblen TT (1993) Structure and tree-fall gaps dynamics of old-growth *Nothofagus* forests in Tierra del Fuego, Argentina. J Veg Sci 4:641–654

Reque JA, Sarasola M, Gyenge J, Fernández ME (2007) Caracterización silvícola de ñirantales del norte de la Patagonia para la gestión forestal sostenible. Bosque 28:33–45

Rusch V, Roveta R, Peralta C, Márques B, Vila A, Sarasola M, Todaro C, Barrios D (2004) Indicadores de sustentabilidad en sistemas silvopastoriles. In: Alternativas de Manejo Sustentable para el Manejo Forestal Integral de los bosques de Patagonia. Informe Final del Proyecto de Investigación Aplicada a los Recursos Forestales Nativos (PIARFON), Dirección de Bosques de la Secretaría de Ambiente y Desarrollo Sustentable de Nación (SAyDS), Buenos Aires, Argentina, pp 681–797

Rusch V, Sarasola M, Schlichter T (2005) Indicadores de Biodiversidad en Bosques *Nothofagus*. IDIA XXI For 5:8–14

Rusch V, Sarasola M, Hansen N, Roveta R (2009a) Criterios e indicadores como método para guiar la sustentabilidad: Principios generales y Criterios e Indicadores ambientales em sistemas silvopastoriles com ñire (*Nothofagus antarctica*). In: Proceedings of the 1st National Congress of Silvopastoral Systems, Misiones, Argentina. INTA Editions, pp 109–119

Rusch V, Sarasola M, Hansen N, Roveta R (2009b) Criterios e indicadores como método para guiar la sustentabilidad: Aspectos productivos y socioeconómicos en sistemas silvopastoriles con ñire (*Nothofagus antarctica*). In: Proceedings of the 1st National Congress of Silvopastoral Systems, Misiones, Argentina. INTA Editions, pp 368–379

Santa Regina I (2000) Biomass estimation and nutrient pools in four Quercus pyrenaica in Sierra de Gata Mountains, Salamanca, Spain. For Ecol Manag 132:127–141

Sarasola M, López D, Gaitán J, Siffredi G (2008a) Productividad de sistemas silvopastoriles en bosques de ñire en la cuenca del río Foyel. Actas de la Segunda Reunión sobre *Nothofagus* en la Patagonia – EcoNothofagus 2008, Esquel, Chubut, pp 156–164

Sarasola M, Fernández ME, Gyenge J, Peyrou C (2008b) Respuesta de los ñire al raleo en la cuenca del río Foyel. Actas de la Segunda Reunión sobre *Nothofagus* en la Patagonia – EcoNothofagus 2008. Esquel, Chubut, p 47

SAyDS (2005) Primer Inventario Nacional de Bosques Nativos. Ministerio de Salud y Ambiente de la Nación. Secretaría de Ambiente y Desarrollo Sustentable, Buenos Aires, p 86

Senanayake SGJN (1995) The effects of different light levels on the nutritive quality of four natural tropical grasses. Trop Grasslands 29:111–114

Simmons MT, Archer SR, Teague WR, Ansley RJ (2008) Tree (*Prosopis glandulosa*) effects on grass growth: an experimental assessment of above- and belowground interactions in a temperate savanna. J Arid Environ 72:314–325

Soler R, Martínez Pastur G, Lencinas MV, Ivancich H, Peri PL (2012) Regeneración natural de *Nothofagus antarctica* bajo distintos niveles de dosel y usos del bosque. In: Proceedings of the 2nd National Congress of Silvopastoral Systems, Santiago del Estero, Argentina. INTA Editions, pp 283–288

Soler R, Martínez Pastur G, Peri PL, Lencinas MV, Pulido F (2013) Are silvopastoral systems compatible with forest regeneration? An integrative approach in southern Patagonia. Agrofor Syst 87:1213–1227

Somlo R, Bonvissuto G, Schlichter T, Laclau P, Peri PL, Allogia M (1997) Silvopastoral use of Argentine Patagonian forest. In: Gordon AM, Newman SM (eds) Temperate agroforestry system. Editorial CAB International, Wallingford, pp 237–250

Speziale K, Ezcurra C (2008) ¿Es importante conservar el ñire? Proceedings II Reunión sobre *Nothofagus* en la Patagonia. Esquel, Chubut

Steinke LR, Premoli AC, Souto CP, Hedrén M (2008) Adaptive and neutral variation of the resprouter *Nothofagus antarctica* growing in distinct habitats in northwestern Patagonia. Silva Fenn 42:177–188

Swift MJ, Heal OW, Anderson JM (1979) Decomposition in terrestrial ecosystems. Blackwell Scientific Publications, Oxford

Tejera L, Hansen N, Fertig M (2005) Efecto de la cobertura arbórea y del pastoreo vacuno sobre el establecimiento de la regeneración de *Nothofagus antarctica*

(G.Forst.) Oerst. In: Proceedings III Congreso Forestal Argentino y Latinoamericano, Corrientes, Argentina
Thompson KF, Poppi DP (1990) Livestock production from pasture. In: Langer RHM (ed) Pastures: their ecology and management. Oxford University Press, Auckland, pp 263–283
Veblen TT, Donoso C, Kitzberger T, Rebertus AJ (1996) Ecology of Southern Chilean and Argentinean *Nothofagus* forests. In: Veblen T, Hill R, Read J (eds) The ecology and biogeography of *Nothofagus* forests. Yale University Press, New Haven, pp 293–353
Vesk PA (2006) Plant size and resprouting ability: trading tolerance and avoidance of damage? J Ecol 94:1027–1034
von Müller A, Lloyd CE, Hansen NE (2014) Disponibilidad forrajera en el Bosque Andino Patagónico. Cartilla Técnica INTA EEA Esquel. Sistemas Silvopastoriles, pp 215–218

# 7 Silvopastoral Systems in Arid and Semiarid Zones of Chile

Patricio Rojas, Marlene González, Susana Benedetti, Peter Yates, Alvaro Sotomayor, and Francis Dube

**Abstract**

In the semiarid Coquimbo Region, Chile, *Acacia saligna* is used particularly where reforestation has been promoted with the objective of recovery of degraded soils, production of fodder for livestock, fuel wood and control erosion. This exotic species also has potential use as an important source of human food, because the seeds of the trees could be harvested and processed for the production of breads and biscuits with nutraceutical properties. *Prosopis* is another genus, six members of which are common in Chile; they are adapted to areas with very low rainfall, infertile soils, and high variations between day and night temperatures. Because of their exceptional adaptability to such difficult conditions where most other species do not survive, *Prosopis* spp. are an important resource for the local peasants and indigenous communities, providing them with food and shelter for their livestock.

**Keywords**

*Acacia saligna* • Fodder • *Prosopis* sp. • Small landowners • Traditional practices

P. Rojas (✉) • M. González • S. Benedetti
Instituto Forestal (INFOR), Sede Región Metropolitana, Sucre 2397, Casilla 3085, Santiago, Chile
e-mail: parojas@infor.cl

P. Yates
World Vision Consultant, 99 Bridgewater Road, Maldon, VIC 3463, Australia

A. Sotomayor
Instituto Forestal (INFOR), Sede Biobío, Camino a Coronel Km. 7,5, Casilla 109-C, Concepción, Chile

F. Dube
Department of Silviculture, Faculty of Forest Sciences, University of Concepción, Victoria 631, Casilla 160-C, VIII - Concepción, Chile



P.L. Peri et al. (eds.), *Silvopastoral Systems in Southern South America*, Advances in Agroforestry 11, DOI 10.1007/978-3-319-24109-8_7

## 7.1 *Acacia saligna in Chile:* A Forage Resource with Potential for Food Production

### 7.1.1 Introduction

*Acacia saligna,* an exotic species, mainly occurs in the semi-arid Coquimbo Region, Chile particularly where reforestation has been promoted with the objective of recovery of degraded soils, production of fodder for livestock, fuel wood and control erosion (Perret and Mora 2001) as seen in the work done at the Hacienda Huentelauquén of the Province of Choapa (Fig. 7.1).

Plantations were established in the semi-arid zone as a result of Decree Law 701 enacted the State of Chile in 1974 and implemented by CONAF,[1] that subsidized up to 75 % the total cost of establishing forest plantations. Currently it is estimated that an area of 7,500 ha of *Acacia saligna* plantations exists in the Coquimbo Region (González 2014). This law was intended to stimulate forestry activities on soil suitable for forestry, on degraded soils and to encourage afforestation, especially by small landowners. *A. saligna* come from the semiarid Western Australian "Wheatbelt" and is adapted to extremely dry conditions through their physiological mechanisms: form and distribution of stomata and a deep root system that allows the tree to access subsurface aquifers and withstand long periods of drought (Maslin 2011).

Plantations near the coast in the Region of Coquimbo allow the species minimize evapotranspiration and survive on coastal moisture. As a pioneer species in plant succession, *A. saligna* can fix atmospheric nitrogen improving the physical and chemical properties of soils, allowing increased fertility, restoration of degraded soils and possible reforestation with native species in a second rotation (Maslin 2011). The *A. saligna* resource represents an important potential food for humans because the seeds of the trees could be harvested and processed for the production of breads and biscuits with functional and nutraceutical properties (Yates 2014).

### 7.1.2 Diversity and Adaptation

*A. saligna* is extremely polymorphic in phenotypic characteristics such as phyllodes, growth habit cortex and also in ecological and biological

**Fig. 7.1** Dunes control work in Huenteauquén farm. Province of Choapa, Coquimbo Region. Lat 31°37′38″ S and Long 71°33′03″ W (Source: Patricio Rojas 2004)

[1] Corporación Nacional Forestal.

attributes. It is endemic to south-western Australia where it occurs in areas with annual rainfall between 250 and 1200 mm. Taxonomic research indicates that polymorphic variation in *A. saligna* fits four sub species referred to McDonald (2007) and in other publications as: (1) the variety *saligna* "cyanophylla", (2) the variety *pruinescens* "Tweed River", (3) the variety *lindleyi* as "typical" and (4) the variety *stolonifera* "forest". This taxonomy offers a range of morphological and growth habits among subspecies and provenances that may yield specific advantages that are adaptable to specific production purposes. Intraspecific variation and growth habits associated are shown in Fig. 7.2.

As an example, those subspecies with a tendency to vegetative propagation such as stolonation (suckering) may not be suited to wood production since they will produce multiple stems of small diameter, however this characteristic may be highly desirable if the goal of production is fodder for animals, and might further the available phyllode biomass. Field observations in native distribution show that the subspecies *saligna pruinescens* have low occurrence of propagation by stolon, however this feature is very common in *saligna subspecies stolonifera.*

### 7.1.3 Forage Resource

*Acacia saligna* is considered a fast growing species, reaching 8 m high at 4–5 years at planting sites with limitations like low rainfall and poor fertility. In dry lands of northern Chile annual increases in height between 30 and 71 cm have been observed. Tree growth is lower in prolonged drought conditions, so that production varies between 1.5 and 10 $m^3$ $ha^{-1}$ according to site conditions in rotations of 5–10 years and coppice management. In rural arid and semi-arid production systems, *Acacia saligna* are generally planted to provide supplemental or emergency food for prolonged periods of drought, shade for small livestock and protection and stabilization of degraded soils (Perret and Mora 2001). However, there are large variations in the nutritional value of this legume species (Maslin 2011), probably because of their genetic variability and/or ignorance of this subspecies in plantations in Chile (W. O'Sullivan, pers. comm. 2005). Typical shape and vegetative propagation of plantations in the Coquimbo Region can be seen in Fig. 7.3.

Normally farmers use this resource to feed sheep and goats, especially during summer and autumn. They harvest acacia leaves and stems from young and mature trees during the dry season to provide a daily supplement to grazing (Meneses 2012). Other farmers permit that sheep and goats also eat directly on the shrubs without any silviculture management.

The recommended planting density for livestock systems is 4×3 m (833 trees $ha^{-1}$). A greater distance between rows, such as 6–10 m may allow for greater production of perennial grasses, herbs and crops under some conditions. The incorporation of silvicultural treatments such as pruning and thinning improves crop productivity. In arid areas the primary purpose is not to produce high quality wood, but rather to create and adapt the tree architecture to maximize the production of forage for livestock (Vita 1996). Pruning and tree management differ from traditional forestry methods. Crop management for fodder purposes may be accomplished by in situ grazing, managed to create a stump of 25–50 cm high after the third year, or by topping trees when they reach 2 m in height. The cutting intervention should be done in the time before the summer growth (Serra 1997). Bratti (1996) concluded that the trees cut to 50 cm height significantly differed better in vigor and growth of other cutting heights made.

The fresh foliage is palatable to animals and can be used as a food supplement for livestock (sheep and goats) containing up to 21 % crude protein in dry weight. Serra (1997) reported that phyllode material contains 10–19 % protein by dry weight, 24–27 % crude fiber and 20–26.48 % digestible organic matter in vitro. Plantations established by INFOR in the arid interior of the province of Choapa showed that forage production can reach a value between 0.8 and 2.2 Mg DM (dry material) $ha^{-1}$ of dry forage at 3 and 4 years after planting (Perret and Mora 2001).

Meneses et al. (2012) concluded that *Acacia saligna* fodder has limitations as feed for goats. In

**Fig. 7.2** Natural geographic distribution of the subspecies of Acacia saligna in Western Australia and differentiation of flower buds (**a**), juveniles (**b**) clusters and (**c**) flowers at anthesis in the subspecies saligna (Adapted from Maslin 2011)

pregnancy its acceptability as the sole forage, represented 65 % of that of alfalfa hay, but during lactation the intake was higher than alfalfa hay. Blood urea is affected with 50 % of acacia inclusion. The response of albumin is inconsistent and the other blood components did not present effects from acacia consumption. For this Acacia fodder should not represent more than 26 % of diet during the last third of pregnancy, according to body and birth weight. During lactation, acacia should not represent more than 25 % of diet to avoid affecting milk production, although the regression equation determined that it should not represent more than 7.3 %. A higher percentage in the diet would affect animal productivity and more than 50 % would affect body weight. Maximum DM intake is obtained with 24.8 % of acacia in the diet (Meneses et al. 2012).

### 7.1.4 The Food Potential of the Seeds, "Wattle Seeds"

The seeds of some species of the genus *Acacia* are a traditional food in Australia and these are processed

**Fig. 7.3** *A. saligna* in plantations in the Region of Coquimbo **a** "typical" growth habit, **b** plants propagated vegetatively, by stolons (Source: Patricio Rojas 2010)

in the "bushfood" industry to create flavoring agents in confectionery, sauces and ice cream and flour for bread, pasta and biscuits. There is interest in the "functional food" potential of acacia seed in Australia due to the high-protein, low glycemic index meal that can be produced (Yates 2014). The glycemic index (GI) is a system for comparing the relative rapidity with which an energy containing food enters the bloodstream as sugars after consumption. A high glycemic index indicates rapid sugar (and thus energy) availability, which requires a relatively robust response from the insulin-endocrine system. Excessive, long-term exposure to foods with a high glycemic index has been implicated in diseases such as type II diabetes. Analyses of *A. saligna* seed from five farms in the Coquimbo region confirm the seed as being high in protein and that their fiber content is suitable for use in baking in combination with other flours (Quitral 2012). Studies are currently underway to ascertain whether *A. saligna* seed may have any beneficial effects on diabetes related illness (Yates 2014).

## 7.2 Silvopastoral Systems with *Prosopis* sp. in Chile

### 7.2.1 Silvopastoral Systems with *Prosopis tamarugo* Phil.

#### 7.2.1.1 Introduction

The species *Prosopis tamarugo* (Tamarugo) grows endemically in the North of Chile and today is mostly present in an area called "Pampa del Tamarugal", 70 km east of the city of Iquique, in the Region of Tarapacá, but in the past it was found growing abundantly between the Regions of Arica and Copiapó (FAO 1984). It was used as fuel wood for the mining activities in the region, due to which it has become practically extinct, and today is considered threatened. In the 1960s, it was planted 20,550 ha of the species in national reserves under the system of Chilean Protected Areas (FAO 1984).

This tree mainly grows in the desert ecosystem of Pampa del Tamarugal under very particular conditions. The climate of the region corresponds to a normal desert climate with the major biological connotations of high temperatures during the day, great temperature variations, almost complete lack of precipitation, occasional occurrence of fog, low relative humidity and high sun radiation (FAO 1981). The mean rainfall varies from 0 to 80 mm per year, depending on weather events in the Los Andes Mountain. The soil is fluvial alluvium material from the Andes mountain range covered with a superficial salt crust of 0.1–0.6 m (FAO 1981).

The Tamarugo receives water from occasional fog and underground water, which typically is found at 1–10 m but can be as deep as 60 m or more (FAO 1984), and which the Tamarugo tree can extract thanks to its special root system up to 8 m (FAO 1984). Although the Tamarugo often grows on soils covered with a thin or sometimes rather thick salt crust with underground water at between 2 and 40 m deep, it has been observed that naturally it grows more frequently in areas where the underground water is at a depth of 20–40 m.

Mature trees have a root system consisting of a dense mass of shallow lateral roots, poorly lignified, of 40–80 cm thick, and one or several tap roots, which grow horizontally until a depth of 7–8 m (FAO 1984). These taproots then grow vertically down without ramifying, generally reaching depths of approximately 4 m. When the plant grows on very sandy soils, as is the case in the Canchones area, the taproots can grow as deep as 7 or 8 m. The moisture-absorbing root mass on the surface typically has approximately the same perimeter as the tree crown.

This species is used as forage by wild and domestic animals such as sheep and goats, feeding on its leaves and fruits. Its wood is also used as firewood, in handcraft and for floors. It grows on saline grounds where no other tree can grow. This makes this species a valuable source of charcoal and firewood for the local peasants.

Under the conditions in Pampa del Tamarugal (Fig. 7.4), a national reserve protected by the National Forest Corporation (CONAF), controlled grazing of goats and sheep is occurring during certain periods of the year supporting small-scale farmers of the adjacent communities.

#### 7.2.1.2 Forage Productivity for Livestock Management with Tamarugo

This tree produces abundant forage mainly consumed by sheep and goats, occasionally also by cattle. Its productivity depends on factors such as tree age and development, density, underground water depth and quality. Its fruits and leaves fall off the tree and spread on the ground, mostly directly under the tree crown, providing a forage material that contains 10.5 and 10.9 % from fruit and leaves of raw protein respectively, 29.7–15.2 % of fibre and 0.46–1.46 % ether extract (CORFO 1984).

Based on measurements of 19 trees, Oyarzún (1967) describes that he obtained an average weight of 2.1 kg $m^{-2}$ seedpods under the tree crown. With 30-year-old trees he observed an average yield of 3.4 kg $m^{-2}$ dry matter of fruits and leaves per square meter under the tree crown. Latrille and García (1968) estimated the yield of leaves and fruits of this tree at 1 kg $m^{-2}$ under the crown of adult trees, which could exclusively feed 3.5 sheep $ha^{-1}$. Furthermore foraging experiments with French Merino sheep have been carried out in Pampa del Tamarugal, comparing different diets with very interesting results as shown in Table 7.1.

It can be observed from data in Table 7.2, that free foraging in Tamarugo forests allows higher weight gains with yearlings and lambs compared to feeding with alfalfa hay and dried Tamarugo leaves and fruits, and lambing percentages are considerably higher under a free foraging system.

Additionally, in the National Reserve "Los Flamencos" in the Antofagasta Region, located 27 km south of San Pedro de Atacama at an approximate altitude of 2,300 m above sea level, there is a Tamarugo plantation implemented by CORFO in 1970 on a surface of 805 ha, known as Sector Tambillo. This area of the Reserve has historically been used for silvopastoral activities by the local communities, being this the only way to provide their livestock with the necessary food and shade under the harsh local conditions (González and Maraboli 1997). Moreover, as of 1994 six families from nearby Talabre have been authorized to make use of this sector, as they had been doing historically before that date (CONAF 2008). This plantation was implemented with a distance between trees of 15 × 15 m, in an area without trees, banning any livestock during the first 6 years, allowing the trees to grow free of animal impact in order to safeguard their survival, achieving an average of 4 ft per plantation cell (González and Maraboli 1997). Although no evaluation on the capacity of animal load has been made until the date, nor has any management been applied to the tree crowns, it has been possible to maintain a certain amount of sheep,

**Fig. 7.4** (**a**) Isolated tamarugo (*Prosopis tamarugo*) tree, growing in the Pampa of Tamarugo, located 70 km east of the city of Iquique, Tarapacá Region, Chile; and (**b**) Tamarugo plantation, Pampa of Tamarugal, Chile (Source: Alvaro Sotomayor)

**Table 7.1** Comparison of results obtained with French Merino sheep in Pampa del Tamarugal Reserve, located 70 km east of the city of Iquique, Tarapacá Region, Chile, fed with Tamarugo leaves and fruits (kg day$^{-1}$), alfalfa hay (kg day$^{-1}$) and free foraging in Tamarugal (Lanino 1966)

| | Feedlot production of sheep according to type of forage (weight gains in kg live weight per animal) | | |
|---|---|---|---|
| Variables | Dried Tamarugo leaves and fruits, 2 kg day$^{-1}$ | Alfalfa hay, 1.5 kg day$^{-1}$ | Free foraging of leaves and fruits in the Tamarugo forest |
| Lamb weight at weaning: | | | |
| Yearlings (kg) | 15.0 | 21.5 | 26.7 |
| Lambs (kg) | 14.0 | 25.0 | 29.6 |
| Fleece weight (kg) | 2.81 | 3.18 | 3.77 |
| Fiber length (mm) | 5.90 | 6.44 | – |
| Lambing (%) | 48 | 63 | 111 |

Note: Fleece, set of wool that is cut from a sheep at shearing time

goats, horses and lamas in the place during short time of the year, in spring, to browse leaves and eat fruits, but supplemented with hay.

#### 7.2.1.3 Conclusions

The results presented above indicate that this type of native forest, with adequate management and maintaining a certain limit of animal load based on the forage production of leaves and fruits per season, would allow sustainable livestock management for the benefit of the local peasants and indigenous communities of this area in Chile. Furthermore, it is recommended to generate public policies allowing the recovery of this species, both for the benefit of the environment and the livestock economy of the local communities, both related to domestic and wild animals.

### 7.2.2 Silvopastoral System with *Prosopis chilensis* (Mol.) Stuntz

#### 7.2.2.1 Introduction

The Chilean Algarrobo tree (*Prosopis chilensis* (Mol.) Stuntz) is a woody leguminous species widely distributed in arid and semi-arid regions in Latin America (Pinto and Riveros 1989). According to Peralta and Serra (1987), its wide distribution area reaches from the South of Peru to the northern and central part of Chile, the South-West of Bolivia to the northwestern, western and central part of Argentina. As to its geographical distribution in Chile, FAO (1997) points out that the Algarrobo tree grows in the arid and semi-arid regions of the northern and

**Table 7.2** Estimation of total production of Algarrobo fruits (kg of seed pods), in the Metropolitan Region, Chile, considering individual trees growing under different conditions, according to the current soil use, surface (ha), production (kg pods tree$^{-1}$) and density (trees ha$^{-1}$)

| Type of formation | Surface (ha) | Production (kg pods tree$^{-1}$) | Density (trees ha$^{-1}$) | Total annual production according to formation type (kg pods)[a] |
|---|---|---|---|---|
| Natural formations | 7,419 | 15 | 28 | 720,524 |
| Associated to agricultural production | 3,213 | 61 | 1 | 195,214 |
| Protection areas | 609 | 15 | 12 | 25,330 |
| Plantations (including wind curtains) | 66 | 2 | 254 | 8,035 |
| Urban | 521 | 5 | 16 | 9,397 |
| Total | 11,828 | | | 958,499 |

[a]Considering that 23 % of the trees are under production each year, except for formations in association with agricultural production

central part of the country in the provinces of Copiapó, Elqui, Limarí, Choapa, San Felipe, Los Andes and is particularly abundant in the northern part of the Santiago watershed. Algarrobo is the most abundant *Prosopis* species in Chile, covering more than 50 % of the national surface in the Metropolitan Region (INFOR 1986). Altamirano (2012) distinguishes it as the southernmost *Prosopis* genus in South America, growing in Chile between latitudes 22°46′ and 33°40′ S (from the Metropolitan Region to San Pedro de Atacama in the Antofagasta Region).

This species occupies a very specific habitat with a well-defined ecological position, it can be found in valleys or in areas with relatively superficial underground water and on the bottom of gullies with more effective water accumulation. It occupies limestone and salty soils, where the species develops under conditions of high aridity (Peralta and Serra 1987). It can be found in the Central Valley and the lower parts of the coastal mountains, as disperse individual trees or in small patches of forest (Altamirano 2012).

The Chilean Algarrobo grows in environments with an average annual precipitation rate of 28–356 mm, concentrated mainly in winter, reaching zero precipitation in dry years, with torrential rainfalls occurring in few occasions, but a dry period of 8–12 months as a rule. The natural habitat of the species is characterized by very high temperature variations between January and July, as the hottest and the coldest month respectively, as well as extreme temperature changes between night and day. The species can endure occasional frosts with temperatures of −5 °C. Annual average temperatures vary between 14.3 and 14.9 °C, absolute maximum temperatures reach 27.4–30.6 °C. In the North as well as in the distribution area further south, relative humidity is always below 78 %, being these very dry environments with high potential evapotranspiration, intense sun radiation and high luminosity. Algarrobo prefers thick, often stony and normally alkaline, secondary soils of volcanic origin with a sandy or loamy-sandy texture. It is very resistant to salinity and can grow on soils with a pH value between 7.6 and 8.9, rich in sodium. It occupies well-drained soils, on gentle to steep slopes in valleys and gullies between 500 and 1,500 m above sea level (FAO 1997). It is never found above 1,800 m altitude and when growing on grounds that occasionally receive irrigation, its development and lushness can be quite impressive (Ortiz 1966).

#### 7.2.2.2 Productivity of *Prosopis chilensis*

According to González (2013), density of natural formations (number of trees ha$^{-1}$) of Algarrobo s in the Metropolitan Region is very low due to its devastation and over-exploitation in the past, being used as firewood, construction wood, production of charcoal and as fuel wood for metal foundries. The surface covered with Algarrobo in

**Fig. 7.5** Cattle using adult Algarrobo tree (*Prosopis chilensis*) for shelter from the sun, Fundo Quilapilun, Province of Chacabuco, Metropolitan Region, Chile (UTM (19H) 345412–6337171, Altitude 804 m) (Source: Marlene González 2013)

2013 was determined to be 11,828 ha, located in Chacabuco Province and differentiated into five types of formation considering different uses, densities and productivity: natural for silvopastoral system, associated to agricultural production (annual crops and fruit production) (Fig. 7.5), protection areas, plantations and urban environments (parks and residential areas) (Table 7.2).

In relation to fruit production, FAO (1997) indicates that an adult Algarrobo tree can produce 100 kg of seedpods; however this is not the rule, as it can be affected by factors such as sun radiation, tree age and growing site. Vásquez et al. (1991) mention that average annual fruit production varies between 30 and 40 kg of pods per plant. However, on sites with less favourable conditions it can lie below 10 kg. On the other hand Serra (1997) points out that some trees show variations in production between 10 and 200 kg per tree. Studies carried out in Pampa del Tamarugal by CORFO (1984) show that forage obtained from fruits and leaves of *Prosopis chilensis* contain between 7.6 and 13.5 % protein and 26.0 and 19.9 % raw fibre respectively, indicating that it would have to be complemented in order to guarantee a balanced diet for sheep and goats, especially during pregnancy and lactation. When analysing its energy input it can be compared to medium-quality (CORFO 1984).

For the Region of Valparaíso, in the commune of Calle Larga, an average of 2.8 trees $ha^{-1}$ can be found in a situation of dry slopes mainly integrated into mixed agricultural/forestry systems, where the average productivity per tree reaches 6.13 kg of seedpods per square meter under the tree crown. It is important to bear in mind that

Algarrobos growing inside silvopastoral system fields have the highest fruit production per tree, but the lowest density (number of trees $ha^{-1}$).

#### 7.2.2.3 Conclusions

*Prosopis chilensis* continues to be a good alternative as forage for livestock management, in particular on small-scale farms in the Metropolitan Region, where trees are well managed in order to provide shelter for the cattle held in this area. Its importance becomes evident through its traditional use as complementary forage, especially in critical periods, for 3 weeks to 1.5 months each year, when the fruits are collected and stored to be used when needed (G. Trivelli, pers. comm. 2012). The same situation was also reported by Contreras (1983), who by quoting the National Academy of Sciences (1979), indicated that the fruits or pods were collected and stored to be used later as feed for the cattle during 1 or 2 months of the year, with a nutritional value comparable to barley and corn. Hence it is recommendable that this species be included in the regional development silvopastoralism strategies and promotion policies, destined to support its recovery for the purposes of soil protection and livestock shelter and forage for small-scale farms and local communities in the central part of Chile.

## 7.3 Traditional Agroforestry Practices in Chile: From Practice to Science

### 7.3.1 Introduction

Agricultural activities associated with trees have existed in Europe since the Middle Age (King 1987). A variety of agroforestry practices have also existed in the Americas since pre-hispanic times (Reynel and Morales 1987; Carlson and Añazco 1990; Budowsky 1994). Unfortunately, most of them have been replaced by the high-tech productive models seen today, with a high degree of artificialization to ensure greater productivity. However, it has become clear that even if these models are successful in economic terms, one of the greatest challenges is their environmental and social viability, that is why it has not been possible to replicate these models on a small scale, and this has generated an enormous gap between large-scale production and family farming.

On the other hand, when one observes rural smallholdings, it is possible to identify different uses and distribution in the field where multiple activities and diversity of products can be found, which responds to a strategy for survival, where the productive rationale is to guarantee the livelihood of the family with minimal risk. As a matter of fact, according to several studies, the hypothesis that would explain the extensive diversification present in small-scale farms in Latin America is the aversion to risk (Reca and Echeverría 1998). Therefore, rural production systems involve a combination of a certain amount of labor and various means and factors of production within a given space carried out by the producer, based on available resources, in order to obtain certain crop and animal goods, hence, the rural production system chiefly consists of agricultural, livestock and forestry subsystems.

The diversified uses of the land come from the "*how it is done*" transmitted from one generation to the next, with the incorporation of new technologies, only if the economic resources are available, but always using the space in a variety of ways and with the interrelation of productive sectors. Farmers, whether they possess the technical knowledge or not, and regardless of the size of their property, plan their activities and the use of the space and resources they have to satisfy their production or consumption needs, and to generate income. Farmers determine how they will combine uses in one plot and how they will make the best use of it; this is nothing more than planning for and diversifying the use of farmlands, in other words, agroforestry (Benedetti 2012).

Therefore, this interrelationship between the sectors stems from historic and empirical knowledge that is not based on scientific studies; therefore, these forms of "doing" are classified as practices. In this sense, agroforestry is a technique known and commonly used by farmers since ancient times, with an enormous variety of combinations since they have to understand and combine the potential and limitations of the ecosystems they inhabit and adapt them to their social and economic conditions and culture (Benedetti and Valdés 1996). However, it was only a few decades ago that agroforestry began to emerge in the scientific-technical scenario as a science that can underpin the failure of the prevailing agroforestry model to meet the basic demands of the rural population. This is also the case in Chile where issues such as integrated production systems, land and property use management, productive diversification and multipurpose species are becoming increasingly important, but have been the subject of very limited research and developments that have focused primarily on forest grazing models.

In this context, it is suggested that Chile's rich traditional agroforestry practices, which are a result of the country's environmental and cultural diversity, contribute enormously to the development of agroforestry models that take into account cultural and social elements to facilitate their adoption. This is why it is so necessary to recover, analyze and understand them.

### 7.3.2 Traditional Silvopastoral Practices in Chile

Family farming in Chile is represented by various types of producers from different ethnic and cultural backgrounds and environmental conditions, namely from highland communities in the north, farm communities in the center-north region, Mapuche communities in the south and small landholders throughout the country. Their productive practices are based on crop and livestock production models in which trees can play either a passive or active role, depending on the agro-climatic conditions. In areas with limited soil and water, there is a low proportion of arboreal elements, and these are mostly associated with crops and fodder. In sectors with better environmental conditions, trees play a more prominent role in maintaining productivity and in contributing to family income. Even if these traditional practices are common to all groups, they often differ in terms of the interrelation and space arrangements, which depend on the environmental conditions and socio-cultural characteristics of each group. The following are some of the most common agroforestry practices identified in Chile.

*Protein banks*: these are shrub species used for fodder grown in soft and steep slopes, on platforms or terraces where they are used to demarcate and protect plots. The fodder is harvested and fed to the livestock in their barns.

*Grazing in croplands*: on soft slopes and plains, in structures known as "melgas" or windrows. These windrows resemble terraces but are larger in size and are used for crop or fodder production. Each sector is separated by rocks, cobblestone fences, or live fences made of shrubs. Animals are brought in to each sector post-harvest to graze for a limited period of time on harvest residue. This is done to control grazing and replenish the soil with organic fertilizer.

*Family gardens*: one of the most common practices in the Chilean highlands. These gardens are located on land close to the farmer's home or outside their family farms in small plots known as *hijuelas* or *tablones*, which are generally square or triangular in shape. Windrows are built using mud or a combination of mud and rocks. These structures are used as entrances and irrigation channels. Fruit trees are planted over them to be used for feed or wood, and to provide shade and protection from low temperatures or frost. The areas used to grow crops are located between the windrows. These areas are demarcated using live fences with trees or bushes. This practice involves fallowing, crop rotation and organic compost and animal guano to cover the earth.

#### 7.3.2.1 Agroforestry Practices in Valleys and the Desert in Chile's Norte Grande Region

*Windbreaks and live fences*: trees on land used for crop or fodder production are a common sight in the desert valley. Trees are situated along the boundaries of the farms as a form of demarcation and to protect crops from the wind. For this, the trees are planted perpendicularly to the direction of the wind. This practice is classified as agroforestry if it involves crop production, and silvopastoral farming if forage crops are being produced, or consumed on site by livestock.

*Mixed tree and grassland farming*: Considers grassland cultivation in sectors of valleys and oasis with irrigation possibilities, using trees or fodder shrubs along the grassland boundaries. The farming areas are rectangular-shaped pasturelands. The trees, normally leguminous, are used for firewood and their fruit for animal or human consumption. The tradition is to plant two or three trees for each tree that has been harvested.

#### 7.3.2.2 Agroforestry Practices in Farming Communities

Farming communities are a form of community-based land tenure that began in colonial times in the Coquimbo and Valparaíso regions, primarily in areas with severe environmental limitations. The following agroforestry practices have been identified in this group:

*Rainfed croplands*: these are plots that the community supplies to a member of the community for dryland farming. The amount of land provided depends on the farmer's and their family's capacity to work the land. The plots are demarcated with live fences with cacti that form an impenetrable live barrier. This fence is essential since the main productive activity is goat farming. Traditionally, these rain fed croplands have been used for growing coriander, anise and wheat for a period of several years until productivity dwindles, after which they are used for grazing. Currently, as a consequence of the drought, these lands have been converted to forests for fodder using species such as *Atriplex nummularia* and, or *Acacia saligna*.

*Protective and productive forest areas*: generally found in shared community fields, especially in ravines where natural forests already exist or trees are planted to shelter animals from the sun or rain, for firewood or to protect water sources.

#### 7.3.2.3 Agroforestry Practices in the Central Region of Chile

*Grazing in the Espinal*: grazing in espino (*Acacia caven*) steppes is a practice that has existed throughout history in the Central Region. Espino is a nitrogen-fixing leguminous tree that provides a natural canopy under which goats and cattle can graze on the wide variety of high quality fodder that grows beneath it. Espino is also used to produce charcoal, which, together with the extensive grazing, has caused serious degradation in this ecosystem.

### 7.3.3 Conclusions

The traditional agroforestry practices represent an important knowledge-base and contribution to developing promising agroforestry technologies.

## References


Altamirano G (2012) Variedad de frutos y semillas en las especies del género *Prosopis* presentes en Chile. Corporación Nacional Forestal. Documento Técnico N 208, 12p (in Spanish)

Benedetti S (2012) Ordenamiento predial y la contribución de la componente forestal al sistema productivo del pequeño productor. En Incorporación y contribución de la componente forestal al sistema predial de la pequeña agricultura. INFOR, Santiago, pp 35–56 (in Spanish)

Benedetti S, Valdés J (1996) Prácticas agroforestales tradicionales en la zona árida y semiárida de Chile. CONAF, Santiago, 65p (in Spanish)

Bratti JM (1996) Efecto de la altura de corte en el rebrote de Acacia saligna (Labill.) Wendl. Universidad de Chile, Facultad de Ciencias Forestales, Santiago, 61p (in Spanish)

Budowsky G (1994) El alcance y potencial de la agroforestería con énfasis en Centroamérica. En: Krishnamusthy L (1995). Agroforesteria en desarrollo: Educación, investigación y extensión. U.A. Chapingo, Mexico, pp 2–17

Carlson P, Añazco M (1990) Establecimiento y manejo de prácticas agroforestales en la sierra ecuatoriana. Red Agroforestal Ecuatoriana, Quito, 187p (in Spanish)

CONAF (2008) Actualización Plan de Manejo Participativo Reserva Nacional Los Flamencos. Corporación Nacional Forestal, Región de Antofagasta, Chile, 142p (in Spanish)

Contreras B (1983) Diversidad morfológica en poblaciones de Algarrobo (*Prosopis chilensis* (Mol.) Stuntz) en la IV Región. Tesis para optar al Título de Ingeniero Forestal. Universidad de Chile, Facultad de Ciencias Agrarias, Veterinarias y Forestales, Escuela de Ciencias Forestales, Santiago de Chile, 100p (in Spanish)

CORFO (1984) Valoración nutricional de tamarugo y Algarrobo y perfiles metabólicos de ovinos y caprinos en la Pampa del Tamarugal. In: Estado Actual del Conocimiento sobre *Prosopis tamarugo*. Arica, pp 75–121 (in Spanish)

FAO (1981) *Prosopis tamarugo*: arbusto forrajero para zonas áridas. Oficina Regional de la FAO para América Latina, Santiago, 143p (in Spanish)

FAO (1984) Estado Actual del Conocimiento sobre *Prosopis tamarugo*. Oficina Regional de la FAO para América Latina y el Caribe, Dirección de Production y Protección Vegetal, Santiago, 483p (in Spanish)

FAO (1997) *Prosopis chilensis*. In: Especies arbóreas y arbustivas para las zonas áridas de América Latina. Serie: Zonas áridas y semiáridas N 12. Santiago, 347p (in Spanish)

González GM (2013). Algarrobo. Rescatando una Especie de Alto Valor Histórico. Una Contribución a una Alimentación Sana y al Combate contra la Desertificación. Informe Técnico N 195, Instituto Forestal, Santiago, 128p (in Spanish)

González GM, Maraboli FF (1997) Actividades Proyecto 2402 Agroforestería Sector Tambillo. Departamento Desarrollo y Fomento Forestal, CONAF, Región de Antofagasta, Chile, p 51p

INFOR (1986) Estadísticas Forestales 1985. Serie Informática 34. División Estudios Económicos, Corporación de Fomento de la Production, Santiago, 98p (in Spanish)

King K (1987) The history of agroforestry. In: Steppled N (ed) Agroforestry: a decade of development. ICRAF, Nairobi, pp 1–11

Lanino R (1966) Comparación en tres razas ovinas alimentadas con tamarugo (Prosopis tamarugo Phil.) (Pampa del Tamarugal). Tesis de Grado, Universidad de Chile, Facultad de Agronomía, Santiago, 80p (in Spanish)

Latrille L, García X (1968) Evaluación nutritiva del tamarugo (*Prosopis tamarugo* Phil.) como forraje para rumiantes. Universidad de Chile, Departamento de Production Animal, Memoria Anual, pp 257–273 (in Spanish)

Maslin BR (2011) Understanding Acacia saligna. Edible Wattle Seeds of Southern Australia. CSIRO. Wattle we eat for dinner. Proceeding of Workshop on Australian Acacias for food security. Alice Spring, 9p.

Mc Donald et al (2007) Improved prospects for the domestication of Acacia saligna in region de Coquimbo. Chile. In: Ciencia e Investigacion Forestal, Instituto Forestal. Infor, Chile, 91–101p

Meneses R, Varela G, Flores H (2012) Evaluating the use of *Atriplex nummularia* hay on feed intake, growth and carcass characteristics of creole kids. Chil J Agric Res 72:74–79

National Academy of Sciences (1979) Tropical legumes: resources for the future. National Academy of Sciences, Washington, DC, 331p

Ortiz GJ (1966) Algunos forestales chilenos de la estepa septentrional. Boletín Técnico N 23. Ministerio de Agricultura, Dirección y Agricultura y Pesca, Departamento de Extensión Agrícola. Santiago, 44p (in Spanish)

Oyarzún S (1967) Algunas variaciones lanimétricas con ovejas merino encastadas en dos épocas diferentes, Canchones (Pampa del Tamarugal). Tesis de Grado, Universidad de Chile, Facultad de Agronomía, Santiago, 91p (in Spanish)

Peralta M, Serra M (1987) Caracterización del hábitat natural de las especies del género *Prosopis*, en las provincias de Huasco y Copiapó, III Región, Chile. Programa Investigación y Desarrollo de Áreas Silvestres Zonas Áridas y Semiáridas de Chile. FO:DO/CHI/83/017. Documento de Trabajo N°9. Santiago, 121p (in Spanish)

Perret S, Mora F (2001) Acacia saligna, leguminosa arbórea de uso múltiple para las zonas áridas y semiáridas de Chile. INFOR, Santiago, 14p (in Spanish)

Pinto M, Riveros E (1989) Variation of fruit production and quality of different ecotypes of Chilean Algarrobo ((*Prosopis chilensis*) (Mol.) Stuntz). In: Wickens G, Hag N, Day P (eds) New crops for food and industry. Chapman and Hall, London, pp 280–287

Quitral V (2012) Análisis proximales desarrollados con muestras de semillas de Acacia saligna de cinco predios de la Región de Coquimbo. Facultad de Medicina, Departamento de Nutrición, Universidad de Chile. Proyecto FIA PYT-2013-0010. Productos Alimentarios Funcionales con semillas de Acacia saligna. Informe Técnico N 2, Santiago, 17p (in Spanish)

Reca L, Echeverria R (1998) Situación actual y propuestas En: Agricultura, Medioambiente y Pobreza Rural en América Latina. IFPRI-BID, Washington, DC, pp 85–120

Reynel C, Morales C (1987) Agroforestería tradicional en Los Andes del Perú. Un inventario de tecnologías para la integración de la vegetación leñosa a la agricultura. FAO/Ministerio de Agricultura, Lima, 154p (in Spanish)

Serra MT (1997) *Prosopis chilensis*. In: Especies arbóreas y arbustivas para las zonas áridas y semiáridas de América Latina. Serie Zonas áridas y semiáridas N°12 FAO/PNUD, Santiago, pp 215–225 (in Spanish)

Vásquez M, Zacarías I, Escobar B, Yáñez E (1991) Calidad biológica de la proteína de los cotiledones de Algarrobo tratadas por calor seco y calor húmedo. Alimentos 1(16):5–8 (in Spanish)

Vita A (1996) Los tratamientos silviculturales. Universidad de Chile. Facultad de Ciencias Agrarias y Forestales, Escuela de Ciencias Forestales, Departamento de Silvicultura. Santiago, 150p (in Spanish)

Yates P (2014) Chile Visit Report. Proyecto FIA PYT-2013-0010. Productos Alimentarios Funcionales con semillas de Acacia saligna. Informe Técnico N 2, Santiago, 2p (in Spanish)

# Silvopastoral Systems in Temperate Zones of Chile

8

Francis Dube, Alvaro Sotomayor, Veronica Loewe, Burkhard Müller-Using, Neal Stolpe, Erick Zagal, and Marcelo Doussoulin

**Abstract**

In Chile, indigenous forests (mostly *Nothofagus* sp. and *Acacia caven*) and plantations of fast-growing exotic species cover 13.5 and 2.4 million ha, respectively. The latter consists principally of *Pinus radiata* and *Eucalyptus* sp., but also include cherry (*Prunus avium*), poplar (*Populus* sp.), and walnut (*Juglans regia*). The main silvopastoral systems that have been established in the temperate zone of the country, excluding arid, semiarid and Patagonian regions, are in: (1) an old *Nothofagus* forest in the foothills of the Andes, (2) a second-growth "Roble" (*Nothofagus obliqua*) forests of the Andes, with the objectives of rejuvenating the over-mature forests, and evaluating the quantity and quality of pasture that is sown under different tree coverages. The *Pinus radiata*-based silvopastoral systems are the most common system adopted by smallscale agroforestry producers in the Mediterranean Chile in diverse environments from semiarid to humid zones. Furthermore, because of the expanding markets for pine nuts and timber of several hardwood species, several experimental trials were established in the past 20 years between the Valparaiso and Los Lagos regions, including exotic species such as cherry, walnut, poplar and stone pine (*Pinus pinea*). Through regular income from the sale of diverse woody and non-woody forest products, these innovative silvopastoral

F. Dube (✉) • B. Müller-Using
Department of Silviculture, Faculty of Forest Sciences, University of Concepción, Victoria 631, Casilla 160-C, VIII - Concepción, Chile
e-mail: fdube@udec.cl

A. Sotomayor
Instituto Forestal (INFOR), Sede Biobío, Camino a Coronel Km 7.5, Casilla 109-C, Concepción, Chile

V. Loewe
Instituto Forestal (INFOR), Sede Región Metropolitana, Av. Sucre 2397, Casilla 3085, Santiago, Chile

N. Stolpe • E. Zagal
Department of Soils and Natural Resources, Faculty of Agronomy, University of Concepción, Vicente Méndez 595, Chillan, Chile

M. Doussoulin
Department of Animal Production, Faculty of Agronomy, University of Concepción, Vicente Méndez 595, Casilla 537, Chillan, Chile


P.L. Peri et al. (eds.), *Silvopastoral Systems in Southern South America*, Advances in Agroforestry 11, DOI 10.1007/978-3-319-24109-8_8

systems help improve the quality of life and wellbeing of small farm owners in temperate Chile.

**Keywords**
Forage crops • Over-mature forest • Sustainability • Economic analysis • "Roble" • Small producers

## 8.1 Introduction

Forests in the Chilean territory cover an area of 15.9 million ha, of which 13.5 are native forests and 2.4 exotic plantations (INFOR 2013). The latter consists mostly of *Pinus radiata* (1.47 million ha) and *Eucalyptus* spp. (774,000 ha) that have been established mainly for industrial purposes. The principal products obtained are pulp and paper, sawn wood, particleboard, and plywood, which in 2012 represented a USD 5.3 billion share of total exports. Other planted species, although in a minor proportion include poplar, acacia, chestnut, Atriplex, ponderosa pine, lodgepole pine (*Pinus contorta*), and Douglas fir (*Pseudotsuga menziesii*).

The native forests of the temperate zones of Chile are distributed in two broad bands, one along the Andes at altitudes between 600 and 900 m above sea level, and the other along the Coastal mountain range between 400 and 800 m above sea level. The latter band, however, is discontinuous as the Coastal mountain range is an area where large areas of native forest have been replaced by faster growing plantations (e.g., *Pinus radiata*). Most of the remaining spots belong to small landowners who use their forests not only as a source of fuelwood and building material but also as pasture for cattle grazing and shelter for animal protection in winter.

In many cases, this multiple use is not in equilibrium with the limited potential of these forest ecosystems. A recent study from INFOR (2012) shows that in the Biobío Region for example, 50 % of the *Nothofagus* forests are so thin that they should not be intervened within the next 10–15 years in order to recover normal stocking levels. One of the causes is the presence of cattle that cause stagnation of natural regeneration processes.

In the dry lands of the Mediterranean region of Chile (also known as "Secano Interior"), most of the landowners use a large proportion of their properties to raise livestock (cattle and sheep) and produce annual crops (barley and wheat), without concern about the long-term sustainability of their lands. As a result, large areas are found with different degrees of erosion, mostly from overgrazing, but also because of insufficient soil protection that would normally be provided by trees and shrubs if these were present at higher densities. According to CIREN (2010), more than 50 % of Chilean soils are eroded, an alarming situation that may have irreversible consequences if not soon addressed.

As a consequence, improved management of rural properties should include innovation and incorporation of appropriate technologies in order to increase the efficiency of traditional activities, and take into account the overall sustainability of the systems in place. With this in mind, forestry and livestock activities can be integrated into efficient and productive silvopastoral systems, resulting in a symbiosis, or a synergy between forests and grazing components (Dube et al. 2012; Sotomayor et al. 2008).

From the technical point of view, the overall productivity of agricultural resources found in silvopastoral systems can be significantly increased (Anderson et al. 1988) because of the better environmental conditions created within the area of influence of the established trees (Dube et al. 2012). These include reduced wind speed, higher ambient temperatures, better soil moisture, fewer periods of water deficit, and protection of livestock against rain, wind and cold temperatures. Trees also provide an efficient soil protection against wind and rain erosion (Dube et al. 2012, 2013; Sotomayor 1990; Quam and Johnson 1999; Sotomayor and Teuber 2011). Simultaneously, the introduced trees benefit from (i) the presence of livestock that may adversely affect weed growth, especially in the early years after establishment of the tree component, which in turn reduce the risks of forest fires, (ii) enhanced nutrient cycling into the soil, (iii) application of fertilizer for the pasture and forage crops, and (iv) larger space available for plant growth as a result of the agroforestry arrangement.

From the economic point of view, local producers can maintain a constant cash flow throughout the year coming from the sale of animal products (meat, milk and wool), and periodic cash income from thinning and pruning the forest to produce diversified wood products (e.g., firewood, lumber, posts for fences) (Dube et al. 2002; Sotomayor 1989). These products may help to pay some of the household expenses of landowners and improve their quality of life (Leslie et al. 1998; Polla 1998). Furthermore, high-quality timber products (e.g., premium clear wood, wood with tight knots) can be obtained at the end of the forest rotation. Additionally, barren rural properties also usually gain value when trees are established and appropriately managed therein.

A unique feature of these integrated production systems is their overall sustainability. For instance, they not only permit reclaiming degraded lands through efficient control of erosion, but also afford the protection of waterways and improvement of water quality; they can contribute to increase $CO_2$ sequestration (Dube et al. 2011); they may generate aesthetical landscapes; and they can preserve and increase the amount of wildlife. The following sections describe the principal silvopastoral systems that have been established in the temperate zone of Chile, excluding arid, semi-arid and Patagonian regions.

## 8.2 Silvopastoral Management in Old *Nothofagus obliqua* Forests in the Mediterranean Region of South Central Chile

### 8.2.1 Introduction

Within the remaining indigenous forest areas of the Biobío Region of Chile, second-growth "Roble" (*Nothofagus obliqua*) forests, with or without the presence of "Raulí" (*N. nervosa*), cover approximately 450,000 ha (INFOR 2012). This forest resource has a high potential for timber production, and a series of related products such as fuelwood, particleboard, and lumber. However, the forests are not only used to produce fuel and building materials, but also as pastures for cattle grazing and as shelter to protect the animals in winter. Cattle can severely impair the establishment and regeneration of natural vegetation, as well as cause soil compaction from trampling of the topsoil, and overstocking the cattle on the pastures.

Unfortunately, there is no specific governmental regulation regarding the use of second-growth forests for pastures, although it is known that uncontrolled cattle-raising is one of the factors that substantially contribute to the degradation of forests. The degraded areas, especially those located in sectors with steep slopes can undergo severe erosion that will complicate, in some cases, the reintroduction of forest species (Ortega and Rodríguez 1994). In general, the second-growth "Roble" forests are presently found in areas that are not suitable for agriculture (i.e. with pronounced relief).

Non-systemic grazing will eventually alter the environmental benefits that are provided by forests (e.g., protection of water quality, increased biodiversity, and mitigation of greenhouse gas emissions that are responsible for global warming). Forest degradation due to deforestation, grazing and erosion implies that the damaged tree stands and underlying soils will sequester less carbon, resulting in an increase of net $CO_2$ emissions to the atmosphere (Dube et al. 2011). Given the absence of studies on sustainable silvopastoral systems in "Roble" forests having different amounts of canopy closure that are used in conjunction with improved pastures under different light conditions, there is an urgent need to investigate the effect of pastures and grazing on the natural and artificial regeneration of "Robles" and "Raulíes" forests of south–central Chile. Additionally, it is important to learn how these systems will affect the live weight gain of cattle that graze in them.

Therefore, the general objective of this study is to develop standards of silvopastoral management in old "Roble" forests with different levels of canopy closure as an option for sustainable management in a native forest area (Ranchillo Alto) in the Andes, Biobío Region, Chile. The specific objectives are to: (i) Evaluate the quantity and quality of pasture established under different degrees of tree-canopy coverage, as well as changes in the botanical composition of the herbaceous strata after 3 years of silvopastoralism; (ii) Monitor the establishment of native species (an essential component for the sustainability of silvopastoral systems) that were planted to effectively rejuvenate the over-mature forest resource; (iii) Determine the combined effect of herbaceous legumes (*Trifolium* sp. and *Lolium* sp.) and nitrogen fertilization on the content of total C and N in soil as well as on the growth of the planted trees; (iv) Assess the effect of canopy closure (open, partly open and partly closed) on the quality and quantity of soil organic matter; and (v) Encourage the establishment of productive silvopastoral systems in the properties of neighboring communities, using appropriate rural experiences and traditions combined with scientific innovation to produce a better quality of life.

### 8.2.2 Problem of the Study Area

The "Ranchillo Alto" property is state owned land and includes a large area of native forest, which is under heavy pressure from long-term and ongoing land uses that include cattle grazing, and tree cutting for firewood, charcoal and timber. These processes strongly threaten overall biodiversity, soil quality and the very existence of the forest itself. The "Calabozo" and "Avellano" areas (north and south sectors of the property, respectively) are located in the foothills of the Andes Mountains, and are both occupied by remnants of the native forest with indigenous species such as *Nothofagus antarctica*, *N. dombeyi*, *N. nervosa*, *N. obliqua*, and *N. pumilio*, *Laurelia sempervirens*, *Gevuina avellana*, *Drimys winterii*, *Lomatia hirsuta*, and *Podocarpus saligna*.

The local economy around "Ranchillo Alto", is characterized by a predominance of agricultural and forestry activities. The "Calabozo" northern sector has a population of 30 people, while 112 inhabitants live in the "Avellano" southern sector. Regarding the socio-economic conditions of the households, 93 % are classified as being poor. In this rural economy, women significantly contribute to family nutrition through gardening, horticulture, livestock management (cattle and sheep), and the preparation and preservation of food. Consequently, women play an important role in food security for the family and community. Small farm or "campesino" communities are organized into groups of neighbors who, through regular meetings, take democratic decisions so that they might implement strategies for equitable and profitable rural development. The strategies, however, are often based on short-term considerations rather than long-term benefits to the community.

Livestock raising is a common activity throughout "Ranchillo Alto", except for the areas with higher elevations and peaks. The land that normally provides most of the animal forage is highly valued by the community since the majority of surrounding properties generally do not have the capacity to produce sufficient forage for livestock consumption throughout the year. Unfortunately, continual animal grazing also hinders the natural regeneration of trees in the more open sites, which eventually affects the quality and density of the stands of "Robles" and "Raulíes", and also favors the proliferation of *Chusquea quila*, which is an invasive species of bamboo in the region (AMBAR 2010).

Illegal timber extraction occurs mainly in the western part of the property, where both dead and green woods are harvested. However, intensive logging can be found throughout the site, and a large proportion of the forest has already been intervened and subsequently degraded. Obvious signs of degradation can easily be seen in the plant communities of "Coigüe (*N. dombeyi*)-Roble" and "Roble", where continuous, non-systemic, grazing is used. It is also possible to see large areas where all of the timber has been harvested, beside areas where selective extraction of some individual trees has occurred. In these areas, the forest is relatively open, or thin, and is comprised mainly of small trees, and patches of degraded pastures that support livestock activity. Harvested wood is used primarily for heating and charcoal production, with the firewood being used for family consumption, while charcoal is usually sold in nearby villages and contributes to cash income for families in the area. Thus, one can perceive that the vegetation has been intensively and illegally exploited without proper management, whether commercially or for subsistence, which leads to overexploitation of the forest.

In 1992, the EULA center of the University of Concepción, Chile conducted a study to classify the amount and extent of soil erosion for the Biobío River Watershed (Peña and Carrasco 1993). The authors concluded that 76 % of the Andean Foothills was affected by moderate to severe erosion. In pastures of "Ranchillo Alto", erosion indicators include compaction of topsoil, small gullies, and light-colored surface soils that implies a loss of the original topsoil that contained a higher content of organic matter. Soils with decreased contents of organic matter will have reduced fertility over time, and lower plant vigor and reduced forage production for grazing.

### 8.2.3 Silvopastoral Recommendations for Sustainable Forest Management

In both the north and south areas of the "Ranchillo Alto" property, a portion of each of the forest types present was left without any production management whatsoever and subsequently protected from grazing and illegal logging. However, in other parts of the forests, management was applied that tends to encourage rejuvenation of the over-mature forest. The silvicultural treatment that is mainly recommended here is called "continuous cover". This system uses neither extensive clear cutting, nor broad cuts in small areas, but rather the selective cutting technique of silviculture harvesting of small groups of trees or individual trees. The adequate dosing of incident light is achieved using these rather precise interventions, such that seed regeneration by parent trees can be initiated, or the existing regenerations can be enhanced.

In the specific case of the pure "Roble" forests the situation is different. These areas are more open given that "Roble" as a deciduous species lets in a greater amount of light reaching the ground during the year, thus creating favorable conditions for growth of *Chusquea quila* and *C. culeu*. If these areas were not exposed to livestock grazing, an impenetrable understory would be established in just a few years. In these sectors, we conduct sequential silvopastoral management, with a series of paddocks, where cattle is grazed, while respecting the carrying capacity of the site. "Roble" is a species that is well suited to silvopastoral management given its deciduous leafing, canopy structure and its light colored leaves that allow better light transmission; characteristics that also favor the establishment and growth of grasses. Additionally, as part of the management strategy, cattle are excluded from riparian buffer strips that protect the watercourses. This management system harmonizes sustainability with the traditional raising activities, and respects the cultural tendencies of the local population.

### 8.2.4 Materials and Methods

#### 8.2.4.1 Site Description and Characterization

The "Ranchillo Alto" state property belongs to the Municipality of Yungay, province of Ñuble, Biobío Region (1200–2000 m altitude, 37°03′–37°04′ S and 71°38′–71°39′ W). It is located 120 km east of Concepción and covers approximately 653 ha. The site is located in the biological corridor "Nevados de Chillán – Laguna del Laja", and is technically an area having official protection by the state. It is also part of CIAF (Center for Investigations in Agroforestry) of the University of Concepción, which has the overall mission to support the future of agricultural and forestry farms through the achievement of economic, environmental and social sustainability.

In this part of the Andes, the prominent climate is "warm temperate Mediterranean with a short dry season" (<4 months), with an average annual rainfall of 3000 mm, and mean annual temperature of 13.5 °C (AMBAR 2010). Two series of volcanic soils (Andisols) are mapped (CIREN 1999) in areas of lower to higher slopes and have been classified as Humic Haploxerands and Typic Haploxerands, respectively. Some of the most important features of volcanic soils include high content of organic matter in the topsoil, low bulk density, which varies between 0.6 and 0.9 g $cm^{-3}$, low pH, and high fixation of soil phosphorus that may cause deficiency of this nutrient in plants. The chemical properties from the north and south locations are summarized in Table 8.1.

The contents of soil organic matter, and total C and N are high in both sites, which were expected given the volcanic origin of these soils. The content of soil nitrate-N is slightly higher in the south, but still low in both sectors. In general, nutrient levels are medium to low, with the exception of certain micronutrients (Fe, Mn, Cu). Sulfur is low only in the south while boron is very low in both sectors. Adequate fertilization will therefore be required at the establishment of improved pastures.

Two areas for silvopastoralism were defined in "Ranchillo Alto" (Fig. 8.1): north and south. In the southern part of the forest, three degrees of canopy closure (or treatments) were identified: open canopy, partly open canopy and partly closed canopy. All treatments have similar slopes (1–3 %) and uniform exposure. Each treatment covers 4 ha and has three plots (or replicates) of 1.33 ha each, all geo-referenced by GPS. In the northern area, two degrees of tree canopy closure were identified: partly open and partly closed. In this sector, each treatment covers an area of 6 ha and both have three plots of 2.0 ha each. The silvopastoral systems in each area (north or south) represent 12 ha. The aforementioned treatments, however, correspond only to the first year of the study, since improved pastures will be established the second year of the study in the same areas. Table 8.2 provides a description of the different treatments that have been applied in each area.

#### 8.2.4.2 Conditioning of Tree Covers to the Desired Levels of Luminosity

To evaluate the efficiency and yield of the silvopastoral systems (for cattle grazing), it was first necessary to define the levels of luminosity under the tree stratum and, once defined, establish uniform sections of luminosity in plots through dosed interventions in the treetop stratum. Ranchillo Alto's over-mature "Roble" forest has areas of diverse tree coverages, and it was a challenge to find adjacent sites having

**Table 8.1** Chemical properties (soil organic matter, organic C and N, pH, effective cation exchange capacity, and contents of aluminum, nitrate, phosphorus, potassium, sulfur and boron) of the volcanic soil (Haploxerands) (0–20 cm) at two temperate silvopastoral locations (north and south plateaus). Measurements were taken in January 2014 in the Ranchillo Alto state property, Chile (north site at 1300 m altitude, Lat 37°03′30.86″ S, and Long 71°38′50.60″ W; south site at 1200 m altitude, Lat 37°04′52.10″ S, and Long 71°39′12.39″ W)

| | SOM | C | N | C/N | pH | ECEC | Exch. Al | $NO_3$-N | Avail. P | Avail. K | Avail. S | B |
|---|---|---|---|---|---|---|---|---|---|---|---|---|
| Site | (%) | (%) | (%) | | | (cmol $kg^{-1}$) | (cmol $kg^{-1}$) | (mg $kg^{-1}$) | (mg $kg^{-1}$) | (mg $kg^{-1}$) | (mg $kg^{-1}$) | (mg $kg^{-1}$) |
| North | 18.7 | 10.81 | 0.47 | 23.0 | 6.0 | 5.38 | 0.10 | 7.6 | 2.7 | 118.8 | 12.3 | 0.1 |
| South | 24.6 | 14.22 | 0.79 | 18.0 | 5.4 | 1.56 | 0.34 | 9.6 | 3.0 | 87.9 | 8.7 | 0.1 |

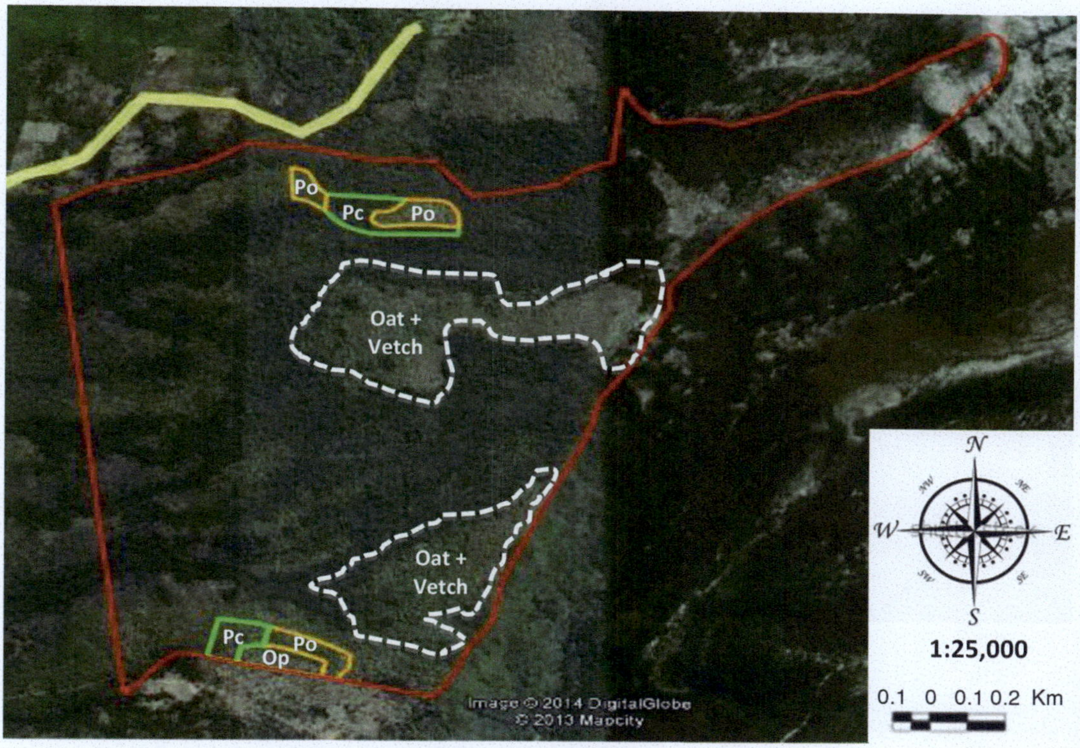


**Fig. 8.1** Satellite photograph of the "Ranchillo Alto" state property, Chile (north site at 1300 m altitude, Lat 37°03′30.86″ S, and Long 71°38′50.60″ W; south site at 1200 m altitude, Lat 37°04′52.10″ S, and Long 71°39′12.39″ W), with the north and south sectors designated for silvopastoralism (*orange* and *green lines*), and central plateaus for producing hay bales of oat and vetch (*white dotted lines*). In each polygon, the following letters correspond to different levels of canopy openings: opened (*Op*), partly opened (*Po*), and partly closed canopy (*Pc*) (map by Google Earth 2014) (Color figure online)

the three approximate degrees of luminosity at ground level. Once the general sites were located, the next task was to intervene in smaller plots to represent the partly opened and partly closed canopies to achieve the desired levels of luminosity. In the southern area of the property, these goals were accomplished fairly easily, with a minimal adjustment of the respective tree covers. In the northern area, it has been a little more difficult to establish a partly closed canopy, since the zone adjacent to the partly opened area was too closed to sunlight because of the predominance of *N. dombeyi* stands of poor quality and the over maturity of the trees. In this case, 1/3 of the total basal area had to be cut and extracted. Figures 8.2a–d illustrate the degrees of canopy closure in the silvopastoral sites.

#### 8.2.4.3 Establishment of Improved Pasture for Cattle Grazing

In the silvopastoral areas, initial site preparation consisted in removing old stumps and dead wood debris, and eliminating weeds that may compete with future improved pastures. Barbed wire fences were constructed to delineate the paddocks in each treatment. The fences are also intended to prevent livestock access to watercourses, which ensure adequate protection of surface water resources. In each treatment, two 80-l water troughs were built for the cattle. It is worth mentioning that currently, there is cattle present in the area (belonging to the neighbors) with a stocking density of 1 cow ha$^{-1}$, that is fed from the existing grass which is mostly *Festuca* sp. In this current undeveloped system, the silvopasture areas provide basic feeding and shelter to the ani-

**Table 8.2** Description of treatments (location, degree of canopy opening, and amount of solar radiation at ground level) that were established in the Ranchillo Alto state-owned property, Chile (north site at 1300 m altitude, Lat 37°03′30.86″ S, and Long 71°38′50.60″ W; south site at 1200 m altitude, Lat 37°04′52.10″ S, and Long 71°39′12.39″ W)

| Treatment | Location | Canopy opening | Code | General description |
|---|---|---|---|---|
| T1 | South | Open | S-Op | Gap of several treetop widths between treetops |
| | | | | Ground with 85–95 % of external light (average of area) |
| T2 | South | Partly open | S-Po | Gap of one treetop width between treetops |
| | | | | Ground with 65–75 % of external light (average of area) |
| T3 | South | Partly closed | S-Pc | Gap of half treetop width between treetops |
| | | | | Ground with 45–55 % of external light (average of area) |
| T4 | North | Partly open | N-Po | Gap of one treetop width between treetops |
| | | | | Ground with 65–75 % of external light (average of area) |
| T5 | North | Partly closed | N-Pc | Gap of half treetop width between treetops |
| | | | | Ground with 45–55 % of external light (average of area) |

mals, and the cattle effectively control the growth of herbaceous and shrubby weeds.

In May 2016, improved pasture will be sown under the different tree canopies, for each of the five treatments described in the previous section (Table 8.2). To achieve this, farm tractors will first perform a chisel plowing followed by "debris dragging" (discarded tires used as a tire drag) to smooth the soil surface. Pasture seeding will consist of a mixture of forage grasses and legumes that will be able to adapt to the different levels of luminosity in the understory. The forage species to be sown include Winter Star (*Lolium multiflorum westerwoldicum*) (4 kg ha$^{-1}$), Phalaris Holdfast (*Phalaris acuatica*) (2 kg ha$^{-1}$), Nutrapack Super 9 (*Lolium perenne*, *Festuca arundinacea* and *Dactylys glomerata*) (15 kg ha$^{-1}$) and Med 700 (*Trifolium incarnatum*, *T. subterraneum* and *T. vesiculosum*) (6 kg ha$^{-1}$). Application of NPK fertilizer will include Supernitro (100 kg ha$^{-1}$), Triple Superphosphate (88 kg ha$^{-1}$) and Muriatic potash (25 kg ha$^{-1}$).

### 8.2.4.4 Establishment of Pasture for Bale Production

Two additional areas (within the white dotted-line polygons) for pasture bales were defined in the "Ranchillo Alto" property (Fig. 8.1). In order to establish hay bale production systems, all remaining dead wood and weeds will first be removed from the sites. A chisel plowing followed by debris dragging will then be performed using a standard farm tractor. A total of 10 ha (5 ha per polygon) of oat (*Avena sativa*) (90 kg ha$^{-1}$) with vetch (*Vicia atropurpurea*) (20 kg ha$^{-1}$) will be sown annually in order to produce enough hay for baling, according to the stocking rates that are described below. Barbed wire fences will be constructed to prevent livestock access. The application of NPK fertilizer will consist of Supernitro (35 kg ha$^{-1}$), Triple Superphosphate (25 kg ha$^{-1}$) and Muriatic potash (12 kg ha$^{-1}$). Upon completion of this process and after sufficient plant growth, hay bales will be harvested and stored in a nearby barn for feeding the cattle during the 3-month winter. Figure 8.3 shows one of the sites where oat and vetch will be sown for bale production.

### 8.2.4.5 Establishment of Enrichment Tree Plantations within the Cattle Grazing Systems

The commitment to develop sustainable silvopastoral systems requires additional tree plantations in those parts of the property where the woodland

**Fig. 8.2** Silvopastoral sites in southern "Roble" (*Nothofagus obliqua*) forest areas with different levels of canopy openings in Ranchillo Alto state property, Chile (south site at 1200 m altitude, Lat 37°04′52.10″ S, and Long 71°39′12.39″ W): (**a**) open: 85–95 % of solar radiation at ground level, (**b**) partly open: 65–75 % of solar radiation at ground level, (**c**) partly open with black Angus and Hereford cattle being grazed, and (**d**) partly closed canopy: 45–55 % of solar radiation at ground level, before sowing improved pasture

is over-mature or dying out and its tree density is too low, or in areas with old trees that are poorly distributed across the landscape. Unlike a regular plantation for timber production purposes, one must also consider its simultaneous use for cattle grazing, for which taller planting trees (>1.5 m) were placed in the ground and individually protected so that the cows could not damage them by grazing or trampling. However, the tree densities used in these enrichment plantations are not the same as in traditional forest plantations, since the objective here is to re-establish the appropriate density of trees that will permit perpetual silvopastoral use of the forest at a level for optimal yield, but which is also compatible from the ecological point of view (i.e. without permanently altering the forest environment while keeping open the option of 1 day returning to a timber use only system). With this in mind, trees were planted at a density of 50 stems ha$^{-1}$ (35 "Robles" and 15 "Raulíes" per hectare). The protection of individual trees consisted of triangle shelters made of three wooden stakes and chicken wires. During the summer, each tree is watered monthly with 10 l of water.

#### 8.2.4.6 Grazing System

Given the size of the plots, the recommended system is rotational grazing of improved pasture so that all animals are grazed sequentially in paddocks, or plots, of the silvopastoral areas. In the southern sector, 12 animals will be grazed for a period of 7–10 days in each 1.33 ha-plot, while in the northern sector, 12 animals will be grazed for 11–15 days in each 2 ha-plot. Upon completion of the grazing period in each plot, the herds are

**Fig. 8.3** Portion of the central plateaus where oat and vetch will be sown among "Robles" (*Nothofagus obliqua*) for hay bale production for livestock feed during the winter. Photo taken in fall 2014 after leafdrop, Ranchillo Alto state property, Chile (1250 m altitude, Lat 37°03′–37°04′ S, and Long 71°38′–71°39′ W)

transferred to the other plots. The average annual stocking density is 1 cow ha$^{-1}$.

#### 8.2.4.7 Air and Soil Temperature and Soil Moisture

The soil moisture and temperature (0–20 cm depth), and air temperature above the soil (+5 cm) are currently being measured every 2 h throughout the year using Decagon Devices EM-5B Data Loggers, EC-20 soil moisture sensors and ECT temperature sensors, respectively (Decagon Devices Inc., Pullman, WA, USA) as done by Dube et al. in a previous research in Chilean Patagonia (2012 and 2013). Table 8.3 presents the values of the readings thus far obtained from the established silvopastoral systems in Ranchillo Alto. Soil moisture at 0–20 cm depth increases with canopy closure in both locations, but is always higher in the southern than northern area when the same degrees of tree canopy coverage are compared between sectors. However, soil temperature decreases as the canopy closure increases, being on average more than 2.5 °C higher in the northern than the southern sector for the same level of canopy closure. This inverse relationship between soil temperature and moisture has been consistently observed in the treatments (Table 8.3). As for the superficial air temperature (+5 cm), the trend is similar to that of the soil temperature, except that it is slightly higher (0.2–0.4° C) for a given treatment.

### 8.2.5 Key Results and Final Remarks

Given the current silvopastoral reality in south-central Chile, the systems established in the Ranchillo Alto state property shall deliver the following results:

1. Development of standard guidelines on the production and quality of forage pastures that are established under different degrees of canopy closure of the tree component of the system.
2. Evaluation of the benefits of rotational grazing for the pasture and soil, as well as the benefits of the annual harvest of oat and

vetch for the production of hay bales, and the confinement of livestock in protected enclosures or barns during the winter.
3. Improvement of the quality of life of small farmers, optimizing forage production for their livestock and getting regular income from the sale of wood, meat and dairy products.
4. Provision of a greater stability to the small farm or "campesino" families, by training them to manage a technically feasible system that is sustainable under the given conditions.
5. Dissemination and implementation of the main elements of the silvopastoral systems established in "Ranchillo Alto" in the properties of neighboring communities, taking into account the different types of forests and levels of tree coverage; in addition to providing advice and strategies on how to adapt the results of the study to the particular socio-economic conditions that are prevalent in the daily lives of "campesinos".
6. Evaluation of the benefits of the introduction of herbaceous legumes in pastures, on the improvement of soil fertility and growth of the tree component in the system.
7. Evaluation of the benefits of good management practices on soil organic matter in order to enhance the biological processes of soil, which are important parameters to assess soil quality.
8. Monitoring of forest degradation from the loss of biodiversity caused by alteration of the natural regeneration process, and lower soil quality due to soil compaction and increased erosion. This will fill a knowledge gap concerning the adequate animal stocking rates for the various forest ecosystems.
9. Proposal of a series of guidelines on silviculture in old *Nothofagus* forests under sustainable silvopastoral systems in Chile.
10. Provision of new methodologies to the Ministry of Agriculture, in particular CONAF to assess the global sustainability of silvopastoral activities in Chile in order to regulate the misuse of natural forest resources through non-timber uses (i.e. cattle grazing).

All of the knowledge that has been acquired to date is gradually being disseminated within the Municipality of Yungay, the Biobío Region and southern Chile, and is contributing to rural development and environmental improvement in the natural range of the "Roble-Raulí-Coigüe" forest type, which is highly threatened by overexploitation from logging, and cattle grazing.

**Table 8.3** Mean annual soil moisture and temperature (0–20 cm depth), and air temperature above the soil (+5 cm) measured between August 2013 and 2014 in the north and south sectors of Ranchillo Alto, Chile (north site at 1300 m altitude, Lat 37°03′30.86″ S, and Long 71°38′50.60″ W; south site at 1200 m altitude, Lat 37°04′52.10″ S, and Long 71°39′12.39″ W)

| Location and level of canopy opening | Soil moisture 0–20 cm (% VWC) | Soil temperature 0–20 cm (°C) | Air temperature +5 cm (°C) |
|---|---|---|---|
| Northern sector (partly open canopy) | 9.9 | 13.5 | 13.9 |
| Northern sector (partly closed canopy) | 10.7 | 11.8 | 12.0 |
| Southern sector (open canopy) | 10.9 | 12.9 | 13.2 |
| Southern sector (partly open canopy) | 11.8 | 10.6 | 10.9 |
| Southern sector (partly closed canopy) | 12.5 | 9.1 | 9.4 |

*VWC* volumetric water content

## 8.3 Experiences on the Establishment of Silvopastoral Systems by the National Agroforestry Program (NAP) of the Forestry Institute of Chile (INFOR)

### 8.3.1 Introduction

As mentioned before, *Pinus radiata* (radiata pine), due to its plasticity, is the most planted tree in Chile, growing from Valparaiso to Los Lagos region of Chile. It is grown from semi-arid zones with 450 mm of precipitation in the central zone, to humid zones with ≥2000 mm in the southern zone of Chile. It supplies the main logs for pulp and paper mills, sawmills, panel board industries and finished timber products (INFOR 2013). Government has promoted the species by subsidies on small and medium sized farms particularly where soils are better fitted to plantations or silvopastoral systems.

The Chilean Forestry Institute (INFOR) developed the National Agroforestry Program (NAP), to evaluate the potential for agroforestry systems in Chile. Research included different potential agroforestry systems (Table 8.1), including suitable species. Further, demonstrative radiata pine agroforestry units were established as extension tools, aimed at small farmers and livestock producers. The results of this extension program are discussed below.

### 8.3.2 The Chilean National Agroforestry Program (NAP), to Promote Agroforestry Systems in Chile, 2006–2013 Period

Following the research conducted throughout Chile, the use of silvopastoral systems, amongst others, was promoted by working with 1600 producers between 2006 and 2013. An area of 1114 ha between Coquimbo in the north to the Magallanes Region in the south was planted (Table 8.4).

The farmers involved with NAP preferred agroforestry systems to planting trees as it was closer to their farming traditions and allowed them to continue with crop and animal production as well as producing wood. A survey of growers found that they were resistant to normal industrial plantations as they reduced their cattle farming. On average participants planted 0.7 ha of trees.

Of the total program established during the 2006–2013 period, the most accepted systems were the silvopastoral and windbreaks systems, each of which accounted for 44 % of the total planted, with the objective of producing animal fodder and wood (Table 8.4). This preference for silvopastoral systems is primarily due to the most viable land for planting trees is located on slopes without irrigation, where there are natural prairies with a low productive value for cattle farming purposes, or for the cultivation of dry land cereal crops. Windbreaks are used for crops and cattle farming affected by the wind, especially in insular zones or close to the sea. Producers recognize their importance for increased crop and livestock yield.

These results suggest that agroforestry systems offer a new way of mixed system farming, that is helpful to both small farmers and larger forestry owners. These agroforestry systems permit:

- Income in the short term through animals and agricultural crops.
- Timber revenue in the medium term through commercial thinning or direct usage of forest products.
- Higher production from the prairies or crops from the protective effect of the trees, such as in silvopastoral systems and windbreaks, due to reduced evapotranspiration of the understory.
- Wood production at the end of the rotation.
- Improved farm valuations and rural landscape improvement.

### 8.3.3 Experimental and Demonstrative Experience

As a way of generating information and also to serve as demonstrative units several silvopastoral pilot schemes with radiata pine were established in the temperate zone with 700–1300 mm rainfall between the Maule and Biobío Regions of Chile.

#### 8.3.3.1 Los Aromos Agroforestry Demonstrative Unit, Cauquenes Commune, Maule Region

This unit was situated on a northeast-facing slope in the Municipality of Cauquenes in the Maule Region, which belongs to an agro-ecological

**Table 8.4** Agroforestry systems established on small farms, 2006–2013, National Agroforestry Program of Chile (INFOR 2013)

| Agroforestry system | Agroforestry surface established per year and system (ha) | | | | | | Total (ha) |
|---|---|---|---|---|---|---|---|
| | 2006 | 2007 | 2008 | 2011 | 2012 | 2013 | |
| Silvoagricultural | 6.7 | 26.3 | 10.0 | 20.5 | – | – | 63.5 |
| Silvopastoral | 162.2 | 112.5 | 66.6 | 78.9 | 74.9 | 24.0 | 495.1 |
| Windbreaks | 12.5 | 97.7 | 35.0 | 201.3 | 140.3 | 21.5 | 486.8 |
| Riparian buffers | – | 3.5 | 10.0 | 0.5 | 1.3 | – | 15.3 |
| Bioenergy | – | 10.0 | 5.0 | – | 16.9 | 0.3 | 31.9 |
| Beekeeping | – | – | – | 16.8 | 4.3 | – | 21.1 |
| Total | 181.4 | 250.0 | 126.6 | 318.0 | 237.7 | 45.8 | 1113.7 |

zone of semi-arid land. Temperatures fluctuated between a mean annual maximum of 29 °C in January to minimum of 4.9 °C in July, with a mean annual temperature of 14.1 °C. The period free of frost was of 259 days, with an average of 6 frosts with temperatures lower than −5 °C each year. The average annual rainfall is 696 mm, with an annual water deficit of 931 mm and a dry period of 7 months (Peralta 1976).

In 2003 four plots were established, T1: 500 radiata pine trees $ha^{-1}$ with natural praire the first 5 years and oversown the last 2 years; T2: 1000 trees $ha^{-1}$ of radiata pine, where the prairie was oversown the first 2 years and the last 2 years with *Trifolium subterraneum* (subclover), *Trifolium michelianu* (bigflower clover) and *Medicago polymorpha* (burclover), and only natural grasses between 2008 and 2010 (Fig. 8.4 and Table 8.6). These silvopastoral treatments were compared with T3: a normal forestry system with 1250 trees $ha^{-1}$ as forestry control, and with T4: an area of natural prairie that the local farmer used for livestock production, as a control area). The silvopastoral layout was in strips, with two plantation rows at 2×3 m spacing and with a distance between them of 14 m for a 500 tree $ha^{-1}$ density, and of 7 m between strips for a 1000 tree $ha^{-1}$ density. The influence of the trees on the prairie production was analyzed and compared with a natural prairie area without trees.

For the evaluation of tree growth, total height (H), diameter at breast height (DBH) and the Basal Area (BA) were measured in three permanent plots of 1000 $m^2$ was used, randomly distributed within the treatments. The data were statistically analysed by a mixed linear model because of the absence of assumptions required by a traditional variance analysis: a normal distribution, the independence of the data, and variance heterogeneity. A longitudinal analysis was carried out, evaluating treatment, time and treatment * time. The variance model used to analyze the forestry, prairie and animals variables was:

$$Y = \mu + T + t + (T\,x\,t) + E,$$

where μ = constant, T = treatment, t = Time and E = error.

The analysis used PROC MIXED and LSMEANS in SAS version 9.3 (SAS Institute 2003).

For the prairie evaluation a randomly distributed plot design was used, with locations changing every year. For each treatment three exclusion cages of 0.5 $m^2$ were used to estimate dry matter (DM) production per hectare. Samples were taken at the end of the vegetative period, between December and January of each year, weighed to determine their fresh biomass and then dried at 60 °C in an oven to get DW. The results were statistically analyzed and averages compared with the LSD (Least Significant Difference) test at a 5 % significance level.

The silvopastoral treatment T2, with 1000 trees $ha^{-1}$, was thinned down at age 10 years to 427 trees $ha^{-1}$. It was pruned twice: at 5 years and 8 years to of 2.1 m and 3.2 m height, respectively. The forestry plot (T3) was thinned to 520 trees $ha^{-1}$ at age 10 and pruned to 2.1 m at 5 years.

**Fig. 8.4** (**a**) Radiata pine-based silvopastoral system in 2006, at 2 years of age, established in two lines of planting, in forest strips, with densities of 500 and 1000 trees ha$^{-1}$; (**b**) silvopastoral system with 1000 trees ha$^{-1}$, pruned to 2.1 m in year 2012, at the age of 8 years, with pasture established with subterranean clover and Harding grass in Los Aromos Agroforestry Unit, Commune of Cauquenes, Maule Region, Chile

**Table 8.5** Forest parameters in silvopastoral system with 1000 trees ha$^{-1}$, and forest system with 1250 trees ha$^{-1}$ in 2013, Los Aromos Agroforestry Unit, Maule Region, Chile

| Treatments | DBH (cm) | H (m) | Final density (trees ha$^{-1}$) | BA (m$^2$ ha$^{-1}$) |
|---|---|---|---|---|
| T2: silvopastoral system with 1000 trees ha$^{-1}$, thinned to 427 trees and pruned to 3.2 m | 16.6$^a$ | 12.0$^a$ | 427 | 9.38$^a$ |
| T3: forest system with 1250 trees ha$^{-1}$, thinned to 520 trees and pruned to 2.1 m | 13.7$^b$ | 10.8$^b$ | 520 | 7.84$^b$ |

aDifferent letters indicate significant differences, LSMEANS ($p<0.05$), for analysis over time, by year of evaluation and treatment

b*DBH* diameter at breast height, *H* total height, *BA* basal area

The results presented in Table 8.5 indicate that tree growth in DBH, H and BA were all superior in the silvopastoral system. For DBH and BA this was probably due to the wider original spacing in the silvopastoral treatment as compared to the forestry control, where trees have more competition. For H, reasons could be little differences in site, and also because use on fertilizer for oversown pastures.

Prairie production between 2006 and 2013 (tree ages 3–10 years) is given in Table 8.6. In 2006, 2007, 2009 and 2010 the silvopastoral plot with improved pastures (T2) was better than the either the silvopastoral plot with natural pasture (T1) or the control of natural prairie (T4); T1 and T4 had a similar low productivity. In 2008 there was no difference between treatments, which may be due to the use of natural pastures. In the latter years, 2012 and 2013, pasture productivity was higher in both the silvopastoral treatments compared to the control treatment.

#### 8.3.3.2 Los Alamos Demonstrative Unit, Arauco Province, Biobío Region

This unit was established in the Los Alamos Commune, Arauco Province, Biobío Region of Chile, on an area of 1.3 ha. The unit was on flat land, with gentle hills of gradients between 3 and 5 %, characteristic of the marine terraces in the Arauco Province. The soil is deep, clay loam on the surface to dense clay at depth (Peralta 1976).

The Los Alamos Commune is located within the Concepción agro-climate zone, a coastal area, with influence from the ocean. The thermal regime is characterized by an annual temperature of 13.2 °C, with a monthly mean maximum of 25.1 °C in January and mean minimum of 5 °C in July. This climate has a period of 7 months free of frost. The hydrous regime has an average annual rainfall of 1330 mm (Peralta 1976).

This silvopastoral trial was established using radiata pine, with a random block design with

**Table 8.6** Production of prairie, both sown and natural, in kg DM ha$^{-1}$, per season in silvopastoral systems and control area treatment without tree influence, Los Aromos Agroforestry Unit, Maule Region, Chile

| | Production of prairie, sown and natural (kg DM ha$^{-1}$) per season | | | | | | |
|---|---|---|---|---|---|---|---|
| Treatment | 2006–2007 | 2007–2008 | 2008–2009 | 2009–2010 | 2010–2011 | 2012–2013 | 2013–2014 |
| T1: silvopastoral, 500 trees ha$^{-1}$ | 1333$^{a\#}$ | 1600$^{a\#}$ | 756$^{a\#}$ | 930$^{a\#}$ | 839$^{a\#}$ | 4306$^{b*}$ | 1035$^{c*}$ |
| T2: silvopastoral, 1000 trees ha$^{-1}$ | 3920$^{b*}$ | 4760$^{b*}$ | 851$^{a\#}$ | 1531$^{b\#}$ | 1189$^{b\#}$ | 3781$^{b*}$ | 2473$^{b*}$ |
| T4: natural prairie (Control) | 1813$^{a\#}$ | 1810$^{a\#}$ | 648$^{a\#}$ | 835$^{a\#}$ | 835$^{a\#}$ | 837$^{a\#}$ | 565$^{a\#}$ |

Different letters indicate significant differences, LSD ($p<0.05$), between treatments in the season
*Sown prairie, with Trifolium subterraneum, T. michelianu and Medicago polymorpha, and natural grasses. #: Natural prairie, without fertilization

three replicates, testing five densities to analyze their effect has on the production of fodder. Trees were planted in two rows spaced 2 m in the row and 3 m between rows, and between double rows from 6 to 18 m, forming five treatments, T1: with 1111 trees ha$^{-1}$, T2: 833 trees ha$^{-1}$, T3: 666 trees ha$^{-1}$, T4: 555 trees ha$^{-1}$, and T5: 454 trees ha$^{-1}$ (Fig. 8.5). The prairie between the double rows was both natural and sown with *Trifolium pratense* L. and local grasses.

The method used, including the statistical analysis procedure, was the same as described in Sect. 8.3.1.3.1 for the experiment at the Cauquenes Commune in the Maule Region.

The natural prairie was not grazed for the first 3 years after planting to prevent browsing of the young pine trees. The prairie, established during the spring of 2012, comprised of 15 kg ha$^{-1}$ of red clover (*Trifolium pratense* L.) and natural grasses of the zone, fertilized with 200 kg ha$^{-1}$ of triple superphosphate. The prairie was not fertilized in the second growing season (2013–2014) and this resulted in a natural grass prairie, a loss of clover, and reduced pasture productivity (Table 8.7).

In the first 2012–2013 season of production, the highest prairie production was at the wider spacings of 12–18 m between bands of trees. In the second season, no clear tendency was observed, which can be attributed to poor prairie management (no fertilization) and associated loss of clover.

These studies demonstrated it was possible to establish silvopastoral systems on small farms with *Pinus radiata* in two environmental situations in the central zone of Chile. Prairie production was improved when enriched with herbaceous species and fertilized and similar to prairies without trees. According to INIA (2001) natural prairies without fertilization have low production and there are no clear advantages with trees. With silvopastoral systems, in addition to prairie production for farm animals, there is the possibility of obtaining timber, which can be viewed as a savings account when harvested at the end of the rotation.

### 8.3.4 Economic Analysis of a *Pinus radiata* Silvopastoral System with Sheep in the Central Mediterranean Coastal Zone of Chile

The small scale farm producers, located in the coastal drylands of the Libertador Bernardo O'Higgins Region of Chile (VI Region of Chile), use a great part of their land, with gentle to strong slopes, for extensive cattle farming, principally with sheep, without considering that the land may be more suitable for forestry. Thus, natural prairies are medium to bad quality, with low meat and/or wool productivity, and growing in eroded soils due to over- grazing and with little protection from perennial vegetation (CIREN 2010). However, this practice allows owners to receive an annual income, supplemented by annual crops such as wheat, barley and lentils both for home consumption and for sale. Another established usage for this coastal area is radiata pine timber

**Fig. 8.5** Radiata pine-based silvopastoral system established in randomized blocks with densities from 454 to 1111 trees ha$^{-1}$, Los Alamos Silvopastoral Unit, Arauco Province, Biobío Region, Chile

**Table 8.7** Annual prairie production in the first two growing seasons in the *Pinus radiata* silvopastoral system, Los Alamos Silvopastoral Unit, Biobío Region, Chile

| Treatments and distance between tree strips (m) | Productivity of prairie (kg DM ha$^{-1}$) | |
|---|---|---|
| | 2012–2013 | 2013–2014 |
| T1: 6 m | 3203[a] | 1641[a] |
| T2: 9 m | 3409[a] | 2060[a] |
| T3: 12 m | 4247[b] | 1221[b] |
| T4: 15 m | 4733[b] | 1866[a] |
| T5: 18 m | 3876[b] | 1180[b] |
| Average production | 3894 | 1594 |

Different letters indicate significant differences, LSD ($p < 0.05$), between treatments in the season

plantations. Because of their long rotations without short-term returns plantation growing by small-scale farmers have declined, and many of these are harvested early with minimal returns.

To improve this situation a study was established in the mid-1980s to compare the productivity of silvopastoral systems using radiata pine associated to three types of prairie with sheep, versus traditional sheep production, and typical forest management.

This study was in the Tanumé Forestry Experimental Center (CEF) owned by the National Forest Corporation (CONAF), located in the coastal area of Pichilemu Municipality, VI Region of Chile, (geographic location 34°15′ Latitude South and 74°49′ Longitude West). This marine influenced zone is characterized by eight dry months, with 704 mm precipitation per year and mean monthly minimum and maximum temperatures of 8.6 and 25 °C, respectively, and a mean annual temperature of 11.6 °C.

The study had seven treatments:

- T1: A silvopastoral system (PS 625), with radiata pine planted at 625 trees ha$^{-1}$ in groups of four plants at 2×2 m, spaced 6 m apart. A new pasture was sown with subterranean clover (*Trifolium subterranean*) and hardinggrass (*Phallaris aquatica* cv Sirosa and cv Steptanera), fertilized annually with N, P, K, with rates according with annual soil evaluation, and grazed with sheep. The forestry component was thinned and pruned in three occasions to a final density of 200 trees ha$^{-1}$ and to a height of 7 m free of branches;
- T2: Silvopastoral System (PM 625), with 625 trees ha$^{-1}$, with the same design and management as T1, with natural prairie (PN) fertilized

each 3 years with N, P, K, with rates according with annual soil evaluation, and grazed with sheep;

- T3: Silvopastoral System (PN 625), with 625 trees ha[1], with the same design and management as T1, with natural prairie (PN) without fertilizer and grazed with sheep;
- T4: Forestry system (Forestry), consisting in a forestry plantation established at 1600 trees $ha^{-1}$, at a $2 \times 3$ m spacing. There were two thinnings and two prunings at ages 7 and 12, to reach an expected final density of 500 trees $ha^{-1}$, and a pruned to a height of 4.1 m in two times at ages 7 and 12. There was no grazing;
- T5: Traditional sown pastoral system (PS) grazed with sheep. This system was established in 1985 with a 24 year projected life. The area was sown with subterranean clover (*Trifolium subterraneum*) and phalaris (*Phallaris. aquatica*), associated with wheat crop the first year; fertilized annually with N, P, K, with rates according with annual soil evaluation, and grazed with sheep.
- T6: Sheep pastoral system with improved prairie (PM). This system was established in 1986, with a projected life of 23 years. It corresponded to a natural improved prairie for sheep usage, fertilized each 3 years with N, P, K;
- T7: Sheep pastoral system with natural prairie (PN). In this system no costs were incurred for prairie establishment and management, as it was worked as a natural prairie without additional inputs to improve the pasture growth.
- For all treatments with pasture, the management of sheep consisted of monthly weighing, health checks, and the application of medicines, vitamins and supplements. The sheep were introduced in 1986, and the pastures grazed from beginning of the growing season in October through to February–March, depending on the fodder availability.

The economic evaluation of the treatments included all forestry and livestock incomes, as well as the incentives the Government gives on the basis of subsidies for forestation. The costs of the various treatments were also accounted for.

The production and value of the radiata pine trees at 24 years was estimated using the Radiata© Simulator and utilized the products descriptions and values given in Table 8.8. To evaluate the various treatments the net present value (NPV) at a discount rate of 10 %, and the internal rate of return (IRR) were calculated.

As with many forestry studies the trees established at a lower density, had greater diameters than those at a high density (Table 8.9). The largest diameters were obtained in the silvopastoral systems, with mean diameters of 50–52 cm, versus 41 cm for the forestry system. Basal area, which is strongly influenced by stand density, was greatest with the forestry treatment. Mean total height was not influenced by stand density, indicating that the treatments were installed on a uniform site.

In terms of total volume, the highest growth was in the forestry treatment with 480 $m^3$ $ha^{-1}$, due to the higher stocking at the end of the rotation (Table 8.9). The silvopastoral treatments with about 390 $m^3$ $ha^{-1}$ were 23 % lower. The valuable pruned products (P1 and P2 – see Table 8.8), made up 37–38 % of the total volume in the silvopastoral treatments, versus only 3 % in the forestry control. On the other hand, the forestry system had a greater volume of intermediate

**Table 8.8** Description of forest products used in the simulation, using Radiata Simulator ©, and prices of forest products in 2008

| Type of product | Quality of forest product | Length of logs (m) | Minimum diameter of logs (cm) | Price (US$ $m^{-3}$) |
|---|---|---|---|---|
| P1 | Pruned logs | 2.80 | 36 | 60 |
| P2 | Pruned logs | 2.80 | 32 | 55 |
| P3 | Sawtimber logs, w/p | 3.25 | 32 | 50 |
| P4 | Sawtimber logs, w/p | 3.25 | 24 | 45 |
| P5 | Sawtimber logs, w/p | 3.25 | 18 | 35 |
| P6 | Sawtimber logs, w/p | 3.25 | 14 | 30 |
| P7 | Pulp logs | 2.44 | 10 | 20 |
| P8 | Firewood | 2.44 | 6 | 10 |

Prices at year 2008, value placed on industrial plant
*w/p* without pruning

**Table 8.9** Forest parameters results of *Pinus radiata* at age 24 for silvopastoral and forest treatments, Tanumé Experimental Center, Cardenal Caro Province of Chile

| Treatment | Final density (trees $ha^{-1}$) | Forest volume ($m^3$) | DBH (cm) | BA ($m^2$) | H (m) |
|---|---|---|---|---|---|
| T1. Silvopastoral 625 (PS) | 224 | 390 | 52 | 49 | 29.8 |
| T2. Silvopastoral 625 (PM) | 198 | 387 | 50 | 39 | 31.1 |
| T3. Silvopastoral 625 (PN) | 180 | 389 | 52 | 42 | 31.5 |
| T4. Forestry treatment | 435 | 480 | 41 | 57 | 30.7 |

*DBH* diameter at breast height, *BA* basal area, *H* total height

**Table 8.10** Economic evaluation of silvopastoral, forestry and pastoral treatments, using NPV (10 %) and IRR, Tanumé Experimental Center, Cardenal Caro Province, Region of Libertador Bernardo O'Higgins, Chile

| | Results of economic evaluation | |
|---|---|---|
| Treatments | NPV (10 %) | IRR (%) |
| T1. Silvopastoral (PS 625) | 74 | 12.4 |
| T2. Silvopastoral (PM 625) | 148 | 16.2 |
| T3. Silvopastoral (PN 625) | 142 | 16.2 |
| T4. Forestry treatment (1600) | 108 | 15.5 |
| T5. Sheep pastoral system (PS) | −181 | Indet |
| T6. Sheep pastoral system (PM) | −49 | Indet |
| T7. Sheep pastoral system (PN) | 1 | 10.5 |

*Indet* indeterminate value due to high negative value
*PS* Sown prairie, *PM* Improved prairie, *PN* Natural prairie, *NPV* net present value, *IRR* internal rate of return

value P4 and P5 products (74 %), versus 32 % of these products in the silvopastoral systems. The proportion of the lower valued products, such as pulp and firewood, was 10 % of the total in the forestry system, but only 4 % in the silvopastoral ones.

Income from animal production remained until 10 in silvopastoral systems due to low productivity of the pasture from year 11, because of tree canopy closure, when we made the decision to end animal production evaluation. In contrast, treatments T5, T6 and T7 were evaluated during the entire period of the study. Animal production income in T5 was 259, 483 and 836 % better than T1, T2 and T3, respectively, and 341 and 309 % regarding T6 and T7, respectively. The second treatment in animal income was T1, indicating that the sown pasture incremented animal production.

#### 8.3.4.1 Economic Evaluation

The results of the economic evaluation after 24 years found in the T2 and T3 silvopastoral treatments were best followed by the T4 forestry treatment, although there was little difference between them (Table 8.10). In contrast, the least profitable were the pastoral treatments (T5, T6 and T7), especially when sown and fertilized pasture was used (T5 and T6). This is principally due to the high fertilizer cost that the subterranean clover and phalaris sown prairie required for its development in the study zone. The T1 treatment was intermediate between these two groups because, although it was over-sown with clover and phalaris, the income from timber improved the economic evaluation.

The results indicated that the best profitability (IRR ~16 %) was where radiata pine was present as either a silvopastoral or forestry system. In contrast, all solely pastoral systems had low or negative economic returns. The results indicate that the silvopastoral systems are an interesting alternative for the small dryland landowners of Cardenal Caro Province, Libertador Bernardo O'Higgins Region of Chile, from a long-term profitability viewpoint. Silvopastoral systems were found to have a series of social benefits, namely:

- The owners retain yearly incomes from annual crops during the first year when establishing the system and from their sheep.
- Obtain a substantial income at the end of the tree rotation from the sale of high quality forestry products.
- Silvopastoral systems are better adapted to the small farmer cultural systems than the Chilean government afforestation programs aimed solely at timber production.
- Silvopastoral systems can achieve greater soil protection and reduced erosion, especially in hilly area.
- Where silvopastoral systems allow small farmers to remain on their farms this reduces migration to cities and avoids major social breaks in their lives.

On the basis of these outcomes, we recommend that silvopastoral systems be emphasized in the government forestry policies and that incentives be available to assist farmers to engage in them.

## 8.4 Alternatives Species for Use in Silvopastoral Systems in Temperate Areas of Chile: Stone Pine, Walnut, Cherry and Poplar

### 8.4.1 Introduction

The analysis of world trends in forestry indicates the importance of diversifying the forest sector, considering species composition and geographic distribution, because of the wide range of environments occurring in Chile. Diversification also helps to limit biotic and abiotic risks, withstand economic risks and market fluctuations, and maximize sites use. Agroforestry allows diversification of agriculture and forestry, and it can be implemented with different species and in different settings according to specific conditions. A 20-year study conducted by INFOR considered several productive options, some related to arboriculture involving broadleaved species known as "noble", and the results consider adaptations of European technologies and some developments that are suitable to our situation (Loewe 2003). Several experimental trials with pure or mixed plantations were established in the central-southern area of Chile, including the exotic species Walnut (*Juglans regia*), Cherry (*Prunus avium*), Sweet gum (*Liquidambar styraciflua*), Ash (*Fraxinus excelsior*), Black walnut (*J. nigra*), Chestnut (*Castanea sativa*), Yellow-poplar (*Liriodendron tulipifera*), Black alder (*Alnus glutinosa*), Red American Oak (*Quercus rubra*) and European Oak (*Quercus robur*); and the native Chilean hazelnut (*Gevuina avellana*), Raulí (*Nothofagus alpina*) and others. Other studied options are poplar (*Populus* sp.) and Stone pine (*Pinus pinea* L.) because their fruits (pine nuts) are highly demanded and expensive and, therefore, very attractive for land owners. According to the director of FAO's Forestry Division,[1] "agroforestry systems play a crucial role in the livelihoods of rural people by providing employment, energy, nutritious foods and a wide range of other goods and ecosystem services". Mixed systems allow a rural sustainable development and enhance biodiversity whilst preserving landscapes (Eichhorn et al. 2006). Reisner et al. (2007) identified regions within 32 European countries for silvoarable agroforestry, involving Walnut, Cherry, Poplar, Stone pine and *Quercus ilex*. To promote their cultivation we identified potential zones in Chile, analyzed their economic impact and tested and developed production techniques. Some experimental plots were implemented in combination with agriculture and/or cattle-raising in mixed systems. Below we present the results.

[1] BBC. 2014. Eva Mueller: UN urges action on forest diversity. http://m.bbc.com/news/science-environment-27963330. Cited June 23, 2014.

### 8.4.2 Innovative Species for Use in Silvopastoral Systems

#### 8.4.2.1 Stone Pine (*Pinus pinea* L.)

This species is native to the Mediterranean basin, where it stabilizes soils; it is important for its edible seeds, the pine nuts, and is one of the nine most important species that produce dried fruit in the world (Gordo et al. 2011). Cones harvest dates back to the Paleolithic (Prada et al. 1997; Gil 1999; Badal 2001); in France there is evidence of its use in medieval rural settlements (Ruas 2005). In Chile, it was introduced more than a century ago by European immigrants, who used it for dune and soil improvement and livestock shading (Loewe et al. 1998). The species has multiple uses of environmental importance (such as fauna feeding, watershed protection, and dune and erosion control) as well as economic importance (pine nut and resin production) (Loewe and Delard 2012). It is attractive to Chilean producers for its wood and especially for the high value of its nuts (Loewe and González 2007; Soto et al. 2008). It is an option for small and medium owners because of the annual income it provides, and its adaptation to eroded areas.

Agroforestry systems that include this species have been established in Spain intercropping agricultural crops as well as in low density settings in vineyards and pastures (Gordo et al. 2011). It is easily integrated in agroforestry systems; grafting is interesting because it increases pine production, along with its rusticity, beauty and highly demanded fruit. According to Eichhorn et al. (2006), the use of conifers may expand the potential applications of agroforestry. INFOR has investigated the species behavior for more than two decades, and is now conducting a project[2] aimed at developing technological information for its cultivation in Chile, where conditions are suitable for stone pine growth. The species is also useful for soil recovery and erosion control. In two experimental plots located in El Carmen, Biobío region (533 m), we tested agroforestry systems that were established in July 2010, which include agricultural crops and sheep production (Fig. 8.6).

The obtained crop yields are lower than the average in the country (INE 2014). For forage oat, average yield is 5070 kg ha$^{-1}$, so our yield represents 46 %, and the production derived from sheep that fed directly on the field should be added to the estimates. In the case of potatoes, our average yield reaches 31,300 kg ha$^{-1}$, equivalent to 23–30 % of the national average production. In Europe, the main species associated with livestock are chestnuts, poplars, hardwood plantations, Mediterranean oaks, pines and fruit trees (Pardini and Nori 2011); stone pine is usually located in the coastal area. Forage production in these plantations is not enough to sustain grazing, but livestock from nearby farms occasionally graze there; if grazing is periodical, it reduces shrub growth, fire risk and the cost of periodical mechanical clearing. Soil fragmentation and nutrient enrichment through goat defecation was reported during grazing, accelerating litter decomposition and nitrogen incorporation, reducing pine needles accumulation and fire risk (Mancilla-Leyton et al. 2013). Liguori et al. (2011) reported a less favorable effect of stone pine agroforestry systems on phytoseiid mite development (natural enemies of pests).

#### 8.4.2.2 High Value (Noble) Timber Species: Walnut and Cherry

The growth of the Chilean forest sector in the last decades has been based on the cultivation of two species (radiata pine and eucalyptus). Monoculture has generated phytosanitary and environmental problems, evidencing the importance of diversifying forestry both in terms of species and productive models. Among the most studied and successful species are walnut and cherry because their timber and fruit production can be incorporated into rural agricultural productive systems under fruit-forest or agro-fruit-forest models. Other options include chestnut, European oak or others with maize, beans, wheat and peas (INFOR-FDI 2004). Silvoarable systems for furniture timber produc-

[2] "Development of techniques for producing Stone pine (*Pinus pinea* L.) nuts, an attractive commercial option for Chile" funded by FONDEF (CONICYT)

**Fig. 8.6** Stone pine-based agroforestry system. (**a**) Potato harvest at age 2 years; (**b**) general view at age 3.5 years (Source: Verónica Loewe 2012)

tion in the UK include species such as Ash, Black walnut, Cherry, European oak and Sycamore (*Acer pseudoplatanus* L.) with cereal and pulse alley cropping (Eichhorn et al. 2006). In the Netherlands, innovative combinations of trees grown for high-grade timber and ground level flower production have been assessed with several species, including cherry intercropped with hyacinth for flowers and bulbs. With good management it is possible to get high quality timber in the medium-long term can be obtained using good management strategies, while getting short-term income from agricultural crops and other non-timber forest products, such as fruit, fungi, leaves, and honey. Agricultural crop alternatives that can be associated with high value trees are forage species (with the exception of alfalfa (*Medicago sativa*)), cereals and vegetables, with the exception of potatoes and tomatoes, because of the allelopathic effects of crops on trees.

Walnut has been associated with maize (with yields of up to 14,000 kg $ha^{-1}$), beans, carrots, lettuce, and other crops, with the culture with best results being winter wheat, cultivated when walnuts are in vegetative recess. No agricultural cultures have presented negative effects on walnut, except for alfalfa, which reduces walnut growth for the following 3–5 years. Furthermore, maize induces increased height. Walnut growth is superior if associated with nitrogen-fixing species, herbs or shrubs.

Good management practices are necessary to obtain good timber and high crops yield. A high tree density reduces the period in which crops can be grown. Initial plantation density in these systems varies from 278 to 1100 trees $ha^{-1}$, with a final density of 100–250 trees $ha^{-1}$ according to the species. Values for growth parameters of seven experimental trials 11–19 years after the establishment were presented by Loewe et al. (2013a), with ranges in average annual DBH growth of 0.5–1.8 cm and average annual height growth of 0.38–1.07 m. depending on the species and sites.

#### Walnut

This is a well-known species of agricultural and forest interest since it produces nuts and wood that is appreciated for style furniture production and handicrafts. It is one of the most appreciated timbers in Europe, with a stable market for centuries. It is a vigorous tree, 20–25 m in height and up to 1 m of DBH, with a straight trunk and a wide crown (Loewe and González 2001). It is also a fast growing species, contrarily to what is commonly thought (Loewe and González 2003), that can be incorporated into agricultural systems as a complement to the economic unit activity, especially by farmers. The species produces valued fruits and timber that is highly quoted worldwide (US$ 250–2500 $m^{-3}$). Additionally, nut and timber production are also compatible. Throughout the Mediterranean region, small walnut orchards are intercropped with vegetables and cereals (Eichhorn et al. 2006), with walnut being a preferred species in mountains. It is a component of silvoarable systems in Italy, with dual-purpose fruit and timber. In the south it is intercropped with vegetables and often with hazel (*Corylus avellana* L.) grown for wood and nuts, where hazel plays the role of trainer to improve the shape of walnut trunks. The combined production of timber and nut simultaneously can be obtained through the high grafting technique (Loewe and González 2002), which was successfully applied in some experiments in Europe and in Chile. Walnut is relatively demanding from the ecological standpoint (Minotta 1989; Pellegrino and Bassi 1993). Compacted soils favor rotting by *Armillaria mellea* and *Phytophtora* sp. The potential cultivable area with irrigation in Chile reaches 3,109,672 ha.

Regarding timber prices and markets, walnut timber is very much appreciated and has been attractive for centuries for luxury applications, which is reflected in its high prices. It is hard, moderately durable, and responds well to preservation treatments; it is easy to work with and its aesthetics makes it the best timber to make quality furniture, veneers, panels, butts and arms parts, musical instruments and fine crafts and cabinetwork (Loewe and González 2003). First quality timber for veneers must have regular growth rings, attractive dimensions (3 m long and 40 cm diameter or more), no defects and a homogenous and demanded color (clear in general). Prices for elaborated timber range between US$ 830–2800 $m^{-3}$, according to quality; standing trees vary between US$ 330–2200 $m^{-3}$, according to dimensions and quality. Its high prices generate demand for timber substitutes.

##### Walnut Growth and Costs in Temperate Agroecosystem

In Chile, walnut is cultivated traditionally in pure plantations for nuts, and innovative agroecosystem have been implemented: pure walnut plantations for timber and nuts; pure walnut plantations for timber and nuts intercropped with crops (maize, beans and vegetables); mixed walnut plantations for timber and nuts associated with nurse species, trees or shrubs; mixed walnut plantations for timber and nuts associated with secondary species, trees or shrubs, and forage production. Growth obtained in these experiments is good if the required techniques are applied; otherwise, growth and profitability diminish drastically (Loewe and González 2006).

**Table 8.11** Agroforestry systems established with walnut in the central southern area of Chile and walnut growth performance

| Agroforestry systems | Age (y) | Spacing (m) | Height (m) | DBH (cm) | Management intensity | Environment |
|---|---|---|---|---|---|---|
| 1. Pure walnut for timber, nuts and agriculture intercropped | 11 | 3×6 | 11.35 | 16.34 | High | Central valley, agricultural soil, irrigation, VII region |
| 2. Pure walnut for timber, nuts and agriculture intercropped | 10 | 3×3 | 10.02 | 12.72 | High | Central valley, agricultural soil, irrigation, VII region |
| 3. Pure walnut for timber, nuts and agriculture intercropped | 8 | 3×3 | 4.71 | 6.72 | Medium | Central valley, agricultural soil, irrigation, IX region |
| 4. Pure walnut for timber and nuts | 8 | 3×3 | 3.15 | 3.58 | Low | Piedmont, Andes, irrigation, VI region |
| 5. Walnut for timber and nuts associated to *Elaeagnus angustifolia* and intercropping | 9 | 3×3 | 9.79 | 11.86 | High | Central valley, agricultural soil, irrigation, VII region |

*DBH* Diameter at breast height

Growth obtained in experimental units is presented in Table 8.11.

Walnut development depends strongly on environmental conditions and management type and intensity. Units located in appropriate sites and with suitable management present interesting results. Nut production begins at the age of 5 in a quantity enough to afford the costs associated with the mixed management techniques. Average height reached 11.4 m at age 11 in plantation type 1; 10 m at age 10 in type 2; 4.7 m at age 7 in type 3; 3.1 m at age 7 in type 4, and 9.7 m at age 9 in type 5. Average DBH reached 16.3 cm at age 11 in plantation type 1; 12.7 cm at age 10 in type 2; 6.7 cm at age 7 in type 3; 3.5 cm at age 7 in type 4, and 11.8 cm at age 9 in type 5, indicating that average growth rates are higher than those observed in Europe.

Mixed plantation from year 3 onwards overcame the height of pure stands, which is in agreement with findings reported by Mohni et al. (2009). The pure plantation of lower density (3×6 m) grew faster than the denser plantation (3×3 m) due to lower competition in the former. Mixed plantation results are attractive, superior to the pure plantation at the same age, in agreement with results of Loewe and González (2006a). In general, the combination of walnut with nitrogen fixing species as secondary species, such as *E. angustifolia*, presents higher growth than the monoculture. Tall intercropped crops, like maize, have the same effect if introduced from year 2 on, since they induce fewer and thinner lateral branches and long, less expanded crowns, of high photosynthetic efficiency, adequate for high quality timber production. Management in high value timber plantations is important for obtaining the desired products, including timely and suitable formation pruning in winter and summer, epicormic shoot elimination during the growth period, weed control, irrigation in spring and summer, and necks or trunk flares checking, an important activity since for every cm that the stem is covered the rotation increases 1 year and the plant losses vigor or even dies. Management cost of mixed plantations is lower than in pure plantations for several reasons: fewer plants to manage (only walnuts), simpler formative pruning, and elimination of annual fertilization. From year 3 on, when the length of the targeted log is already reached, management costs decrease (Fig. 8.7).

**Fig. 8.7** High value timber and nut walnut plantation at 6 years (**a**) and 15 year of age (**b**), located in Maule Region, Chile (Source: Verónica Loewe 2007 and 2012, respectively)

## Cherry

Cherry produces quality timber (Fig. 8.8), being the most important species of all *Prunus*. It is a noble species of importance in Austria, Belgium, France, Germany, Italy, Portugal, Spain and other countries worldwide. Currently, Germany, France, England and Italy are developing R&D programs to improve their wood production (Loewe et al. 2001). Cherry has a strong apical dominance; it reaches 25–30 m in height, has a straight and cylindrical trunk, with diameter of up to 70–80 cm, and a crown with few, thin, ascending and regularly arranged branches. In Chile, it is cultivated from the Metropolitan to Araucanía regions, mostly in the central regions (Metropolitan to Maule). Potential cultivable area reaches 4,458,719 ha and 3,456,928 ha with and without irrigation, respectively.

Regarding timber prices and markets, the demand for cherry timber depends partially on the consumption trends or fashion and on market fluctuations. Nevertheless, since the First World War, the demand for this timber has exceeded the low offer, so there is low probability of an over-supply of high quality logs in the short or medium term (Loewe and González 2004). Hardwood timber market considers numerous factors in determining log quality and price; if a good size log presents knots, color alterations, wood contraction, spiral grains or cracks, it is disqualified and its price falls. In North American markets, prices differ in relation to defects, piece length and color. In Germany, dark timber is generally more valued than in other European countries; in the UK, most of the sawn cherry timber is destined to the furniture industry. In Italy, in some periods, the price of cherry timber almost reached walnut values due to a fashion trend, stimulated principally by architects and interior designers. Standing forests are sold in US$ 140–1000 $m^{-3}$; veneer logs US$ 850–1000 $m^{-3}$; and saw timber between US$ 300–1350 $m^{-3}$. High cherry timber prices also generate demand for substitute similar timbers.

**Fig. 8.8** High value timber Cherry plantation (**a**) and associated to hazel (**b**), both 6 years old, located in Yacal and Alupenhue, respectively, Maule Region (Source: Verónica Loewe 2007)

#### Cherry Growth and Costs in Temperate Agroecosystem

In Chile, cherry is cultivated traditionally in pure plantations to produce cherries, and some attempts have included the species in agroecosystem, such as pure plantations for timber and fruit, pure cherry plantations for timber and fruit associated with forage production areas, mixed cherry plantations for timber and fruit, associated with nurse species; and mixed cherry plantations for timber and fruit, associated with secondary species (trees or shrubs).

The species has been tested since 1994 with poor to excellent results, depending on the management quality (weed control, irrigation, cattle management and pruning) and plant quality/type. Growth was usually slow during the first year, while root system establishment was developing. Loewe et al. (2013b) indicated annual height growth from 0.50 to 0.92 m in six mixed plantations, whereas the growth rate was 0.37–0.83 m year$^{-1}$ in monoculture in the same sites. Therefore, the observed height growth in Chile was greater than in 8 year-old agroforestry plantations in Languedoc, France, where annual height growth ranged between 0.31 and 0.69 m year$^{-1}$ (Balandier and Dupraz 1999). These findings agree with those of Tani et al. (2006) and Loewe (2003), who reported best performances in cherry trees grown along with accessory trees, especially N-fixing species. Positive effects reflect many co-occurring favorable causes, such as higher nitrogen availability. Mohni et al. (2009) indicated that a dense plantation with companion species has often demonstrated greater height and diameter growth than pure stands. In soils with reduced fertility, the role of companion species is even more important than in fertile soils (Chifflot et al. 2005). Even if growth is high, there are some pests and diseases that affect the species, such as bacterial canker disease (*Pseudomonas siringae pv. Mors-prunorum* Van Hall.) and cherry slug-

**Fig. 8.9** (**a**) 1-year old poplar plantation with land preparation for agricultural crops; (**b**) 12-year old poplar, with natural pasture for grazing (Source: Alvaro Sotomayor 2011)

**Table 8.12** Dry matter production of forage crops (Mg DM ha$^{-1}$) according to canopy cover in silvopastoral poplar plantation, San Jose de la Mariquina, Los Ríos Region, Chile (INIA 2001)

| Forage species | Mean dry matter production of forage crops (Mg DM ha$^{-1}$) per canopy cover (m$^2$ tree$^{-1}$) in silvopastoral poplar plantation | | | | | | | |
|---|---|---|---|---|---|---|---|---|
| Crown cover (m$^2$ tree$^{-1}$) | 3.2 | 4.0 | 4.4 | 5.2 | 5.4 | 6.2 | 6.4 | 7.2 |
| Perennial ryegrass | | 7.5 | | 7.7 | 6.9 | 5.1 | 1.8 | 1.6 |
| Oat | 5.5 | 6.5 | | | 5.9 | | | |
| Barley | 6.3 | 6.8 | 8.9 | 4.1 | 4.8 | 2.3 | | |
| Triticale | | | | | 8.7 | 4.1 | | |
| Natural prairie | 2.5 | 1.9 | 1.8 | 1.8 | 1.8 | 1.1 | 1.1 | 1.0 |

worm (*Caliroa cerasi* L.), which causes losses in fruit yield and even tree death. Canker is one of the major limitations in timber production due to its difficult control; for this reason, the Forestry Commission imposes a 15 % limit on cherry trees in farm woodlands (Roberts 2001).

#### 8.4.2.3 Experiments with Poplar (*Populus* sp.) for Silvopastoral Purposes

The El Alamo Forest Company located in Retiro, Maule Region, Chile, has established poplar plantations for industrial purposes, mainly for manufacturing matches, popsicles and boxes. The rotation considered corn and vegetables production between poplar rows in the first 3 years and then cattle grazing, because the system retains the pastures and, consequently livestock production. When canopy closes the forage production decreases (Fig. 8.9).

**Study of the Effect of Forest Cover with Poplar on Forage Species in Wetlands of Chile**

The Agricultural Research Institute (INIA) has investigated the feasibility of silvopastoral systems with poplar and different forage species in Los Lagos region. The climate in the area is humid, with rainfall of 2000 mm year[1] and average temperatures between 4 and 20 °C.

The following forage species were sown: *Medicago sativa* L. (alfalfa), *Avena strigosa* Schreb (oats), *Lolium perenne* L. (perennial ryegrass), *Hordeum hexastichon* L. (barley), *Triticum aestivum* (triticale) and natural pasture as control for animal production, in poplar plantations of 2, 4 and 6 years. Poplar planting was established at a density of 278 trees ha$^{-1}$, with spacing of 6×6 m. Soil was prepared in a traditional way by type of crop with offset harrow, seedbed, road roller, and drill; crops were fertilized as necessary; natural pasture was not fertilized. Sowing date was in August for alfalfa and oats; March for ryegrass; September for barley, and May for triticale. For statistical analysis, forage crops were evaluated with exclusion cages 0.5 $m^2$ with three replications per crop. Tree growth was evaluated in three 200-$m^2$ plots evaluating height (H), canopy cover (CC) and diameter at breast height (DBH) growth.

Poplar growth under the effect of forage crops showed a higher average increase 2.48, 2.64, 2.51, and 2.56 cm $year^{-1}$ in sectors that were established with alfalfa, biannual ryegrass, hybrid ryegrass and barley, respectively, compared to a condition without crop (2.45 cm $year^{-1}$). Fertilization performed on crops improved growth of trees in diameter, but no differences in overall height were observed.

The canopy cover of poplars affected production of forage crops established under the silvopastoral arrangement (Table 8.12). In ryegrass, an average production of 8.1 Mg DM $ha^{-1}$ was obtained with 4 $m^2$ of canopy cover, and 6.2 Mg DM $ha^{-1}$ with 5.4 $m^2$; triticale reached 8.7 and 4.1 Mg DM $ha^{-1}$ with 5.4 and 6.2 $m^2$ of crown cover, respectively. In oats, a production of 6.5 Mg DM $ha^{-1}$ was obtained with 4 $m^2$ of cover, and the maximum production of forage crops tested was achieved in barley, with 8.9 Mg DM $ha^{-1}$, with 4.4 $m^2$ of canopy cover. According to the results shown in Table 8.12, fodder crops had adequate performance under the influence of established poplar silvopastoral densities, only until poplar trees reached cover of about 6.5 $m^2$, which is achieved at the age of 7–8 years. The evaluation shows that the condition for proper growth of the analyzed crops was achieved at 7–8 years, reaching an average production equivalent to 36 % of the traditional culture without trees, which justifies its sowing under tree canopy considering costs and performance.

## 8.5 Conclusions

Agroforestry and silvopastoral systems are attractive alternatives for many companies and landowners with moderate to high management capacity. When establishing high value timber species, plantation goals must be clearly set (species and markets) for correctly defining the techniques to be applied. Walnut and cherry are productive, economic and socially interesting species in agroforestry due to its rapid growth, noble timber production and present international market and prices. Both species have been naturalized in Chile for centuries. They are well known and belong to the rural culture; they are vigorous and can be integrated to agroforestry by incorporating timber production as a complementary goal to the fruit production or agricultural crops, or to produce timber as main goal. Walnut does not present important diseases, whereas cherry is susceptible to bacterial canker, recommended in mixed plantations to reduce risks. Stone pine turns out to be a highly interesting species for combined productive systems due to its high value fruit, plasticity and rusticity, and because it has been present in Chile for more than a century, exhibiting high vigor and an excellent phytosanitary performance. It is possible to design agroforestry systems adapted to different situations, considering available capital, owner's goals, capacity of operative management, and local, national and international economy.

**Acknowledgements** We acknowledge our sincere thanks to Osvaldo Mardones for looking after the "Roble"-based silvopastoral system being established in "Ranchillo Alto", and to José Morales, Alejandro Lucero, Arnoldo Villarroel, Hernán Soto and Jorge Acevedo for assistance in the field and data collection. In addition, we thank Donald Mead for kindly reviewing part of this paper. Part of this study was funded by a grant (Project No. 001--2014) from the Fundo de Investigación del Bosque Nativo-FIBN of the Chilean Forest Service (CONAF).

## References

AMBAR – Consultoría e Ingeniería Ambiental (2010) Confección y elaboración de una guía de manejo en terreno fiscal Ranchillo Alto en la Región del Biobío, 189p

Anderson GW, Moore RW, Jenkins PJ (1988) The integration of pasture, livestock and widely spaced pine in South West Western Australia. Agroforest Syst 6:195–211

Badal E (2001) La recolección de piñas durante la prehistoria en la Cueva de Nerja (Málaga). In: Villaverde V (ed) De neandertales a cromañones-el inicio del poblamiento humano en las tierras valencianas. Universidad de Valencia, Valencia, pp 101–104 (in Spanish)
Balandier P, Dupraz C (1999) Growth of widely spaced trees. A case study from young agroforestry plantations in France. Agroforest Syst 43:151–167
Chifflot V, Bertoni G, Cabanettes A, Gavaland A (2005) Beneficial effects of intercropping on the growth and nitrogen status of young wild cherry and hybrid walnut trees. Agroforest Syst 66(1):13–26
CIREN (1999) Estudio Agrológico VIII Región de Chile. Centro de Informaciones de Recursos Naturales, Santiago, 555p
CIREN (2010) Determinación de la erosión actual y potencial de los suelos de Chile. Publicación N° 138, CIREN, Santiago, Chile, 281p
Dube F, Couto L, Silva ML, Leite HG, Garcia R, Arauco GA (2002) A simulation model for evaluating technical and economic aspects of an industrial eucalyptus-based agroforestry system in Minas Gerais, Brazil. Agroforest Syst 55:73–80
Dube F, Thevathasan NV, Zagal E, Gordon AM, Stolpe NB, Espinosa M (2011) Carbon sequestration potential of silvopastoral and other land use systems in the Chilean Patagonia. In: Kumar BM, Nair PKR (eds) Carbon sequestration potential of agroforestry systems, vol 8, Advances in agroforestry. Springer, Dordrecht, pp 101–127
Dube F, Espinosa M, Stolpe N, Zagal E, Thevathasan NV, Gordon AM (2012) Productivity and carbon storage in silvopastoral systems with *Pinus ponderosa* and *Trifolium* spp., plantations and pasture on an Andisol in Patagonia, Chile. Agroforest Syst 86(2):113–128
Dube F, Thevathasan NV, Stolpe N, Espinosa M, Zagal E, Gordon AM (2013) Selected carbon fluxes in *Pinus ponderosa*-based silvopastoral systems, exotic plantations and natural pastures on volcanic soils in the Chilean Patagonia. Agroforest Syst 87(3):525–542
Eichhorn MP, Paris P, Hderzog F, Incoll LD, Liagre F, Mantzanas K, Mayus M, Moreno G, Papanastasis VP, Pilbeam DJ, Pisanelli A, Dupraz C (2006) Silvoarable systems in Europe – past, present and future prospects. Agroforest Syst 67:29–50
Gil L (1999) La transformación histórica del paisaje: la permanencia y la extinción local del pino piñonero. In: Marín F, Domingo J, Calzado A (eds) Los montes y su historia, una perspectiva política, económica y social. Spain, pp 151–185 (in Spanish)
Gordo J, Mutke S, Calama R, Gil L (2011) El uso del pino piñonero en sistemas agroforestales. Jornadas de cultivos alternativos con especies forestales. Valladolid, Spain (in Spanish)
INE (2014) Instituto nacional de estadísticas. Agropecuarias. Available at http://www.ine.cl/canales/menu/publicaciones/calendario_de_publicaciones/pdf/agropecuarias_informe_anual_2 014.pdf. Accessed 2 Dec 2014 (in Spanish)
INFOR-FDI (2004) Technical report no. 1, Appendix 1 "Caracterización productiva, socioeconómica y ambiental de pequeños y medianos productores asociados a bosque nativo, entre la VII y XI Región". In: Proyecto FDI-CORFO Generación de modelos de manejo sustentable en bosque nativo y guías expertas de manejo forestal para pequeños y medianos productores. Santiago, Chile (in Spanish)
INFOR (2012) Desarrollo de tecnología silvícola productiva como apoyo para la implementación de la Ley de Bosque Nativo. Anexo 6. Elaboración de pautas silvícolas de manejo para renovales. Proyecto INNOVA/CORFO – 09CN14-5854. Valdivia, Chile, 28p. (in Spanish)
INFOR (2013) El Sector Forestal Chileno 2013. Santiago, Chile, 2013, 32p. (in Spanish)
INIA (2001) Mejoramiento de la rentabilidad del álamo a través de cultivos asociados. Boletín INIA N° 64, Santiago, Chile, 190p. (in Spanish)
Leslie B, Knowles R, Moore R (1998) Silvopastoreo con *Pinus radiata* en Zonas Frías. In: Compilación de Resultados en Diversos Ensayos de Modelos Silvopastorales en Chile y en el Extranjero, Santiago, Chile (in Spanish)
Liguori M, Tixier MS, Hernandes AF, Douin M, Kreiter S (2011) Agroforestry management and phytoseiid communities in vineyards in the south of France. Exp Appl Acarol 55:167–181
Loewe MV (2003) Perspectivas para el desarrollo de la arboricultura para producción de madera de alto valor en Chile. INFOR-FIA, Santiago, 289p (in Spanish)
Loewe MV, Delard RC (2012) Un nuevo cultivo para Chile, el Pino Piñonero (*Pinus pinea* L.). INFOR-CORFO, Santiago, 364p (in Spanish)
Loewe MV, González OM (2001) Producción de madera de alto valor: tecnología productiva apta para un desarrollo rural con equidad. In: Latin American Congress "Retos y perspectivas del desarrollo rural para alcanzar la equidad de género". CATIE, Turrialba, Costa Rica, pp 18–20
Loewe MV, González OM (2002) High grafting, technique that allows producing wood and fruit of quality simultaneously. 17th Symposium of the international farming systems association. Orlando, pp 17–20
Loewe V, González M (2003) Sicomoro, grevillea, roble rojo americano, pino piñonero, castaño, ruil y cerezo americano, alternativas para producir madera de alto valor. INFOR-FIA, Santiago, 320p (in Spanish)
Loewe MV, González OM (2004) Cerezo común: alta cotización mundial. Chile For 303:41–43
Loewe MV, González OM (2006) Plantaciones Mixtas, un modelo productivo con potencial para Chile. INFOR-FIA, Santiago, 299p (in Spanish)
Loewe MV, González M (2007) Pino piñonero: el potencial de su madera y fruto. Chile For 334:49–53
Loewe MV, Toral M, Delard C, López C, Urquieta E (1998) Monografía de pino piñonero (*Pinus pinea*). CONAF-INFOR-FIA, Santiago, 81p (in Spanish)

Loewe MV, Pineda BG, Delard RC (2001) Cerezo común (*Prunus avium* L.), una alternativa para producir madera de alto valor. INFOR-FIA, Santiago, 105p (in Spanish)

Loewe MV, Álvarez DA, Barrales L (2013a) Growth development of hardwood high value timber species in central south Chile, South America. In: Proceedings International Scientific Conference on Hardwood Processing (ISCHP), Florence, Italy, 7–9 October 2013, pp 50–61

Loewe MV, González OM, Balzarini M (2013b) Wild Cherry tree (*Prunus avium* L.) growth in pure and mixed plantations in South America. For Ecol Manage 306:31–41

Mancilla-Leyton JM, Sánchez-Lineros V, Vicente AM (2013) Influence of grazing on the decomposition of *Pinus pinea* L. needles in a silvopastoral system in Doñana, Spain. Plant Soil 373:173–181

Minotta G (1989) Il rimboschimento dei terreni ex-agricoli: primi risultati sul confronto tra diverse tecniche di lavorazione del suolo. L' Ital For Mont XLIII 6:474–483

Mohni C, Pelleri F, Hemery GE (2009) The modern silviculture of *Juglans regia* L.: a literature review. Die Bodenkultur 59:19–32

Ortega R y Rodríguez I (1994) Manual de gestión del medio ambiente. Mapfre, 343p

Pardini A, Nori M (2011) Agro-silvo-pastoral systems in Italy: integration and diversification. Pastoralism: Res Policy and Pract 1(1):26

Pellegrino S, Bassi R (1993) Aspetti della tecnica colturale del noce. Fruticoltura 12:53–59

Peña L y Carrasco P (1993) Clasificación de erosión de suelos de la cuenca del Biobío. Proyecto FONDECYT N° 91/0355. Universidad de Concepción, Chillán, Chile.

Peralta M (1976) Uso, clasificación y conservación de suelos. Servicio Agrícola y Ganadero, Santiago, Chile, 340p (in Spanish)

Polla C (1998) Estrategias de Acción en el Tema Silvopastoreo. In: Actas Seminario "Manejo Silvopastoral", Trabajo N°8, Young, Uruguay (in Spanish)

Prada MA, Gordo J, De Miguel J, Mutke S, Catalán G, Iglesias S, Gil L (1997) Las regiones de procedencia de *Pinus pinea* L. en España. Organismo Autónomo de Parques Naturales, Madrid

Quam V, Johnson L (1999) Windbreaks for livestock operations. University of Nebraska Cooperative Extension EC 94-1766-X

Reisner Y, De Filippi R, Herzog F, Palma J (2007) Target regions for silvoarable agroforestry in Europe. Ecol Eng 29:401–418

Roberts SJ (2001) Bacterial canker of cherry (*Prunus avium*): biology, epidemiology and resistance. Review Report WD0224/WD0234 Horticulture Research International

Ruas MP (2005) Aspects of early medieval farming from sites in Mediterranean France. Veg Hist Archaeobot 14:400–415

SAS Institute Inc (2003) Proceedings of the Twenty-eight annual SAS(r) users group international conference. Cary

Soto D, Gysling J, Loewe V (2008) Antecedentes del Mercado Internacional de Piñones de Pino. Cienc Investig For 14(3):599–623 (in Spanish)

Sotomayor A (1989) Sistemas Silvopastorales y su Manejo. Chile Agríc 156:203–206

Sotomayor A (1990) Sistemas Silvopastorales y su Manejo. Chile Agríc 157:203–206

Sotomayor A, Teuber O (2011) Evaluación del efecto de los arboles manejados bajo ordenación silvopastoral en los parámetros climáticos del sitio, en relación a un manejo ganadero sin árboles. Cienc Investig For 17(1):23–40 (in Spanish)

Sotomayor A, González M, Lucero A, Luengo K (2008) Modelos Agroforestales: Sistema Productivo Integrado para una Agricultura Sustentable. Instituto Forestal (INFOR), Concepción, 32p (in Spanish)

Tani A, Maltoni A, Mariotti B, Buresti Lattes E (2006) *Juglans regia* L. Tree plantations for wood production in mining area of S. Barbara (AR). Evaluation of N-fixing accessory trees effect. Proceedings of the V SISEF Congress "Forests & Society – Changes, Conflicts, Synergies"

# Silvopastoral Systems in the Aysén and Magallanes Regions of the Chilean Patagonia

9

Alvaro Sotomayor, Harald Schmidt, Jaime Salinas, Andreas Schmidt, Laura Sánchez-Jardón, Máximo Alonso, Ivan Moya, and Osvaldo Teuber


**Abstract**

This chapter presents the experiences with silvopastoral systems and windbreaks in the Aysén region, located in the extreme south of Chile. These systems include the native tree species of the genus *Nothofagus* and other cold-tolerant tree species such as *Pinus contorta, Pinus ponderosa and Pseudotsuga menziesii* that were introduced in the 1960 from regions of the United States and Canada, where the weather conditions are similar to those of the Patagonian Region of Chile. The experience so far shows that the establishment of trees in silvopastoral systems and windbreaks has helped improve prairies productivity and therefore animal production, and provided wood for industrial purposes and farm needs. The factor that most influenced increase in prairie productivity in silvopastoral systems was tree crown coverage, which reduced the wind speed by 200 %. The use of windbreaks in this region demonstrated the importance of these systems in farm properties, especially in open areas without any protection, where productivity of different fodder species increased by 47 % in the most protected zone of 0–5 H (where H is the height of the windbreak tree) to 14 % in the least protected zone (15 H). In addition, because of



A. Sotomayor (✉)
Instituto Forestal (INFOR),
Sede Biobío, Camino a Coronel Km. 7,5, Casilla 109-C, Concepción, Chile
e-mail: alvaro.sotomayor@infor.cl

H. Schmidt
Department of Silviculture, Faculty of Forest Sciences, University of Chile,
Santa Rosa, Santiago, Chile

J. Salinas • I. Moya
Instituto Forestal (INFOR),
Sede Patagonia, Camino a Coyhaique Alto Km. 4,0, Coyhaique, Chile

A. Schmidt
Technological Consortium on Biomass and Bioenergy, Santiago, Chile

L. Sánchez-Jardón
University Center Coyhaique, University of Magallanes, José Miguel Carrera 485, Coyhaique, Chile

M. Alonso
Department of Environmental Sciences and Natural Resources, University of Chile,
Santa Rosa 11315, Santiago, Chile

O. Teuber
Instituto de Investigaciones Agropecuarias, INIA-TamelAike, Las Lengas 1450, Coyhaique, Chile




P.L. Peri et al. (eds.), *Silvopastoral Systems in Southern South America*,
Advances in Agroforestry 11, DOI 10.1007/978-3-319-24109-8_9

wind reduction by windbreak and trees established in silvopastoral systems arrangement, and management of native forests for this purpose, would reduce soil erosion which is a major problem in the Patagonian region of Chile.

**Keywords**
Windbreaks • *Pinus contorta* • *Dactilys glomerata* • *Nothofagus antarctica* • *Nothofagus pumilio*

## 9.1 Introduction

The principal uses of the 4.9 million ha of agricultural lands of the Chilean Patagonia, which includes the Aysén and Magallanes Regions of Chile, are as natural prairies for extensive cattle farming (Teuber and Ganderatz 2009). Almost 90 % of the agricultural area in the Aysén Region is used for cattle farming, of which 57 % belongs to big owners (INE 1997); a similar situation occurs in the Magallanes region. The other principal landuse is forestry and the Aysén Region, with 4.8 million ha, is the largest area of native forest in Chile. The Magallanes Region has 2.7 million ha of native forest. In contrast there are only 43,000 and 250 ha of plantations in the Aysén and Magallanes regions, respectively (INFOR 2013).

The low rates of afforestation are due to low acceptance by cattle farmers for traditional forestry activities. They perceive it as competition to their farming practices and general culture (Arnold 1983; Longhurst 1983; Sotomayor 1989). Furthermore, there is very little industrial forestry activity in these regions. Thus it has been a slow process to introduce farm forestry concepts. The extensive low-productivity cattle farming in the region is primarily due to climate limitations, such as low winter temperatures and the strong spring and summer winds. Windbreaks and other silvopastoral systems could help solve both the soil erosion and low animal productivity on farms (Sotomayor et al.2009a). Another problem in the area are farmers putting their animals into the native Nothofagus forests. Animal browsing leads to their degradation by impeding natural regeneration (Larraín et al. 2007). This chapter covers silvopastoral systems using cold-tolerant conifers and also their use in native forests. Twelve years of research has shown that silvopastoral systems with introduced conifers are feasible and beneficial. Silvopastoral systems in native forests can also be used provided forest sustainability requirements are emphasized.

## 9.2 Experiences with Silvopastoral Systems with Conifers in the Aysén Region of Chile

### 9.2.1 Silvopastoral System with *Pinus contorta* and Cattle

The silvopastoral systems with conifers adapted to cold areas have great potential for application in the Aysén Region, Chile but a lack of information had prevented this. Thus in 2002 the Forest Research Institute of Chile (INFOR) and Agricultural Research Insitute (INIA) began a series of research projects to fill this gap.

In 2003, a research project was started in a 12 year-old *Pinus contorta* Doug. Ex. Loud. (lodgepole pine) plantation. Two silvopastoral systems were compared to cattle farming without trees and to a normal plantation forest. In one system alternate bands of trees was used and in the second the trees were evenly distributed over the area. The results demonstrated that the silvopastoral systems had enormous potential application.

In 2003 a study, designed to evaluate the feasibility of silvopastoral systems implementation in the Aysén Region were installed in the San Gabriel Agroforestry Unit, situated 30 km north of Coyhaique. The two silvopastoral and the forestry system were installed in a 12-year-old *Pinus contorta* plantation, at an altitude of 650 m above the sea level, Lat S 45°25′55″ and Long W 72°00′41″. At the beginning of the trial the stand height was 4.9 m and density 1,514 trees $ha^{-1}$. The cattle farming system was installed in a

**Fig. 9.1** (**a**) SST, traditional silvopastoral system; (**b**) SSF, strip silvopastoral system; (**c**) SG, cattle system, in the San Gabriel Agroforestry Unit, Aysén Region, Chile (Source: Sotomayor and Teuber (2011)

prairie area without trees, adjacent to the silvopastoral and forestry systems (Fig. 9.1). For more information, the full study details are given by Sotomayor et al. (2010), Sotomayor and Teuber (2011) and Sotomayor et al. (2011b).

The four treatments in the 15.5 ha experimental area were:

- (T1) Managed forestry treatment (TF). In 2003 the stand was thinning from 1514 to 800 trees $ha^{-1}$ and pruned to a height of 1.97 m;
- (T2) Traditional silvopastoral system (SST). The pine stand was thinned to 400 trees $ha^{-1}$, with the trees distributed homogeneously in the site and pruned to a height of 2.01 m;

residues from thinning and pruning were ordered in bands every 21 m;

- (T3) Strip silvopastoral system (SSF). The pine stand was thinned to 400 trees ha$^{-1}$ (Fig. 9.1). There were three rows of tree along the contour, with a 21 m space between them. The trees were pruned to a height of 2.09 m, and the residues from thinning and pruning were placed under the trees in lines;
- (T4) Cattle farming system (SG), with natural prairie and without trees.

For treatments T2, T3, and T4, 5 ha plots were used for the evaluation of the natural prairie and animal production. The managed forestry treatment had an area of 0.5 ha.

The natural prairie, that started to growth in silvopastoral systems after thinned from seedbed, and improved with fertilization, was composed with white clover (*Trifolium repens* L.) and orchard grass (*Dactylis glomerata* L.), and fertilizer was applied annually based on soil analysis made at the beginning of each season. In the spring of 2004, 16 kg of nitrogen ha$^{-1}$ (N), 30 kg of magnesium ha$^{-1}$ (MgO), 24 kg of sulfur ha$^{-1}$ ($SO_4$) and 55 kg of sulfur ha$^{-1}$ (S) were applied.

For tree productivity evaluation three permanent plots of 1,000 m$^2$ were randomly located in the T2 and T3 treatments while in the denser T1 treatment plots of 200 m$^2$ were used. Total height (H), diameter at breast level (DBH), crown cover (CC), and basal area (BA) were measured annually from 2004 to 2008. For statistical analysis, to attend the absence of some of the classic assumptions that require a traditional variance analysis, that are that the data must present a normal distribution, data independency, and heterogeneity of variance, we worked with Mixed Linear Statistics Models. As measurements were repeated in time a longitudinal analysis was used to, evaluate treatment, time and time*treatment effect. The variance model used was: $Y = \mu + T + t + (T^*t) + P + E$, where $\mu$ = constant, T = treatment, t = time, P = plot and E = error. For the statistical evaluation of the tree productivity variables the PROC MIXED version 9.3 of SAS (SAS Institute 2003) and the LSMEANS adjustment were used for the comparison between treatments.

#### 9.2.1.1 Prairie and cattle productivity evaluation

In T2, T3 and T4 treatments seven exclusion cages of 0.5 m$^2$ were distributed randomly to measure prairie dry matter (DM) per hectare (kg DM ha$^{-1}$) between 2004 and 2008. The cages were clipped monthly over the growing season (December–May) and the green pasture dried at 60 °C in a drying oven to a constant weight. The results on prairie and animal production were statistically analyzed through by analysis of variance and LSD Fisher (Least Significant Difference) at 5 % used to compare specific treatments.

The cattle used in treatments T2, T3, and T4, were locally adapted bovines creoles (hybrid crossbreeds, principally based on hereford, angus, red piebald and holstein). To measure their growth from 2004 to 2008 an animal feedlot system was used. Cattle for fattening (300–350 kg initial weight) were grazed over the growing season from December to April–May. Each animal was identified with an ear tag and weighed monthly.

#### 9.2.1.2 Cilmate Evaluation

In 2007, three climate stations were installed, one each in treatments T2, T3 and T4. The stations were situated at the same altitude, and for silvopastoral treatment were midway between trees (T2, the middle distance was 2.5 m and for T3 it was 10.5 m between tree strips). Computers powered by solar panels recorded measurements every half an hour. We measured air temperature (t°C), mean wind speed (vv), mean maximum speed, relative humidity (HR) and precipitation (pp) from January 2007 to June 2008.

The following linear regression was used to explore the effects of wind speed on pine, praire and animal growth parameters (Sotomayor and Teuber 2011): $Vv = \mu + p^*F + E$, where Vv = wind speed, u = constant, p = factor, F = forestry parameter and E = error.

To analyze the effect of the trees on prairie and animal productivity the regression model used was: $Pp(a) = \mu + F + t + (F^*t) + C + t + (C^*t) + P + E$, where Pp = prairie (kg DM ha$^{-1}$), Pa = animal production (kg live weight ha$^{-1}$), $\mu$ = constant, F = forestry parameter, t = time, C = climate

**Table 9.1** Results of total height (H), diameter at breast height (DBH), canopy cover (CC) and basal area (BA) in silvopastoral and forest treatments with *Pinus contorta*, from 2004 to 2008, in San Gabriel Agroforestry Unit, Aysén Region, Chile

| | Mean DBH (cm) | | | Mean BA (m$^2$ ha$^{-1}$) | | | Mean H (m) | | | Mean CC (%) | | |
|---|---|---|---|---|---|---|---|---|---|---|---|---|
| Year | TF | SST | SSF | TF | SST | SSF | TF | SST | SSF | TF | SST | SSF |
| 2004 | 12.7$^a$ | 12.9$^a$ | 13.0$^a$ | 10.6$^b$ | 4.8$^a$ | 5.4$^a$ | 5.3$^a$ | 6.0$^a$ | 5.6$^a$ | 26.9$^a$ | 14.5$^b$ | 24.2$^a$ |
| 2005 | 13.6$^a$ | 14.2$^a$ | 14.1$^a$ | 12.0$^b$ | 5.8$^a$ | 6.4$^a$ | 5.7$^a$ | 6.4$^a$ | 6.1$^a$ | – | – | – |
| 2006 | 14.8$^b$ | 16.0$^a$ | 15.6$^a$ | 14.2$^b$ | 7.3$^a$ | 8.3$^a$ | 6.1$^a$ | 6.8$^a$ | 6.5$^a$ | 36.0$^b$ | 21.7$^a$ | 27.1$^a$ |
| 2007 | 16.0$^b$ | 17.8$^a$ | 17.1$^a$ | 16.6$^b$ | 9.1$^a$ | 9.9$^a$ | 6.6$^a$ | 7.2$^a$ | 7.0$^a$ | 55.1$^b$ | 31.1$^a$ | 28.9$^a$ |
| 2008 | 16.9$^b$ | 19.1$^a$ | 18.0$^{a,b}$ | 18.3$^b$ | 10.3$^a$ | 11.0$^a$ | 7.0$^a$ | 7.6$^a$ | 7.5$^a$ | 60.2$^b$ | 32.2$^a$ | 30.7$^a$ |

*TF* forest treatment, *SST* traditional silvopastoral treatment, *SSF* strip silvopastoral treatment
*Different letters indicate significant differences, LSMEANS ($p \leq 0.05$), for analysis over time, by year and treatment

**Table 9.2** Prairie production, growth season 2004–2005 to 2007–2008 (kg DM ha$^{-1}$), per treatment in San Gabriel Agroforestry Unit, Aysén Region, Chile

| | Prairie production per season and treatment (kg MS ha$^{-1}$) | | | |
|---|---|---|---|---|
| Treatment | 2004–2005 | 2005–2006 | 2006–2007 | 2007–2008 |
| T2: SST | 1486$^a$ | 6110$^a$ | 4153$^b$ | 4331$^b$ |
| T3: SSF | 2685$^a$ | 7182$^a$ | 6395$^a$ | 5360$^a$ |
| T4: SG | 2452$^a$ | 3832$^b$ | 3874$^b$ | 3514$^b$ |
| Average production per season | 2208 | 5709 | 4807 | 4401 |

*SST* traditional silvopastoral system treatment, *SSF* strip silvopastoral system treatment, *SG* cattle treatment
*Different letters indicate significant differences, LSD Fisher ($p \leq 0.05$), between treatments in the season

parameter, P = plot, E = error. Similarly, in silvopastoral systems crown cover (CC) and wind speed (vv) were included in the analysis of prairie and animal growth.

To confirm that the experimental plots were installed in similar sites, we assessed whether there were differences in mean heights (H) between the plots in 2003. Furthermore, after thinning in 2004, the height of the 100 dominant trees (H100) was evaluated. To see if there were differences in soil fertility between treatments T2, T3 and T4, we analysed 0–15 cm soil samples for nitrogen (N), phosphorus (P), potassium (K), organic matter (OM) and acidity (pH) (Sotomayor et al. 2009b). Analysis of variance, performed using the INFOSTAT(c) tool, found no significant differences; this indicated the sites were similar.

#### 9.2.1.3 Results and discussion

DBH varied significantly between the treatments as a function of time from 2006 on with larger and similar diameters in the lower stocked silvopastoral treatments; these differences increased with time (Table 9.1). However, in 2008 the normal forestry and strip silvopastoral systems were not significantly different. With BA there was a significant treatment*time interaction due to the increasing growth difference between the higher stocked forestry treatment and the silvopastoral treatments (Table 9.1). The silvopoastoral treatments had similar basal areas, as their stocking was the same.

There were no significant differences in height between treatments or with time despite the greater tree selection at thinning with the silvopastoral treatments (Table 9.1). For crown cover treatment, time and treatment*time effect were all statistically significant. The main cause of this was the greater stocking of the forestry treatment with the CC between treatments increasing with time (Table 9.1).

Related to prairie productivity, there was no significant effect of the treatment*time interaction in the first production season (Table 9.2). From the second sampling season, the two silvopastoral treatments, especially the strip system,

**Table 9.3** Cattle production per hectare (kg live weight gain ha$^{-1}$) by treatment, seasons 2004–2005 to 2007–2008, San Gabriel Agroforestry Unit, Aysén Región, Chile

| | Live weight gain per hectare of prairie and season (kg live weight ha$^{-1}$) | | | | Total production 2004–2008 (kg live weight ha$^{-1}$) |
|---|---|---|---|---|---|
| Treatment | 2004–2005 | 2005–2006 | 2006–2007 | 2007–2008 | |
| T2: SST | 114[a] | 239[a] | 306[a] | 159[a] | 817 |
| T3: SSF | 110[a] | 256[a] | 318[a] | 172[a] | 855 |
| T4: SG | 96[a] | 228[a] | 349[a] | 144[a] | 817 |
| Average production per season and total | 107 | 241 | 324 | 159 | 830 |

*SST* traditional silvopastoral system treatment, *SSF* strip silvopastoral system treatment, *SG* cattle treatment
*Different letters indicate significant differences LSD Fisher ($p \leq 0.05$), between treatments in the season

began to show a greater productivity, with significant differences in relation to the open prairie treatment. For the last two seasons SSF showed a greater pasture production, with significant differences in relation to the other two treatments. In general, in all seasons, the best production was from the strip system. The worst production season for the treatments was the first evaluation period (2004–2005), because it was the first year after opening of the canopy in the silvopastoral treatments and improvement of the prairie. Note that the pasture productivity in the silvopastoral systems decreased after the second growing season as the canopy closed.

In animal production the treatment and treatment*time effects were not significant but productivity was significantly different over time with the best season being the 2006–2007 growing season (Table 9.3). In the last growing season the effects of increasing crown cover was reducing animal productivity.

Regarding microclimate evaluation, the average wind speed was higher in the cattle system (SG) than the silvopastoral systems throughout the evaluation period (Fig. 9.2a). Further, wind speeds were greater in spring-summer (August–January) than in winter (April–July), with greater differences between the open prairie and the silvopastoral treatments at high wind speeds. The maximum difference was 10 km h$^{-1}$ in October 2007. Further, the traditional silvopastoral treatment had lower wind speed than the strip system, because the former has a more uniform distribution of the trees (Sotomayor and Teuber 2011). The mean maximum speed, increased from August (end of winter) until April (beginning of autumn). The open prairie had the highest mean maximum speeds of 29 km h$^{-1}$ in October. For that same month, the strip system registered 18 km h$^{-1}$ and the traditional silvopastoral system slightly below 10 km h$^{-1}$.

Wind Chill is defined a still-air temperature that would have the same cooling effect on exposed human skin, or animal skin, as a given combination of temperature and wind speed. It called also as chill factor, windchill factor, and windchill index. The lowest wind chill values were presented in the cattle treatment and the higher values, with similar results, in the silvopastoral treatments (Fig. 9.2b). This value is important for animal management, because the animals must use more energy to regulate their body temperature when there are low temperatures and high winds and consequently low values of wind chill (Quam and Johnson 1999). For this reason, the silvopastoral treatments provide them greater protection, reducing animal energy consumption (Sotomayor and Teuber 2011).

In relation to air temperature, no significant differences between treatments were found. Between October 2007 and February 2008, the strip system had an average value of 10.2 °C compared to 9.9 °C for the than the open prairie. In the February to March 2008 period the environmental temperature had dropped to 7.9 – 7.7 – 7.9 °C for the strip, traditional and cattle systems, respectively (Sotomayor and Teuber 2011).

Relative humidity did not differ between treatments. They ranged from 85 % for winter month to 62 % in January 2008, during the summer period. Average monthly rainfall between July 2007 and April 2008 was 260 and 229 mm in the

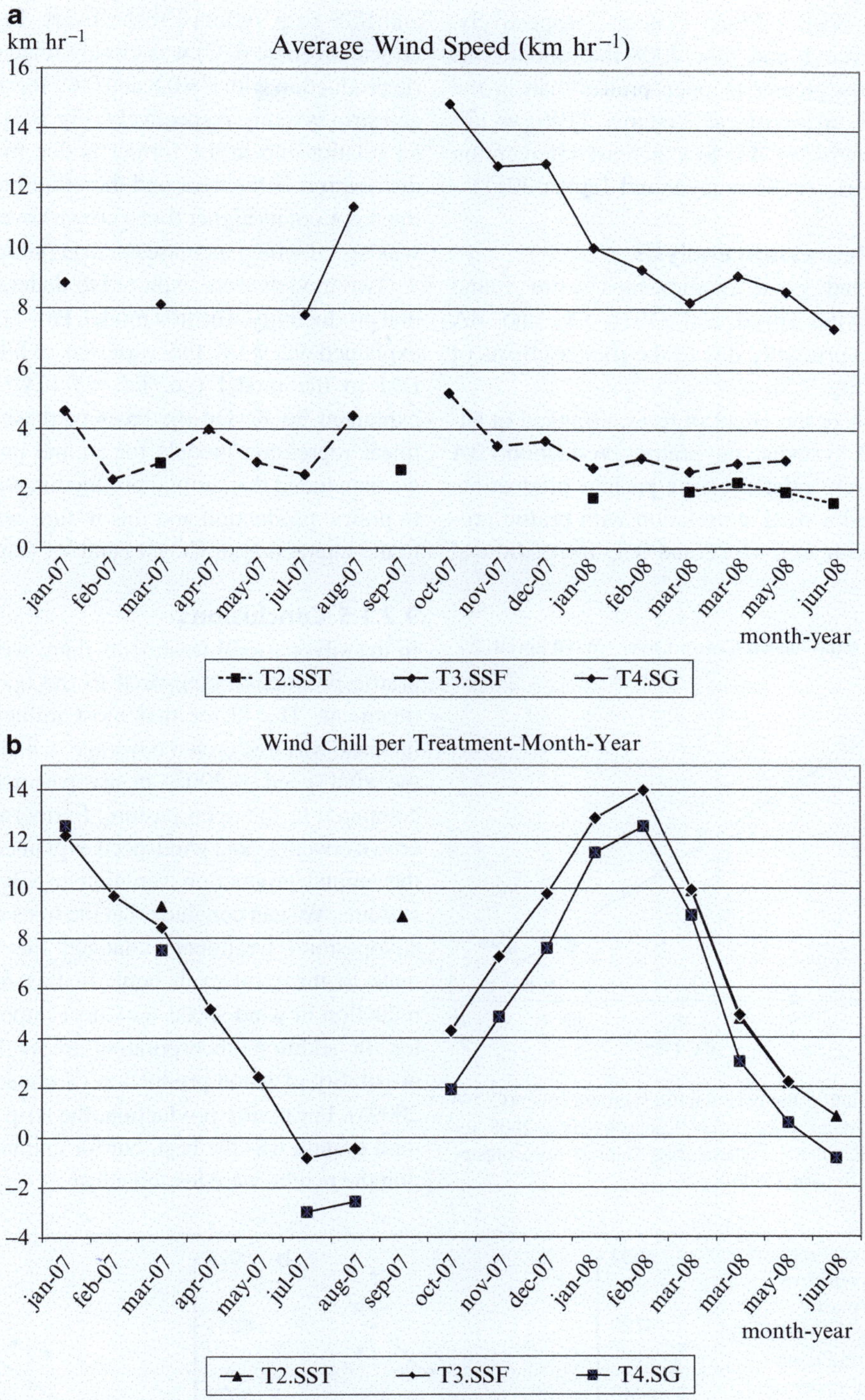


**Fig. 9.2** (**a**) Average monthly wind speed (km h$^{-1}$), January 2007–June 2008; (**b**) average monthly wind chill by treatment and month, January 2007–June 2008, at San Gabriel Agroforestry Unit, Aysén Region, Chile (Source: Sotomayor and Teuber 2011)

strip and open pasture systems, respectively. Between March and June 2008 the rainfall was significantly greater in open prairie than in the traditional silvopastoral systems (196 vs.179 mm, respectively), due to a increase interception of the tree cover (Sotomayor and Teuber 2011).

#### 9.2.1.4 Regression analysis

As expected good relationships were found between wind speed and DBH, CC and BA ($r^2 \geq 0.95$) primarily due to the shelter effects of the trees (Fig. 9.3).

To analyze the effect of trees, arranged in silvopastoral systems, on prairie production, BA and CC were related to the pasture production. BA showed a weak correlation with prairie production, with an $r^2=0.22$ and 0.35 for traditional and SSF strip systems, respectively. In contrast, crown coverage (CC) was closely related to prairie production with $r^2=0.8$ and 0.69 for traditional and strip systems, respectively (Fig. 9.4). The better relationship in the former is due to the even distribution of the trees and therefore better wind interception; its higher forest crown cover reduces windspeed which increases prairie productivity.

Two models were explored to understand animal productivity. The first model, Pa = f (CC*Vv), explained 92 % of the variation, while adding DM to the model (i.e. Pa = f (CC*Vv*DM) explained 96 %. On the basis of these multiple linear regression models for animal production, we concluded that animal production was related to prairie production and this in turn was related to the interaction of CC (%) and Vv (km $h^{-1}$).

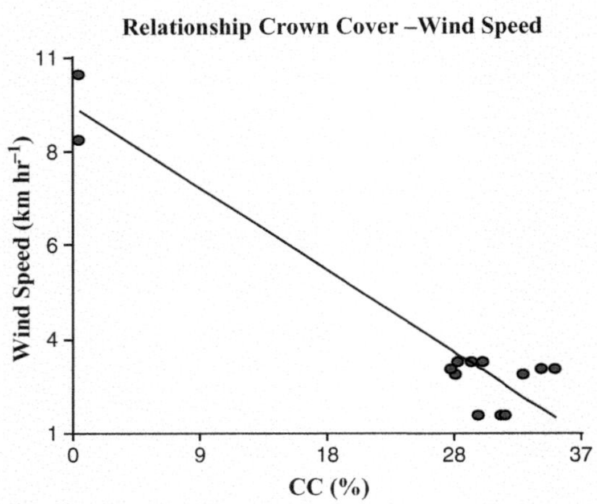


**Fig. 9.3** Relationship between forest crown cover (CC%) and average wind speed (km $h^{-1}$), seasons 2006–2008, for silvopastoral a cattle systems, San Gabriel Agroforestry Unit, Aysén Region, Chile

#### 9.2.1.5 Conclusions

In the silvopastoral treatments there were higher prairie production compared to the open prairie treatment. The factor that most influenced this increase was tree crown coverage, which reduced the wind speed by 200 % in silvopastoral systems compared to the open pasture. Furthermore, tree crown coverage and windspeed explained 92 % of the animal production variation in silvopastoral systems. We can conclude that the trees present in silvopastoral treatments changed the microclimate to the farmer's benefit. In this region the reduction in wind speed was most important factor. In addition, incorporating trees allows the possibility of wood production (Sotomayor et al. 2009a). For prairie production, the strip silvopastoral system was the best, but for animal production the results were less conclusive.

**Fig. 9.4** Lineal regression of forest crown cover (CC%) and prairie productivity (kg DM $ha^{-1}$), in (**a**) SST, traditional silvopastoral system and (**b**) SSF, strip silvopastoral system, seasons 2004–2008, San Gabriel Agroforestry Unit, Aysén Region, Chile

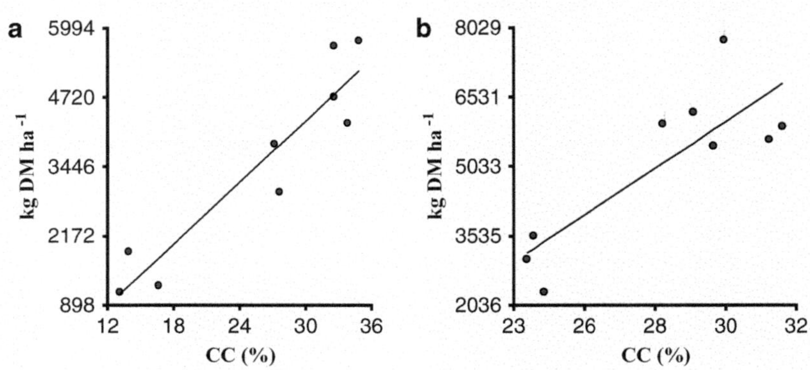

### 9.2.2 Results and Experiences in the Production of Fodder Crops Production with Windbreaks

In the Aysén region of Chile the use of planted tree windbreaks for soil and prairie protection is still scarce, although the detrimental effect of the wind on plant productivity and soil erosion, has been documented (Sotomayor et al. 2011a). From 2002, INIA and INFOR studied the use of windbreaks to enhance the growth of animal fodder and pasture in Patagonia. The details below at taken from the book "Sistemas Agroforestales para la Region de Aysén: Cortinas Cortaviento y Silvopastoreo" (Teuber 2009).

The Los Ñires Farm was selected to establish the windbreak study. The farm is in the lower part of the Simpson Valley, Coyhaique Municipality, at an approximate altitude of 325 m above sea level and is known for strong winds from the northeast. The soil and climate characteristics are typical of the intermediate zone of the Aysén Region.

On this flat site the shelter was provided with a 32 year-old mature tree windbreak having three rows of *Pinus contorta*, *Pinus ponderosa* and *Pinus sylvestris*, with a mean height of 14 m. This shelterbelt was oriented perpendicular to the main damaging wind. The study included two fodder species and one fodder mix, of great importance to the cattle production in the area:

(a) *Dactilys glomerata* (orchard grass);
(b) *Medicago sativa* (lucerne); and
(c) A fodder mixture comprised of *Lolium perenne*, *Festuca arundinacea* and *Trifolium repens* (perennial ryegrass, fescue and white clover).

A randomized complete block design, with three replications was used with the crop productivity being measured at different distances from both the leeward and windward sides of the windbreak. The three treatments were sown perpendicular to the windbreak, during November 2005 and samples of 0.5 $m^2$ were harvested for dry matter production in January and May 2006. These samples were taken at different distances between 0 and 15 H on the leewards side of the windbreak and from 0 to 5 H on the windward side (where H is the height of the shelterbelt). For this study, the control was the area at 14–15 H.

The results were statistically analyzed by analysis of variance, was and Fisher's LSD test at $P=0.5$ was used to compare specified hypotheses.

## 9.3 Results and discussion

Orchard grass is the principal forage grass in naturalized and sowen prairies in the Intermediate Zone of Aysén, due to its good adaptation to water deficit conditions and its good performance under cattle grazing systems.

Production varied significantly with distance from the shelterbelt with the best production, 18.0 Mg DM $ha^{-1}$, found at 2–3 H (Fig. 9.5). The more protected leeward areas between 0 and 5 H had the best production (14.8–18.0 Mg DM $ha^{-1}$) comparing to between 6 and 15 H sector where average production was 12.7–14.5 Mg DM $ha^{-1}$. On the windward side, the pasture production was 11.9–14.2 Mg DM $ha^{-1}$. The pasture growth in the 0–5 H zone was 22 % greater than at 15 H.

The fodder mixture had a minimum productivity of 12.9 Mg DM $ha^{-1}$ at 14 H and the maximum of 18.7 Mg DM $ha^{-1}$ at 3 H. Again greater production was found on the more protected leeward zone between 0 and 7 H where the production was 15.9 and 18.7 Mg DM $ha^{-1}$. At 8–15 H from the shelterbelt the average production was between 13.5 and 15.6 Mg DM $ha^{-1}$. In the case of windward side production ranged from 12.6 to 14.5 Mg MS $ha^{-1}$.

Lucerne is a pasture species of major potential in the Aysén Intermediate Zone, due to its large and deep root system. As it can extract water from depth, it is more resistant to water deficit periods which occurred in summer. When dry matter production was analyzed in the second cut in May, (the first cut had problems with weeds and was not evaluated) it was found that 0–5 H zone had an average production of 3.5–5.1 Mg DM $ha^{-1}$, which was 47 % of greater than the 15 H control zone. On the windward side, the production fluctuated between 2.9 and 3.4 Mg DM $ha^{-1}$.

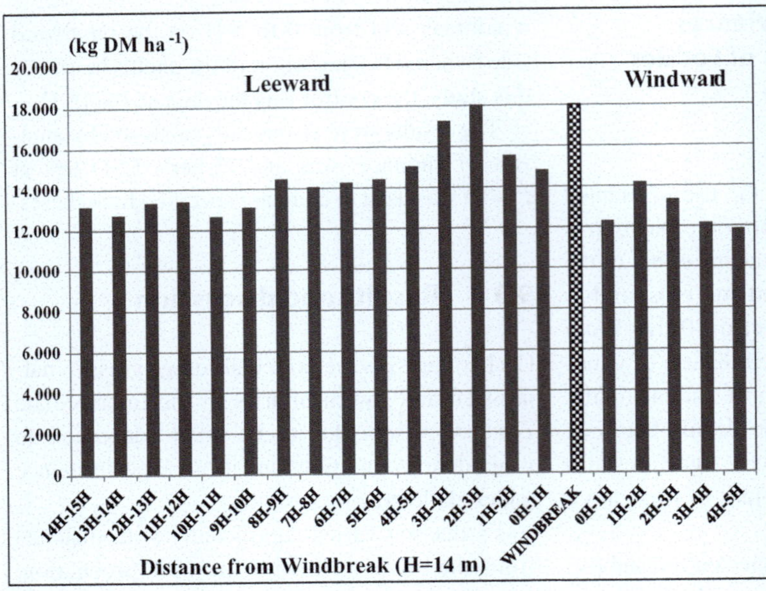


| Distance from Windbreak | Production (kg DM ha$^{-1}$) | Significance ($p \leq 0.05$) |
|---|---|---|
| LEEWARD | | |
| 14H-15H | 13149 | defgh |
| 13H-14H | 12737 | efgh |
| 12H-13H | 13326 | defgh |
| 11H-12H | 13384 | cdefgh |
| 10H-11H | 12657 | efgh |
| 9H-10H | 13087 | defgh |
| 8H-9H | 14461 | cdef |
| 7H-8H | 14052 | cdefg |
| 6H-7H | 14271 | cdefg |
| 5H-6H | 14401 | cdef |
| 4H-5H | 15034 | cd |
| 3H-4H | 17238 | ab |
| 2H-3H | 18019 | a |
| 1H-2H | 15547 | bc |
| 0H-1H | 14812 | cde |
| WINDWARD | | |
| 0H-1H | 12284 | fgh |
| 1H-2H | 14189 | cdefg |
| 2H-3H | 13319 | defgh |
| 3H-4H | 12161 | gh |
| 4H-5H | 11859 | h |

**Fig. 9.5** Cumulative production of orchard grass, 2005–2006 summer seasons, in kg DM ha$^{-1}$, in windward and leeward of a tree windbreak, Aysén Region, Chile (Different letters indicate statistically significant difference, $p \leq 0.05$) (Source: Teuber et al. 2008)

## 9.4 Conclusion

These studies confirmed that windbreaks are of great assistance for animal fodder production in the Aysén region. All the species and mixtures studied responded favorably to the protection afforded by the windbreak, with increases ranging from 14 % with the fodder mixture to 47 % with lucerne when comparing the area of most protection between 0–5 H to the least protected area (15 H), considered in this study as control. Finally, we recommend that cattle and agricultural producers of Aysén Region use tall-tree windbreaks to increase the productivity of fodder species as well as for the protection of soils by wind erosion. There may also be the possibility of wood production when the trees are mature and need to be replaced.

## 9.5 Management of Deciduous Species of the *Nothofagus* Genus for Silvopastoral Purposes in the Chilean Regions of Aysén and Magallanes

### 9.5.1 Silvopastoral Management Experiences in Forests of *Nothofagus antarctica* and *N. pumilio* in the Region of Aysén

The region of Aysén represents the highest proportion of native forests in Chile with a total of 4.3 million hectares, of which Ñire forests however only account for 131,593 ha (CONAF-UACH 2011), mostly distributed in the communes of Coyhaique, Lago Verde and Cochrane, where they grow up to a sea level of 1,200 m.

Despite its wide geographic distribution, information on silvicultural practices associated with ñire is rare. When talking about a ñire forest, the image often is the one of a sick forest with high pressure from cattle grazing, subject to insect plagues and forest fires during decades, leading to a very low commercial value of these ecosystems. The shrub-like growth of this species with crooked stems has not allowed its use as timber (Donoso 2006). Ramírez et al. (1985) classify three morphological types of ñire in the central-southern part of Chile, according to the development of the over ground plant system: tree-like, shrub-like and chamaephyte-like. This ecomorphological plasticity makes it possible for the species to tolerate various and often harsh weather conditions with major temperature variations during the day and atmospheric saturation deficits (Weinberger 1973), as is the case for the ñire forests growing near the tree line in the Aysén region. From the botanical point of view ñire forests typically are regarded as belonging to the class of *Nothofageteapumilionisantarcticae* (Oberdorfer 1960).

Experiences of silvopastoral management in Aysén are twofold. The first one relates to a research project currently being carried out by INFOR with the objective of establishing guidelines for the silvopastoral management of ñire forests. This study evaluated ñire forests under two site conditions: a humid condition related to a swamp area and a drier condition on a site without groundwater influence. In each case three management methods were selected. The first method represents a situation without light limitation (100 % total light transmission) corresponding to open areas where the forest has been cut down or is naturally not present (Peri et al. 2005). The second situation corresponds to silvopastoral management after a homogeneously distributed thinning intervention (extracting 40 % of the basal area) (Table 9.4). The third treatment considers an original forest without intervention corresponding to the low-pole development stage.

In order to characterize the prevailing forest structure a conglomerate composed of three permanent round sample areas (200 m$^2$) arranged systematically in an inverted “L” shape (Martin et al. 2009) was evaluated. To define the incorporation of seedlings in each sample site, four permanent sample plots of 1 m$^2$ (Bahamonde et al. 2011) were installed in line with the cardinal points inside each permanent round sample area. Pasture productivity was measured by including randomly distributed exclusion cages (0.5 m$^2$) within each situation (Sotomayor et al. 2007).

vire has a high vegetative reproduction capacity, responding very well to growth-enhancing hormones (Salinas et al. 2011). Thanks to this typical feature forest management of the species has traditionally been based on agamic regeneration (Peri 2005). In this context, for the purpose of silvopastoral management two types of individual trunk protectors were installed (to protect against pressure from herbivores such as livestock and hare) in order to be able to evaluate the resprouting behaviour of ñire. For estimations of canopy-cover, light conditions and totally transmitted photosynthetically active radiation (PAR) were determined using ten hemispherical photographs of the tree cover were taken from 1 m to the ground, these were analyzed in the Gap Light Analyzer (GLA) software.

Although this research study is still under development, there are some preliminary results

**Table 9.4** Dendrometric variables for silvopastoral management under two site conditions, before and after a thinning intervention (extracting 40 % of basal area) with homogeneous distribution, Coyhaique, Aysén Region, Chile

| | Before the intervention | | | | After the intervention | | | | Indicator degree of intervention |
|---|---|---|---|---|---|---|---|---|---|
| | $H_M$ | $D_{MC}$ | N | G | $H_M$ | $D_{MC}$ | N | G | |
| Site | [m] | [cm] | [tree ha$^{-1}$] | [m$^2$ ha$^{-1}$] | [m] | [cm] | [tree ha$^{-1}$] | [m$^2$ ha$^{-1}$] | $D_{MC\ remaining}/D_{MC\ initial}$ |
| WET | 9.3 | 10.3 | 5 263 | 44.16 | 9.1 | 12.1 | 2 350 | 26.80 | 1.1 |
| DRY | 9.9 | 11.8 | 3 600 | 39.08 | 10.1 | 13.8 | 1 467 | 22.01 | 1.1 |

$H_M$ average height, $D_{MC}$ average quadratic diameter, *N* density, *G* basal area

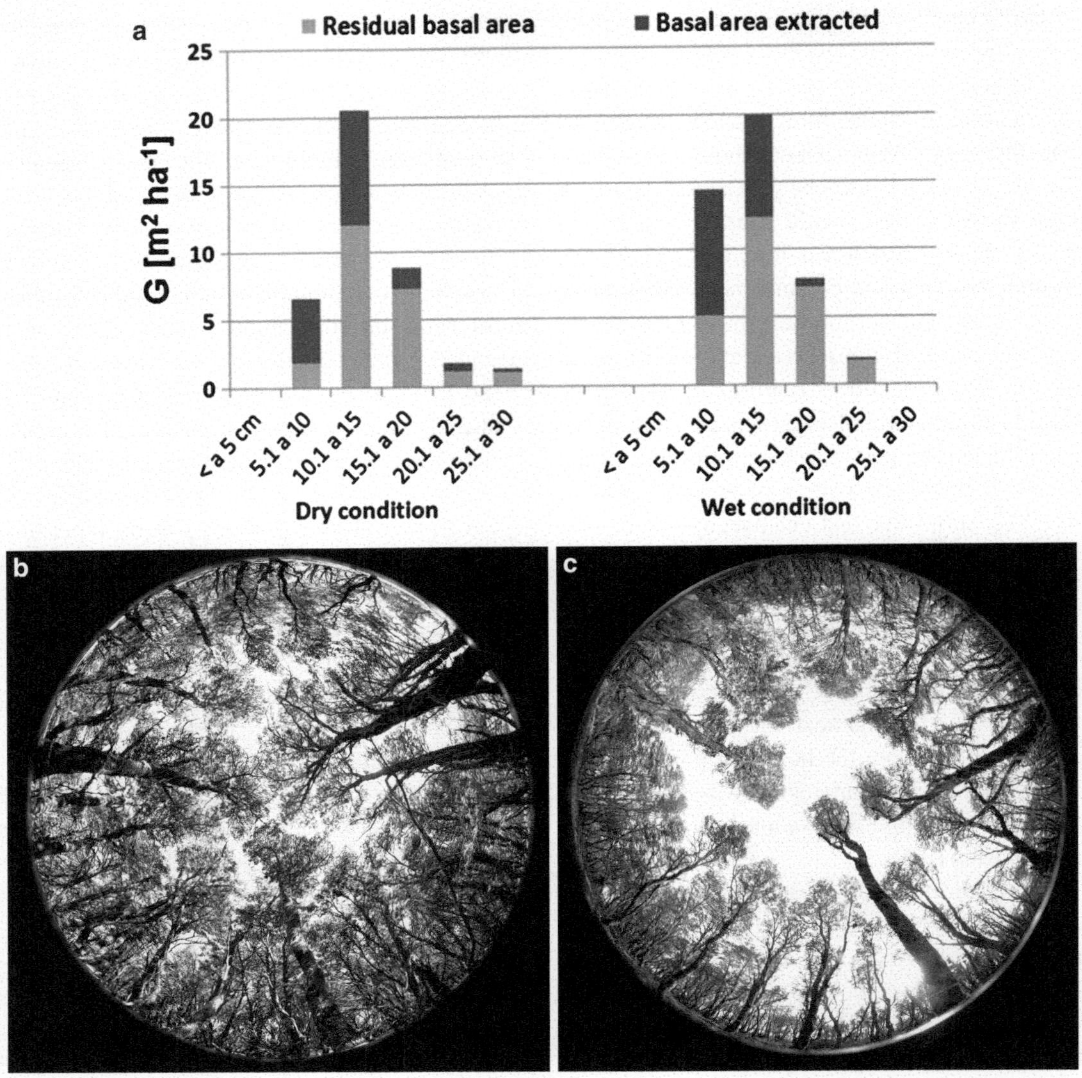


**Fig. 9.6** (**a**) Extracted and total remaining basal area (G) in a silvopastoral system for *N. antarctica*, growing under two site conditions, Aysén Region, Chile; (**b**) Hemispherical photograph taken of the dry site condition before the thinning intervention. **c** Hemispherical photograph taken of the dry site condition after the thinning intervention

on the characterization of forest management (witness areas) and silvopastoral management before and after the intervention available. In the case of silvopastoral management a thinning intervention extracting 40 % of the basal area (Fig. 9.6a) was applied to each study site, with the purpose of leaving an approximate canopy cover of 40 % (Fig. 9.6c).

The second example of silvopastoral management of ñire forests corresponds to an initiative taken by the National Forest Corporation (CONAF) in 2010 (not published). Its objective was to evaluate the reaction in growth and regeneration of a 60-year-old ñire forest in the commune of Río Ibáñez under different thinning schemes (IR) and with different canopy cover (CC). The area was subdivided into four plots of approximately 3800 $m^2$ with a previously defined treatment for each plot; T1 = 50 % of CC and 30 % of IR, T2 = 0 % of CC and 100 % of IR, T3 = 100 % of CC and 0 % of IR, T4 = 30 % of CC and 70 % of IR. The different cover percent-

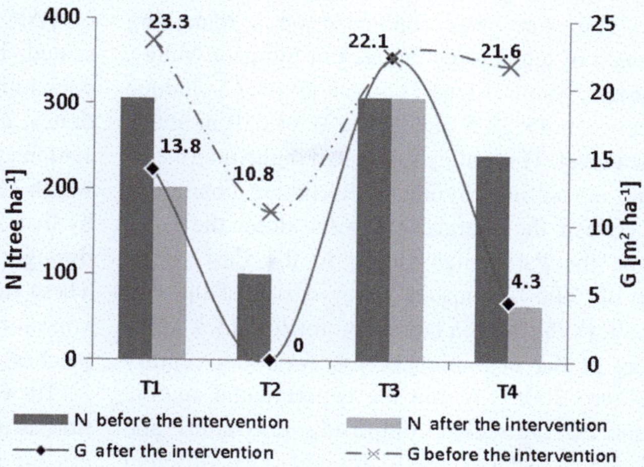


**Fig. 9.7** Comparison of dendrometric variables, density (N) and basal area (G) in a forest of *N. antarctica* before and after a thinning intervention of varying intensity, Aysén Region, Chile

**Table 9.5** Response of natural and agamic regeneration after different intensities of thinning, Bahía Murta, Aysén Region, Chile

| | Regeneration | |
|---|---|---|
| Treatment | Seedlings [n° ha$^{-1}$] | Resprouting stump [%] |
| T1 | 23,000 | 32.5 |
| T2 | 3000 | 21 |
| T3 | 2000 | – |
| T4 | 24,000 | 65.1 |

ages were defined according to Tejera et al. (2006). The thinning intensities were established based on the recommendations of Peri (2005). In order to define the dynamics of regeneration, 10 plots of 1 m$^2$ each were established in each intervention area distributed systematically every 15 m (Fig. 9.7).

Forestry measurement was carried out in September 2011 and April 2012, after which a process of natural (seeds) and vegetative (resprouting from trunk) regeneration was initiated, which was enhanced by excluding livestock.

Although ñire is a shade-intolerant species (Donoso 2006), the highest percentage of regeneration did not occur on the trunks without vegetation cover (T2: clearcut), but under the protection of a higher canopy cover of some 30 % (Table 9.5). The same trend was reported by Tejera et al. (2006) in the province of Chubut, Argentina.

Lenga is represented in Aysén region by 947,685 ha (CONAF-UACH 2011), but has been less studied in relation to silvopastoral management systems. The only study carried out in the Region was initiated within an international project co-financed by the Institute of Agricultural Research (Instituto de Investigaciones Agropecuarias, INIA) and the Spanish Biodiversity Foundation (Fundación Biodiversidad); it was carried out between 2005 and 2008 and new measurements are presently being collected. The project considered silvopastoral systems as an integrated management system, by taking advantage of the synergies resulting from the ecological interaction between forests and pastures. In different climate regions of the world there is empirical evidence that both, production and conservation goals can simultaneously be reached by maintaining a certain tree cover within the agricultural system (Tscharntke et al. 2005; Rivest et al. 2013). For instance, in the Mediterranean basin these ecosystems have existed for more than 5000 years. Based on these experiences, the project aimed at setting the scientific bases for the implementation of silvopastoral systems in the remaining patches of native forests. In order to do so the effect of tree cover on the herbaceous stratum was analysed, considering different agronomical (e.g. production, forage quality) and ecological (e.g. biodiversity, presence of native plant species) variables.

The study was conducted on a remaining patch of lenga forest located in Simpson Valley, sector Santa Elena, commune of Coyhaique between 45°77 S and 72°06 W at 619 m above sea level. This valley is situated in the intermediate agro-climate zone with climate conditions between the evergreen forests along the coast and the Patagonian steppe to the East (Hepp et al. 1988). Simpson Valley is one of the first valleys in the area colonized for livestock farming at the beginning of the twentieth century (Otero 2006). Within the experimental station, one 50 ha sector comprising grasslands and lenga woodlands was delimited. Two different complementary approaches were used: analysing the tree cover continuum between dense forests and open tree-lees grasslands, and the spatial contact between the two ends of this continuum; in the latter four permanent exclusion areas ($10 \times 80$ m$^2$) were established to evaluate the influence of livestock grazing on these ecosystems.

Between October and April of the years 2006 and 2008, the herbaceous biomass inside and outside wire exclourse cages ($70 \times 100 \times 50$ cm), was measured on a monthly basis, cutting the grass at soil level with an electric shearing handpiece. After each sampling date, cages were relocated in order to avoid previously sampled areas. Floristic composition of and soil physicochemical variables were measured as well.

In total, 36 herbaceous species were registered, distributed in 33 genera and 17 families (Sánchez-Jardón et al. 2014a) and 8 shrub species, all of them native; i.e. *Berberis microphylla* G. Forst, *B. serratodentata* Lechler, *Chusquea culeou* E. Desv, *Maytenus disticha* (Hook. f.) Urban, *Ovidia andina* (Poepp. & Endl.) Meisn. *Ribes cucullatum* Hook. & Arn., *R. magellanicum* Poir. and *Senecio magellanicus* Hook. & Arn. Roughly 50 % of the 36 herbaceous species were native. The introduced species *Dactylis glomerata* L., *Trifolium repens* L. and *Taraxacum officinale* (Weber.), were the most abundant ones with a cover of 29, 20 and 12 %, respectively. Net primary production was higher on pastures ($300 \pm 29$ g m$^{-2}$) than on dense forests ($164 \pm 40$ g m$^{-2}$). Some areas with scattered trees (50–70 % of PAR transmitted through the canopy) presented, however, maximum production values, coinciding but lower water-soluble carbohydrates, as expected in shaded plants (Sánchez-Jardón et al. 2010). Also, forest soils were richer, which seem to be associated to the lower intake by livestock and a sustained fertilization process through dung (Sánchez-Jardón et al. 2014b). These results support the implementation of silvopastoral management systems on lenga forest patches.

The effect of tree cover especially in its function as a light filter, and the impact of livestock grazing are the two factors explaining the spatial variation in the structure of the herbaceous plant communities. Grazed-on forests maintain higher native species richness than tree-less pastures, so that the coexistence of both types of patches.

Results from permanent exclusion areas showed that, after 2 years, the absence of livestock grazing changes plant species composition and diversity, and generally favouring native herbaceous species (*unpublished data*). Results suggest that native species diversity could be restored in these grazing systems, thus their conservation value improved.

### 9.5.2 Silvopastoral Management Experiences in Forests of *Nothofagus antarctica* and *N. pumilio* in the Region of Magallanes

Ñire forests do not have an important timber value. However, they are very appreciated by their ecological and scenery services and constitute an important resource for the regional livestock industry. According to CONAF-CONAMA (2006), ñire forests cover 220,045 ha in Magallanes. Most of them are located under 400 m.a.s.l., in sites with slopes up to 30 % and frequently, at the contact zone with the Patagonian rangelands, which make them accessible for grazing most part of the year (Fig. 9.8a).

At silvopastoral trials associated to ñire forests in Magallanes (Schmidt 2008), the highest whole-system productivity results were obtained under

**Fig. 9.8** (**a**) Beef cattle grazing in a Ňire forest, sector el Palenque, Province of Ultima Esperanza, Magallanes Region, altitude 260 m above sea level; (**b**) General view of a silvopastoral system associated to Lenga, sector Monte, Province of Ultima Esperanza, Magallanes Region, altitude 300 m above sea level

tree covers of 41–57 %. At this canopy cover range, the PAR at the ground level, as well as the soil temperature and moisture were the highest during the growing season. These microclimatic conditions allowed for a significant improvement of the understory from a grazing perspective, increasing the presence of species of high nutritional value as *Holcus lanatus* L., *Dactylis glomerata* L. and *Trifolium repens* L. from 15 to 67 %, in a dry weight basis. Pasture productivity also increased, reaching values up to 2 Mg of DM ha$^{-1}$ during the growing season, a significantly higher value than those obtained in sites with a higher tree density or without trees.

Due the favorable microclimatic conditions, pasture growth in silvopastoral systems start earlier and finish later than in pastures without a tree stratum, resulting in higher food availability at the beginning and the ending of the grazing season. Moreover, due the wind protection effect of the trees, animals grazing in silvopastoral systems dedicate less time looking for food, are more selective and increase the total forage consumption, which increases the weight gain (Schmidt 2008).

Under high winds and rain, animals without a tree protection stop eating and group together, whereas animals in silvopastoral systems maintain their normal grazing routine. In snowy conditions, animals in the forest look for food and shelter under the trees. On the contrary, animals in open pastures stop eating and dedicate more time looking for sites of thin snow to graze (Schmidt 2008). Most of the watering points freeze during the winter, but only those in open pastures need for a periodic human intervention to break the ice cover allowing cattle the access to water. All the above reasons have promoted an extensive and permanent grazing of the ñire forests in Magallanes, with great benefits for the regional livestock industry.

Lenga forests are the principal timber resource of the region. According to CONAF-CONAMA (2006), lenga secondary forests cover 120,000 ha in Magallanes. The majority of these secondary forests are not part of the economic activity of the region, and only a small fraction is managed with silvicultural methods. Differently to the silvopastoral systems associated to ñire, silvopastoral systems in lenga secondary forests emerge as a consequence and opportunity when applying silvicultural management methods to tree stands looking to increase the growth rate of selected trees by reducing the competition through thinning once the secondary forest is at pole and sapling stage and 40–60 years of age.

At silvopastoral trials associated to lenga forests in Magallanes (Schmidt et al. 2013a, b), tree density was reduced to 600–1000 individuals $ha^{-1}$. Thinning the stand reduced tree canopy cover from 50 % to 15–30 %, which increased the PAR at ground level from 25 % to 50–65 % and the temperature and moisture of the soil during the growing season (Fig. 9.8b).

The modified microclimatic conditions allowed for the growth of an understory cover of native species associated to the lenga dominated ecosystems. They also allowed for the growth and establishment of naturalized species associated to pastures and of high nutritional value, as *Holcus lanatus* L., *Dactylis glomerata* L., *Lolium perenne* L., *Phleum commutatum* Gardin and *Trifolium repens* L., creating an ecotone between pastures and forests of high biodiversity. Grass cover increased with the PAR at ground level. At secondary forests without thinning, of 1000 and 600 trees $ha^{-1}$ the grass cover was 0.35 and 73 %, respectively. The grass cover in open grasslands was 84 %. The understory yielded up to 2 Mg DM $ha^{-1}$ $y^{-1}$, with 6–8 % of crude protein. Grazing trials demonstrated a live weight gain of 580 g $d^{-1}$ in Hereford steers.

As trees grow, their canopy cover closes and new thinnings are required to keep the productivity of the pasture underneath. When the forest reaches its mature stage at a density of 300–400 tree $ha^{-1}$, the canopy closes definitely and the silvopastoral stage comes to an end favoring the establishment of a lenga seed bank that ensures the forest regeneration.

Results presented here show the high silvopastoral potential of the Aysén and Magallanes forests. However, there is still lack of knowledge in the management of seeding forests with pastures of high fodder value, in the fertilization of the pastures associated to the forests and in the stocking rate and grazing time that maximize the grazing potential of the system without affecting the forest stability and natural regeneration, ensuring their perpetuity. Thus, the implementation of silvopastoral systems adapted to the specific conditions of these regions, should allow to transform these forests in a more productive and diverse ecosystems improving the quality of the timber and cattle products. The design of silvopastoral schemes requires a systemic approach of the components (forest, pairie and cattle).

**Acknowledgements** We sincerely wish to express our sincere thanks to Dr. Donald Mead for kindly reviewing this paper, and to Bernardo Acuña and Luigi Solis for assistance in the field and data collection.

## References

Arnold JEM (1983) Economic considerations in agroforestry. Agrofor Syst 1:229–311

Bahamonde H, Peri P, Monelos L, Martínez Pastur G (2011) Aspectos ecológicos de la regeneración por semillas en bosques nativos de Nothofagus antarctica en Patagonia sur, Argentina. Bosque [online] 32(1):20–29 (in Spanish)

CONAF – CONAMA (2006) Monitoreo y Actualización, Catastro de Uso del Suelo y Vegetación, Región de Magallanes y Antártica Chilena, 11p (in Spanish)

CONAF – UACH (2011) Catastro de los Recursos Vegetacionales Nativos de Chile. Monitoreo de Cambios, Corrección Cartográfica y Actualización del Catastro de Bosque Nativo en la XI Región de Aysén. Valdivia, Chile, 55p (in Spanish)

Donoso C (2006) Capítulo *Nothofagus Antarctica* en: Las Especies Arbóreas de los Bosques Templados de Chile y Argentina. In: Cuneo M (ed) Autoecología. Valdivia, Chile, pp 401–410 (in Spanish)

Hepp C, Thiermann H, Ramírez C (1988) Praderas en la zona austral XI Región (Aysén). In: Ruiz Núñez I (ed) Praderas para Chile. Instituto de Investigaciones Agropecuarias, Santiago, pp 624–638 (in Spanish)

INE (1997) VI censo nacional agropecuario. Santiago, Chile (in Spanish)

INFOR (2013) El sector forestal chileno 2013. Santiago, Chile (in Spanish)

Larraín O, Grosse H, Sotomayor A (2007) Manual de silvicultura en bosques nativos dominados por raulí, roble y coigüe para pequeños y medianos productores. Instituto Forestal (INFOR), Concepción, 21p (in Spanish)

Longhurst WM (1983) Wildlife interactions with domestic animals and forests. In: Hannaway DB (ed) Foothills food forest. Or. State Univ. Timber Press, Beaverton, pp 309–320

Martin M, Bahamóndez C, Quiroz I, Peña O, Barrientos M (2009) Manual N° 40. Inventario de Ecosistemas Forestales: Manual de Operaciones en Terreno. INFOR, Valdivia, 139p (in Spanish)

Oberdorfer E (1960) Planzensoziologische Studien in Chile. Flora et Vegetatio II (III):65–132 (in German)

Otero L (2006) La huella del fuego. Historia de los bosques nativos, poblamiento y cambios en el paisaje del sur de Chile. Pehuén Editores, Chile (in Spanish)

Peri PL (2005) Sistemas silvopastoriles en ñirantales. IDIA XXI Revista de información sobre investigación y desarrollo agropecuario INTA, pp 255–261 (in Spanish)

Peri PL, Sturzenbaum M, Monelos L (2005) Proyecto de Investigación Aplicada a los Recursos Forestales Nativos. Sistemas Silvopastoriles en Santa Cruz. CONICET, pp 417–443

Quam V, Johnson L (1999) Windbreaks for Livestock Operations. University of Nebraska

Ramírez C, Correa M, Figueroa H, San Martín J (1985) Variación de hábito y hábitat de *Nothofagus antarctica* en el centro-sur de Chile. Bosque 6(2):55–73 (in Spanish)

Rivest D, Paquette A, Moreno G, Messier C (2013) A meta-analysis reveals mostly neutral influence of scattered trees on pasture yield along with some contrasted effects depending on functional groups and rainfall conditions. Agric Ecosyst Environ 165:74–79

Salinas J, Ovando V, Acuña B (2011) Estándares de producción vegetativa en plantas de *Nothofagus antarctica* en la Región de Aysén. Instituto Forestal, Chile, 11p (in Spanish)

Sánchez-Jardón L, Acosta B, del Pozo A, Casado MA, Ovalle C, Elizalde HF, Hepp C, de Miguel JM (2010) Grassland productivity and diversity on a tree cover gradient in *Nothofagus pumilio* in NW Patagonia. Agric Ecosyst Environ 137:213–218

Sánchez-Jardón L, Acosta-Gallo B, del Pozo A, Casado MA, Ovalle C, de Miguel JM (2014a) Variability of herbaceous productivity along *Nothofagus pumilio* forest-open grassland boundaries in northern Chilean Patagonia. Agrofor Syst 88:397–411

Sánchez-Jardón L, del Pozo A, Casado MA, Ovalle C, de Miguel JM (2014b) Native and non-native herbaceous species dependence on tree cover in grazing systems from northern Chilean Patagonia. Agric Ecosyst Environ 184:41–50

SAS Institute Inc (2003) Proceedings of the Twenty-eight annual SAS users group international conference. SAS Institute Inc., Cary

Schmidt H (2008) Mejoramiento de la productividad ganadera a través del manejo silvopastoril de los bosques de ñire en la XII región. Informe Final, Proyecto FIA-PI-C-2004-1-F054. Santiago, Chile, 51p (in Spanish)

Schmidt H, Schmidt A, Alonso MF (2013a) Programa de Investigación, Mejoramiento, Productividad y Silvopastoreo en la XII Región. Informe Final, Proyecto BIP-30077120-0, Santiago, Chile, 211p (in Spanish)

Schmidt H, Schmidt A, Alonso MF (2013b) Manejo silvopastoril en los renovales de lenga en la Región de Magallanes. Santiago, Chile, 38p (in Spanish)

Sotomayor A (1989) Sistemas silvopastorales y su manejo. Documento técnico N° 42. Revista Chile Forestal, Diciembre 1989, CONAF, 8p (in Spanish)

Sotomayor A, Teuber O (2011) Evaluación del efecto de los arboles manejados bajo ordenación silvopastoral en los parámetros climáticos del sitio, en relación a un manejo ganadero sin árboles. Cienc Investig For 17(1):23–40 (in Spanish)

Sotomayor A, Moya I, Teuber O (2007) Análisis de un sistema silvopastoral con *Pinus contorta* (dougl. ex loud.) en la XI Región de Chile. Revista Ciencia e Investigación. Inst For 13(3):541–558 (in Spanish)

Sotomayor A, Moya I, Teuber O (2009a) Manual de establecimiento y manejo de sistemas silvopastorales en zonas Patagónicas de Chile. Manual N° 41, Instituto Forestal (INFOR), Coyhaique, Chile (in Spanish)

Sotomayor A, Teuber O, Moya I (2009b) Resultados y experiencia sobre manejo de sistemas silvopastorales en la Región de Aysén. In: Teuber O (ed) Sistemas agroforestales para la Región de Aysén: cortinas cortaviento y Silvopastoreo. Instituto de Investigaciones Agropecuarias, Coyhaique, 211p (in Spanish)

Sotomayor A, Moya I, Acuña B (2010) Comportamiento de las variables dasométricas en plantaciones de *Pinus contorta* Doug. Ex Loud., bajo manejo silvopastoral y forestal en la Región de Aysén, Chile. Cienc Investig For 16(3):265–290 (in Spanish)

Sotomayor A, Moya I, Teuber O (2011a) Diseño, establecimiento y manejo de cortinas cortaviento. Manual N° 43. Instituto Forestal, Santiago (in Spanish)

Sotomayor A, Teuber O, Moya I, Almonacid P (2011b) Productividad animal en un sistema silvopastoral con la especies *Pinus contorta*, en relación a un manejo ganadero sin árboles, en la región de Aysén, Chile. Cienc Investig For 17(2):139–152 (in Spanish)

Tejera L, Hansen N, Ferig M (2006) Efecto de la cobertura arbórea y del pastoreo vacuno sobre el establecimiento de la regeneración de *Nothofagus antarctica* (G.Forst.) Oerst. INTA EEA, Esquel, 7p (in Spanish)

Teuber O, Moya I y Sotomayor A (2008) Efecto de una cortina cortaviento sobre la producción de cultivos forrajeros en Aysén, Patagonia de Chile. Ciencia e Investigación Forestal 14(1):77–90 (in Spanish)

Teuber O (2009) Resultados y experiencia en producción de cultivos forrajeros con cortinas adultas en la Región de Aysén. In: Teuber O (ed) Sistemas agroforestales para la Región de Aysén: cortinas cortaviento y Silvopastoreo. Instituto de Investigaciones Agropecuarias, Coyhaique, 211p (in Spanish)

Teuber O, Ganderatz S (2009) Características geográficas y edafoclimáticas de la Región de Aysén. In: Teuber O (ed) Sistemas agroforestales para la Región de Aysén: cortinas cortaviento y silvopastoreo. Instituto de Investigaciones Agropecuarias, Coyhaique, pp 9–16 (in Spanish)

Tscharntke T, Klein AM, Kruess A, Steffan-Dewenter I, Thies C (2005) Landscape perspectives on agricultural intensification and biodiversity-ecosystem service management. Ecol Lett 8:857–874

Weinberger P (1973) The regeneration of the Araucano-Patagonic *Nothofagus* species in relation to microclimatic condition. Flora 162:157–179

# Silvopastoral Systems in the Cold Zone of Brazil

10

Alexandre Costa Varella, Raquel Santiago Barro, Jamir Luis Silva da Silva, Vanderley Porfírio-da-Silva, and João Carlos de Saibro

**Abstract**

The cold zone of Brazil occupies approximately 6 % of the national territory and is located between latitudes 24° S and 33° S. In this area, extensive cattle and sheep farming systems and conventional cropping and forestry are predominant. With the end of government subsidies by the decade of 1980s, an increase in farming production costs, a decrease of native forest covering, an increase of degraded areas in agriculture and livestock farming systems and a mismatch between timber national supply and demand after 1990s, an opportunity arises for integrate forestry with livestock and agriculture activities in Brazil, particularly in the southern. This chapter initially reports key events over the last three decades that have supported the increasing interest of farmers and enterprises on agroforestry activities, with focus on silvopastoral systems in the cold area of Brazil. Then, relevant advances on silvopastoral systems from research and extension services were reported, highlighting the screening of shaded adapted forage plants and management, trees species screening for silvipastoral systems and animal performance and behaviour under trees.

A.C. Varella, Eng. Agr. (Ph. D.) (✉)
Brazilian Agricultural Research Corporation (EMBRAPA), South Livestock Center (CPPSUL), BR 153, Km 603, 242, 96401-970 Bage, RS, Brazil
e-mail: alexandre.varella@embrapa.br

R.S. Barro, Eng. Agr. (D.Sc.)
J.C. de Saibro, Eng. Agr. (Ph. D.)
Faculty of Agronomy, Plant Forages Department, Federal University of Rio Grande do Sul (UFRGS), 7712 Bento Goncalves Ave., 91540-000 Porto Alegre, RS, Brazil
e-mail: raquelbarro@gmail.com; jnsaibro@terra.com.br

J.L.S. da Silva, Eng. Agr. (D.Sc.)
EMBRAPA Temperate Climate (CPACT), 403, Rodovia BR-392, Km 78, 9º Distrito, Monte Bonito, 96010-971 Pelotas, RS, Brazil
e-mail: jamir.silva@embrapa.br

V. Porfírio-da-Silva, Eng. Agr. (D.Sc.)
EMBRAPA Forestry (CNPF), 319 km 111, Estrada da Ribeira, Bairro Guaraituba, 83411-000 Colombo, PR, Brazil
e-mail: vanderley.porfirio@embrapa.br


P.L. Peri et al. (eds.), *Silvopastoral Systems in Southern South America*, Advances in Agroforestry 11, DOI 10.1007/978-3-319-24109-8_10

Finally, the chapter analyses the existent opportunities to increase silvopastoral areas in southern Brazil and future challenges for research, development and technology transfer.

**Keywords**
Southern Brazil • Agroforestry • Tree • Forage • Cattle • Sheep

## 10.1 Introduction

The "cold zone" in Brazil is located in the southern region of the country, between latitudes 24° S and 33° S, comprising the states of Rio Grande do Sul, Santa Catarina and the south-central region of Paraná, occupying an area of approximately 576,410 km$^2$ (6 % of the national territory). This region has two important Biomes: the Pampa located at the most southern part of Rio Grande do Sul and the Atlantic Forest in southern Brazil territory. The climate is classified as subtropical cold with frequent frosts in winter and hot summers. Average annual temperatures range from 12 to 20 °C with well-defined seasons, and the annual average rainfall varies from 1250 to 2000 mm well distributed throughout the year. In the cold zones of Brazil, adaptation to frosts is the most important plant survival factor. As the frosts may vary each year (Fig. 10.1), the time of occurrence of frosts is critically important with regard to development of vegetation. The most damaging are early frosts in autumn and late spring. They eventually reach plants that are not cold hardy. In southern Brazil, 200,355 km$^2$ are arable land, 48,430 km$^2$ cultivated pastures, 108,426 km$^2$ grasslands, 62,088 km$^2$ natural forests and 25,313 km$^2$ tree plantations. Of the total arable land, 87 % is being used in row crop agriculture (i.e. soybean, maize and rice mainly) and 10 % is devoted to fruits and horticulture and 3 % to cutting forages (IBGE 2006).

Brazil has the second largest forest cover in the world, equivalent to 14.5 % of the global forest area, surpassed only by Russia (FAO 2014). Out of the total area of the country (845.7 M ha = million ha), approximately 54.4 % are covered by native forest and only 0.8 % with plantations. Planted forests in Brazil occupy approximately 7.2 M ha, of which about 5.1 M ha are eucalyptus, 1.6 M ha pines, and 0.5 M ha other species (ABRAF 2010). According to the Brazilian Association of Planted Forest Producers (ABRAF), the southern region currently has 1.9 million ha of eucalyptus (*Eucalyptus* spp.), pine (*Pinus* spp.), and black wattle (*Acacia mearnsii*) plantations and 511,000 ha of agroforestry areas (IBGE 2006). Among the main products of the national forest industry, the cold region of Brazil has a great emphasis on the production of yerba mate (*Ilex paraguariensis*), firewood, sawmill logs and wood for the panel, pulp and paper industries, besides the black wattle bark for tannin extraction.

This puts Brazil in a strategic position in global environmental issues due to the great productive potential of timber and non-timber forest products. From 1967 to 1987, reforestation programs with tax incentives resulted in a staggering growth in the sector, allowing organization of activities and consolidation of the forest sector as of great importance to the country. Investments, which totaled about US$ 10 billion, according to the Brazilian Ministry of Environment (MMA), resulted in a surplus in the supply of wood during this period, as a result of significant technological development attained by the planted forest sector, raising productivity in plantations of pine and eucalyptus from 20 to 40 m$^3$ per hectare per year.

With the end of tax incentives in 1987, there was a drastic reduction in forest plantations, compromising the expansion of the sector and leading to a mismatch between timber supply and demand to meet the needs of projected growth by forest-based industry in the medium and long term. According to estimations of the Federal Government agencies and the Brazilian Society of Silviculture (SBS), Brazil will need about 275

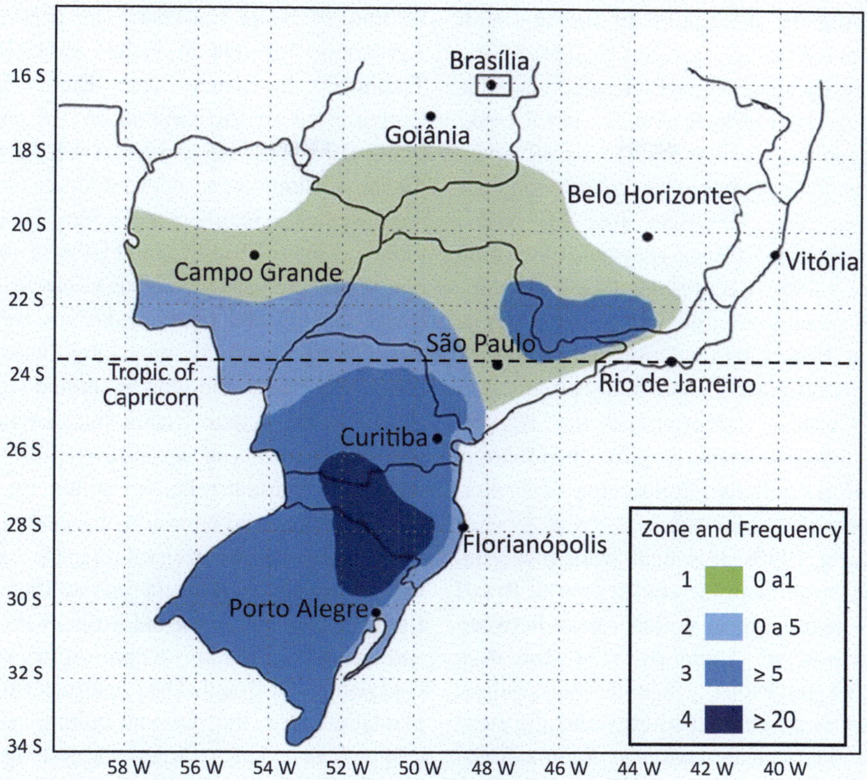


**Fig. 10.1** Cold zones and frequency of frosts in southern Brazil. Data are from historical averages (Adapted from Sentelhas and Angelocci 2012)

million m$^3$ of wood from planted forests by 2020 to meet the national demands of the pulp and paper, lumber, panels, charcoal and firewood industries. This means an estimated area of 8.7 M ha to reach future demand of wood in the country by 2020, according to SBS (2008) and MMA. As southern Brazil concentrates 33 % of the national forest planted forests, then RS, SC and PR states would need to increase from current 1.9 M ha (IBGE 2006) to 2.9 M ha by 2020 or about 71,000 ha per year from 2006 to 2020. The scenario also pointed out a marked imbalance between supply and demand, with imminent risk of a deficit in wood supply and negative consequences to primarily sawmills and the furniture industry. The crisis of wood from commercial forests, known as a national "forest blackout," strongly affected the forest industries. Therefore, since 2004, government new incentives through extensive credit and low interest rates were offered and a new cycle of expansion for planted forests occurred mainly with eucalyptus for the pulp and paper industries and pine for sawmill lumber and veneer in southern Brazil. Due to the favorable climate and soils, the high productive potential, the existence of high-tech production tools, the availability of land at a low acquisition cost and the simultaneous crisis of livestock and agricultural sectors in the early 2000s, national and international companies and producers increased the area of plantations by approximately 331,000 ha in the last 5 years, mainly with eucalyptus and black wattle on the border area with Uruguay, and pine in highland areas of the states of Rio Grande do Sul and Santa Catarina.

From the early 1990s, the increase in production costs (especially agricultural inputs), a decrease of native forest cover, an increase of degraded areas mainly through soil erosion consequent to an increase in agricultural activities and livestock operations, and the phenomenon of rural exodus and concerns about animal welfare, led to the search for new and more sustainable production systems in southern Brazil, particularly involving the integration of forestry with agriculture and livestock systems. As a result, the first agroforestry systems were initiated in southern Brazil. According to Radomski and Ribaski (2010), Paraná was the first State to use silvopastoral systems to any great extent, mainly on beef cattle farms. The Grevillea tree (*Grevillea robusta*) and species of Eucalyptus and Corymbia (*Corymbia citriodora*) represented the most common forest species identified in these systems. Associations between Eucalyptus and Grevillea with native trees such as Canafistula (*Peltophorum dubium*), Gurucaia (*Parapiptadenia rigida*), Guabiroba (*Campomanesia* sp.), Aroeira (*Schinus terebinthifolius*) and the Yellow Ipe (*Tabebuia vellosoi*) were also observed.

In the early 1990s, technical work involving agroforestry systems in the cold region of Brazil was begun, mainly from collaboration between research teams of Embrapa (the Brazilian Research Corporation), federal universities, southern states research institutes and the rural extension and technical assistance organizations. Also in the 1990s, several undergraduate and graduate courses such as Forestry, Agronomy, Animal Husbandry and Veterinary Science, as well as polytechnics and technical schools included in their academic curricula knowledge and training on agroforestry systems. In addition, between 1980 and 1994, 76 technical and scientific articles were published by workers at the southern research institutes, of which 70 % were on agrosilvicultural systems and 30 % on silvopastoral systems (Montoya et al. 1994). Also in the 1990s, the first Theses and Dissertations on agroforestry systems from federal universities in southern Brazil were published, allowing the formation of new knowledge and specialized human resources in this area.

Also since 2000, the technical and scientific experiences in agroforestry systems have multiplied in southern Brazil, especially with the work of scientists at Embrapa National Forestry Center (CNPF) and Sheep and Animal Husbandry Southern Center (CPPSUL), the Federal Universities of Rio Grande do Sul (UFRGS) and Paraná (UFPR), the Agricultural Research Foundation of Rio Grande do Sul (FEPAGRO), the Agricultural Research and Rural Extension Institute of Santa Catarina State (EPAGRI), the Agronomic Institute of Paraná (IAPAR) and at Technical Assistance and Rural Extension Corporations of Rio Grande do Sul and Paraná States (EMATER/RS and EMATER/PR). Among the most common and relevant topics observed in these studies were: tolerance of forage and agricultural species to shading; indication of timber and non-timber tree species for agroforestry systems; forest spacing and planting designs for agroforestry; tree productivity, interaction between trees, understory plants and grazing animals (cattle and sheep) in the systems; economic, environmental and social aspects of agroforestry; microclimate, water and nutrient relations within the systems; carbon balance and emission of greenhouse gases.

Another relevant observation in the past decade was the involvement of the private sector, mainly from the pulp and paper industries, with scientific and technical activities on agroforestry systems in cold regions of Brazil. The opportunity to increase plantation area, the regional culture on singular cropping or extensive livestock systems, and the increasing concerns about environmental issues, motivated forestry companies to initiate technical and financial collaboration with research institutions and extension to conduct applied research on the integration of forest-agriculture-livestock systems. In addition, from 2004, forestry companies started promoting partnerships with the State Government of Rio Grande do Sul and credit financial institutions to expand agroforestry areas using eucalyptus and several crops and forages in southern Brazil. It was during this period that the greatest increase in agroforestry planted area in southern Brazil occurred, estimated at 3 million hectares currently. These areas mainly involved the combination of eucalyptus with various agricultural crops (soybean [*Glycine max*], maize [*Zea mays*], sorghum [*Sorghum bicolor*], sunflower [*Helianthus annuus*], watermelon [*Citrullus lanatus*], wheat [*Triticum* spp.], barley [*Hordeum vulgare*]) with forage species grown (annual ryegrass [*Lolium multiflorum*], oats [*Avena* spp.], white clover [*Trifolium repens*], red clover [*Trifolium pratense*], birdsfoot trefoil [*Lotus cornicuatus*]), and ruminant animals (Fig. 10.2).

Models of integrating trees with agriculture and livestock, promoted by forestry companies in

**Fig. 10.2** Agroforestry systems used in southern Brazil: integration between eucalyptus and sorghum [*Sorghum bicolor*] at VCP Company (*above* and *left*), sunflower [*Helianthus annuus*] (*above* and *right*); integration between eucalyptus (*Eucalyptus grandis*), tropical grass (*Brachairia brizantha*), eucalypt, grass temperate pasture and beef cattle at a smallholder property in Paraná State (*below* and *left*) and integration between black wattle and *Digitaria diversinervis* grass at FEPAGRO (*below* and *right*) (Images from: A.C. Varella; V. Porfirio-da-Silva; Z.M. Castilhos)

southern Brazil, had occurred before in the southeast part of Brazil, mainly in Minas Gerais and São Paulo. However, these models had serious limitations in performing as a long term system in this southern environment, particularly in Rio Grande do Sul, because of the high tree densities used (above 1000 trees per hectare) and radiation conditions with lower photoperiod and a different solar angle (maximum 38° in July) in the winter period than tropical areas. These conditions combined with the rapid early growth of eucalyptus, imposes heavy shading to understory plants at these latitudes. In addition, these agroforestry models failed to prevent excessive shading on understory by pruning and thinning trees at 4–5 years after forest establishment. This resulted in short term agroforestry integration, allowing ruminant grazing under trees only for the first 3 years after trees had been planted. Then the area turned into an exclusive forest plantation, allowing grazing only in the border areas. This fact has discouraged many farmers and producers from Rio Grande do Sul, who considered that this model would turn their traditional animal production systems into conventional forestry rather than long-term integration systems. In 2010, the forest development programs were discontinued in southern Brazil because forest companies and environmentalists disagreed about environmental issues raised by the government of Rio Grande do Sul that were aimed at permitting establishment of more plantations. Also the lack of firm positioning from State Government for the consolidation of forestry sector in southern Brazil discouraged advances in forestry investments. These factors caused forest companies to revise their strategic plans, quit the expansion of forestry and agroforestry areas, particularly in Rio Grande do Sul, and change their investments to the tropical and central region of Brazil.

Conversely, forestry investments were maintained similarly in Santa Catarina and Parana over the last 5 years. In these States, agriculture is predominant and area for expansion of conventional forestry investments is limited, nevertheless there is still land area for the expansion of agroforestry systems.

The situation will likely remain the same in the near future. The southern region of Brazil will experience a decrease in the supply of raw material for forest products, particularly to meet the demands of the furniture industry, sawn wood, medium density fiberboard (MDF) and particleboard, joinery and carpentry business, pulp and paper industry, tannins, resins, and chemicals derived from black wattle, posts and treated wood for construction, and wood for energy and charcoal. New debates for the resumption of forestry investments returned to the public agenda of governments, forestry managers, technicians, producers and more recently companies. This time, trying to establish a new strategy based on territorial and sustainable development, as required by a modern society, with production systems adapted to the environmental, cultural and productive conditions of the southern region. Clear rules and better integration for negotiating parties (government, environmentalists, farmers and companies) are needed to set a new cycle of forestry and agroforestry growth in southern Brazil. Therefore, agroforestry systems may provide a sustainable alternative to all parties, leading to new challenges for stakeholders and encourage the return of investments and expansion of agroforestry area in southern Brazil by private sector. To achieve that objective, some challenges must be addressed:

- Proposition of feasible models for forestry-livestock farming systems in contrast to the traditional agricultural cultivation, the extensive cattle and sheep rangeland systems and the specialization of forestry systems commonly used on farms and by companies in southern Brazil;
- Implementation of new programs promoting agroforestry, integrating public and private sector and encouraging the establishment of low tree density (below 800 trees/ha) systems and offering financial credit compatible with the length of time required to obtain wood in these systems. Forest companies should change their strategy and focus on increasing the planting area (suppliers) by attracting more farmers (mostly rangeland producers) with the proposition to establish integrated systems. Animal producers are generally resistant to conversion of rangeland systems to a long-term profit system such as forestry plantations;
- Reversal of the current situation of little knowledge and appreciation for the role of trees in farming systems (commercial, environmental, animal welfare, landscape) by reinforcing curricula in schools and universities;
- To implement public policies that support a strategy of territorial management using geospatial tools (satellite imagery and georeferenced information system) capable of supporting decisions and rules for forestry and agroforestry business and reducing environmental concerns, particularly avoiding exceeded occupation of natural biomes in southern Brazil;
- Enhance the certification of forest products from sustainable systems such as agroforestry;
- Create massive technical training programs on agroforestry systems in the public and private sectors, associated with consulting firms, official technical assistance, rural extension, and cooperatives.

This chapter includes main results collected from agroforestry experiments in Rio Grande do Sul, Santa Catarina and Parana, including several $C_3$ and $C_4$ pastures (natives and exotics) performances under shade, beef cattle and sheep performance and behaviour under trees, climate-tree-pasture-animal interactions, and other relevant results from the last 20 years. As part of this challenge, this chapter aims to: (i) review the major scientific and technological advances in silvopastoral systems, with emphasis on silvopastoral systems developed by public and private institutions in southern Brazil; (ii) Report success cases from producers and forestry companies; (iii) propose alternatives for expanding the area of silvopastoral systems in southern Brazil in order to promote development with sustainable land management and providing opportunities to many services and multifunctional activities in farms, benefiting producers, industry and consumers involved in this chain.

## 10.2 Potential of Forage Plants for Silvopastoral Systems in Southern Brazil

The focus of research on silvopastoral systems in the late 1990s was to conduct experiments based on reductionist models, aiming to study binary interactions between the components of the integrated system, such as tree-pasture, soil-plant, plant-environment, etc. (Saibro 2001) and rarely contemplated the effect of the grazing animals (Saibro and Barro 2009).

Since then, a series of silvopastoral projects were developed with grasslands, which is an important and remarkable forage resource in the southern region of Brazil, particularly in Rio Grande do Sul State. These studies demonstrated that tree species, age, and density are factors that interfere with the growth of $C_3$ and $C_4$ species of the native grasslands. This occurs because of changes in light quantity and quality that passes through the trees. For example, Varella and Saibro (1999) found no differences on dry matter yield underneath three *Eucalyptus saligna* densities (816, 400 and 204 trees $ha^{-1}$) in the first and second years of a silvopastoral system at the experimental site of UFRGS. According to the author, the incident radiation was not limiting to understory vegetation because trees were approximately 2 m in height in this experiment. In addition, Fucks (1999), working at the same experimental site, reported decreases on grazing days, dry matter yield and stocking rates under the tree densities at years 3 and 4. In another UFRGS experiment, Silva (1998), working with winter pasture under *E. saligna*, reported that tree density influenced animal performance as pasture density and carrying capacity decreased, affecting beef cattle gains after the second year of tree establishment.

In the past, several studies focused on quantifying the effects of reduced radiation on the yield of understory pasture, seeking to determine limits of pasture growth and forage shade tolerance when altering tree densities. In these experiments, the shading levels were usually high after the second and third year as a consequence of the high tree populations established (800–1000 trees $ha^{-1}$), using fast growing trees and simple rows arrangement (Schreiner 1987; Silva et al. 1998; Fucks 1999; Varella and Saibro 1999; Lucas 2004). Despite these important advances, research on silvopastoral systems faced a discontinuation of support by financial agencies in the Brazilian subtropics and studies with a sequence of tree-pasture interactions in experimental sites were missed at that stage. Therefore, there is still lack of scientific information on pasture yield and management of silvopastoral system over the complete productive cycle of silvopastoral systems, particularly underneath fast growing tree species such as *Eucalyptus* sp. and *Pinus* sp.

The next step in silvopastoral research in southern Brazil was the investigation of the changes on pasture morphology, physiology, and nutritive value in response to shading, from the early 2000s (Barro et al. 2008; Soares et al. 2009). More recently, important scientific advances have been made in understanding the mechanisms of pasture adaptation and tolerance to shading from research groups in southern Brazil (Varella et al. 2011; Barro et al. 2012; Pontes et al. 2014; Baldissera et al. 2014). For instance, Pontes et al. (2014) and Baldissera et al. (2014) showed the relationship between soil N supply, plant height and light interception of six perennial tropical forages under trees, leading to important management practices of pasture intensity for silvopastoral systems. In addition, Barro et al. (2012) showed the N nutrition dynamics and forage yield of four important forages from southern grasslands in Brazil under shading and potential to use them in silvopastoral systems. This meant an extra step to better establish and manage pastures underneath trees and consequently, increase pasture productivity and quality in integrated systems.

The preliminary results indicated that more conservative management of pasture under shade has greater chances of success (longevity, yield, and quality) (Baldissera 2014). Overall, the management of understory pastures requires continuous investigation on silvopastoral systems in order to generate growth models as a function of available solar radiation and management applied. This would support pasture and grazing management decisions by farmers as trees develop in the system.

### 10.2.1 Overview of Forage Research in Shaded Environments

Grazing management can be established by controlling the frequency and intensity of defoliation (Pedreira et al. 2007). The combination of these variables greatly affects the structure of the pasture canopy (Da Trindade et al. 2007), which are also influenced by shading. However, only recently basic investigations concerning the influence of grazing management on shaded pastures and morphophysiological changes have been conducted in southern Brazil.

In agroforestry systems, grazing management should explore the phenotypic plasticity of forage species under shade. Scientists have recommended that forages under shade should be managed in order to maintain more residual leaf area or post-grazing heights greater than those in full sun. However, further studies on frequency and intensity of grazing pastures are required in silvopastoral systems, along with studies of pasture nutrient and water efficiencies in order to develop better pasture management strategies.

The interception of photosynthetically active radiation (iPAR) by the pasture canopy is the main variable used for determining the optimal timing for grazing defoliation (Silva and Carvalho 2005). In full sunlight, 95 % light interception by the pasture canopy results in optimum growth (Silva and Carvalho 2005; Da Trindade et al. 2007). Thus, consistent relationships between sward height on pre-grazing and iPAR by the canopy has being determined for several forage species in full sun (e.g. Zanini et al. 2012; Amaral et al. 2013), but rarely under shaded conditions. Finally, it has been pointed that this relationship may be different in silvopastoral systems than in open pastures as canopy photosynthesis decreases rapidly under shade and less carbon reserves are stored in roots (Barro et al. 2012; Varella et al. 2012). Therefore, more research also needs to be conducted to determine the optimum light interception point by pasture canopies under shade for cutting or grazing.

### 10.2.2 Perennial $C_4$ Forages

The adaptation of perennial forages to different levels of shading has been a target of many studies in Brazil for at least 20 years (e.g. Schreiner 1987). In the cold regions of the subtropics, tropical perennial forages stop growing during autumn and winter seasons and, eventually, susceptible genotypes are unable to survive over the frosty period in southern Brazil. The silvopastoral integration allows obtaining productive perennial $C_4$ pastures all year by minimizing climate extremes under the trees (Lucas 2004; Sartor et al. 2006; Kirchner et al. 2010). Even in the coldest areas of southern Brazil, where severe frosts are frequent from April to August, tropical pastures such as *Brachiaria* spp., *Panicum* spp. and *Cynodon* spp. can remain green and productive for longer periods under trees than they would tolerate in an open area (Fig. 10.3) and this represents an extraordinary feeding resource for cattle and sheep during extreme weather on farms.

Several studies have shown the potential yield of tropical forages under moderate shade in southern Brazil (Table 10.1). Research has shown the best growth responses, morphological and physiological adaptation and nutritive value of $C_4$ grasses under moderate shade (up to 40–50 % shading). In some cases, the shading effect may be very beneficial, resulting in higher dry matter yields compared to open pastures, especially in soil water and nitrogen deficit conditions,

**Fig. 10.3** Experiment evaluation of *Panicum maximum* under *Eucalyptus* spp. (*left side*) and in full sun (*right side*) at FEPAGRO Unit in Rio Grande do Sul. Photos were taken at the same moment after a severe frosting event on September, 12th 2003 (Images from J. C. Saibro)

depending on tree density and type of soil. For example, recent studies have shown the unique response of *Paspalum regnellii* (Table 10.1), highlighting its potential yield under severe shading (up to 80 % shade) and the improvement of forage nitrogen content (Barro et al. 2012). As shown in Table 10.1, the study of $C_4$ perennial forages under shade is highlighted in southern Brazil because of the potential of early production in spring after the second year of establishment and high growth rates in summer when native and annual pastures are usually water stressed. Also, perennials are preferred by farmers over annual pastures for silvopastoral systems as the process of germination and emergence are difficult in shaded environments and pasture production costs are higher (Varella et al. 2011).

In addition to *P. regnellii*, other native grasses from Pampa and Atlantic Forest Biomes have shown potential for use in shaded environments. These species were initially identified at the transitional area between native forests and grasslands of southern Brazil, such as *P. regnellii* (Barro et al. 2012), *Axonopus catharinensis* (Soares et al. 2009; Barro et al. 2008) as well as forage species from open sites which showed high plasticity under shade, such as *Paspalum notatum* (Soares et al. 2009), *P. dilatatum* (Barro et al. 2012) and *Arachis pintoi* (Barro et al. 2014). However, further research advances with these perennial native forages, particularly evaluation under grazing, have faced difficulties because of the lack of seed of commercial cultivars in the market. Research institutions are putting great efforts into getting registration of these native forage cultivars by the Brazilian Ministry of Agriculture, Livestock and Food Supply and to establish seed fields of these cultivars to enable farmers to increase silvopastoral systems based on $C_4$ pastures.

### 10.2.3 Annual and Perennial $C_3$ Forages Adapted to Shade

Besides the perennial $C_4$ grasses, the animal production systems in southern Brazil are characterized by the cultivation of winter forage grasses. The most commonly grown pasture species used are annual ryegrass (Silva et al. 1998; Barro et al. 2008; Kirchner et al. 2010), forage oats-(Deiss et al. 2014) and the dual purpose cereals (wheat, barley, triticale and oats). Currently, these species are objects of research in various institutions (Table 10.2), in different production systems, with or without the presence of trees and with the possibility of being used in rotational systems, particularly with cash crops (soybean and maize) over the summer or composing an agroforestry system.

**Table 10.1** Summary of $C_4$ pastures tolerance to shade from several studies in southern Brazil

| Forage cicle | Carbon fixation | Growing season | Species | Shade level tested | Type of shading | Relative yield to full sun | Effect on nutritive value | References |
|---|---|---|---|---|---|---|---|---|
| Perennial | C4 | Summer | *Axonopus catharinensis* | 24 and 56 % | Tree shade | 52 and 50 % | Not determined | Barro et al. (2008) |
| | | | | 17 and 33 % | Tree shade | 79 and 47 % | Increased | Soares et al. (2009) |
| | | | | 40 to 50 % | Tree shade | 74 % | Not determined | Pontes et al. (2012) |
| | | | *Cynodon* sp. (cv.Tifton 85) | 40 to 50 %[a] | Tree shade | 25 and 50 % | Not determined | Pontes et al. (2012) |
| | | | | 24 and 56 % | Tree shade | 61 and 51 % | Not determined | Barro et al. (2008) |
| | | | | 17 and 33 % | Tree shade | 48 and 24 % | Increased | Soares et al. (2009) |
| | | | *Brachiaria brizantha* cv Marandu | 17 and 33 % | Tree shade | 86 and 36 % | Increased or mantained | Soares et al. (2009) |
| | | | | 40 to 50 % | Tree shade | 45 % | Not determined | Pontes et al. (2012) |
| | | | *Brachiaria decumbens* cv. Basilisk | 17 and 33 % | Tree shade | 48 and 24 % | Not determined | Soares et al. (2009) |
| | | | *Brachiaria decumbens* | 25, 50 and 80 % | Artificial shade | 85, 44 and 22,7 % | Not determined | Schreiner (1987) |
| | | | *Digitaria decumbens* | 25, 50 and 80 % | Artificial shade | 88, 60 and 25 % | Not determined | Schreiner (1987) |
| | | | *Panicum maximum* cv. Aruana | 17 and 33 % | Tree shade | 54 and 9 % | Increased | Soares et al. (2009) |
| | | | | 40 to 50 %[a] | Tree shade | 57 and 20 % | Not determined | Pontes et al. (2012) |
| | | | *Panicum maximum* cv. Tanzânia | 17 and 33 % | Tree shade | 61 and 0 % | Not determined | Soares et al. (2009) |
| | | | | not determined | Tree shade | 25 % | Not determined | Ferreira et al. (2006) |
| | | | *Panicum maximum* cv. Mombaça | 17 and 33 % | Tree shade | 47 and 15 % | Not determined | Soares et al. (2009) |
| | | | | not determined | Tree shade | 22 % | Not determined | Lucas (2004) |
| | | | Paspalum notatum cv. Pensacola | 17 and 33 % | Tree shade | 87 and 26 % | Not determined | Soares et al. (2009) |
| | | | | 40 and 50 % | Tree shade | 54 and 10 % | Not determined | Pontes et al. (2012) |
| | | | | 25, 50 and 80 % | Artificial shade | 93, 55 e 13 % | Not determined | Schreiner (1987) |
| | | | *Paspalum notatum* | 50 and 80 % | Artificial shade | 112 and 81 %[a] | Increased | Barro et al. (2012) |
| | | | *Paspalum dilatatum* | 50 and 80 % | Artificial shade | 117 and 81 %[a] | Increased | Barro et al. (2012) |
| | | | *Paspalum regnelii* | 50 and 80 % | Artificial shade | 118 and 99 %[a] | Increased | Barro et al. (2012) |
| | | | *Hemarthria altissima* | 25, 50 and 80 % | Artificial shade | 99 and 22 % | Not determined | Schreiner (1987) |
| | | | Hemarthria altissima cv. Flórida | 17 and 33 % | Tree shade | 66 and 5 % | Not determined | Soares et al. (2009) |
| | | | | 40 to 50 % | Tree shade | 7 % | Not determined | Pontes et al. (2012) |
| | | | | 24 and 56 % | Tree shade | 58 and 43 % | Not determined | Barro et al. (2008) |

[a]Under severe drought

**Table 10.2** Summary of $C_3$ pastures tolerance to shade from several studies in southern Brazil

| Forage cicle | Carbon fixation | Growing season | Species | Shade level | Type of shading | Relative yield to full sun | Effect on nutritive | References |
|---|---|---|---|---|---|---|---|---|
| | | | | 30 e 60 % | Tree shade | 43 and 22 % | Increased | Kirchner et al. (2010) |
| | | | *Lolium multiflorum* | 24 and 56 % | Tree shade | 46 and 36 % | Increased | Barro et al. (2008) |
| | | | | 33 and 58 % | Artificial shade | 87 and 84 | Not determined | Saibro (1992)[a] |
| | | | | 30 e 60 % | Tree shade | 38 and 13 % | Increased | Kirchner et al. (2010) |
| | | | *Avena sativa* | 24 and 56 % | Tree shade | 75 amd 42 % | Increased or mantained | Barro et al. (2008) |
| Annual | C3 | inverno | *Avena strigosa* | 30 and 60 % | Tree shade | 43 and 8 % | Piora | Kirchner et al. (2010) |
| | | | | 24 and 56 % | Tree shade | 75 and 42 % | Decreased | Barro et al. (2008) |
| | | | *Holcus lanatus* | 50 and 80 % | Artificial shade | 234 and 223 %[a] | Not determined | Varella et al. (2008) |
| | | | *Bromus auleticus* | 50 and 80 % | Artificial shade | 737 and 605 %[a] | Not determined | Varella et al. (2008) |
| | | | *Bromus catharticus* | 50 and 80 % | Artificial shade | 205 and 159 %[a] | Not determined | Varella et al. (2008) |
| | | | *Triticum aestivum*-duplo proposito | 30 and 60 % | Tree shade | 46 and 25 % | Decreased | Kirchner et al. (2010) |
| | | | *Vicia villosa* | 30 and 60 % | Tree shade | 48 and 27 % | Mantained | Kirchner et al. (2010) |
| | | | *Arachis Pintoi* (hybrid ecotype) | 50 and 80 % | Artificial shade | 93 and 60 % | Decreased | Barro et al. (2014) |
| Perennial | C3 | verão | *Arachis pintoi* cv. Alqueire | 17 and 33 % | Tree shade | 37 and 18 % | Not determined | Soares et al. (2009) |
| | | | *Arachis pintoi* cv. Amarillo | 17 and 33 % | Artificial shade | 59 and 23 % | Not determined | Soares et al. (2009) |
| Perennial | C3 | inverno | *Trifolium repens* cv. Zapicân | 24 and 56 % | Tree shade | 25 and 40 % | Decreased | Saibro et al. (2008) |
| | | | *Lotus comiculatus cv.* São Gabriel | 24 and 56 % | Tree shade | 27 and 25 % | Decreased | Saibro et al. (2008) |

[a]Under severe drought

For this group of pastures, Porfirio-da-Silva (2012) found a decrease in forage yield in silvopastoral systems compared to full sun. For instance, the winter pastures *Avena strigosa* and *A. sativa* intercropped with *Lolium multiflorum* were evaluated in two integrated systems: with and without the presence of trees. Forage yield in full sun averaged 2524 kg DM ha$^{-1}$ and was higher than under shade which yielded 2210 kg DM ha$^{-1}$. However, beef cattle liveweight gains were similar between the treatments, ranging from 0.55 to 1.10 kg LW day$^{-1}$. In addition, Kirchner et al. (2010) reported yields of several winter forages varying from 38 to 48 % under 15×3 m *Pinus taeda* system (30 % shade) and from 8 to 27 % under 9×3 m (60 % shading) compared to open pastures in Santa Catarina State. Likewise, annual ryegrass performance was highlighted in this study, yielding 3478 kg DM ha$^{-1}$ and showing 18 % of total crude protein under the 15×3 m system.

The responses of $C_3$ legumes on DM yield and nutritive value are individual and dependent on their agronomic performance under shade. Results as to the quantitative and qualitative performance of winter forage legumes in single or mixture pastures and under different shading levels were obtained in the subtropics (Table 10.2). For example, white clover (*Trifolium repens*) and birdsfoot trefoil (*Lotus corniculatus*) yields were similar under 15×3 m *Pinus taeda* system (moderate shading) compared to full sunlight (Barro et al. 2006) growing on a sandy soil of the coastal area in Rio Grande do Sul State. In a different experimental site, Sartor et al. (2006) reported that birdsfoot trefoil yielded 2844 and 2669 kg DM ha$^{-1}$ under the 15×3 m and 9×3 m *Pinus taeda* system, respectively, whereas in an open pasture yielded 8121 kg DM ha$^{-1}$ on clay soil of Santa Catarina State. In sequence at the same experimental site, Kirchner et al. (2010) showed that the annual legume forage *Vicia villosa* yielded 2300 kg DM ha$^{-1}$ under intermediate shade (15×3 m) and 1292 kg DM ha$^{-1}$ under the heavy shade (9×3 m) compared to 4771 kg DM ha$^{-1}$ in full sun. In artificial shade conditions, Barro et al. (2010) reported that *Arachis pintoi* yield was affected by intense shade (80 %), but still showed potential to increase nutritive value of natural grass-legume mixtures under intermediate (50 %) shade level. Under intense shading (80 %), *Arachis pintoi* showed a decrease of 40 % in dry matter yield compared to full sun. However, under moderate shading (50 %), this legume performance was not affected (Table 10.2).

The performance of forage legumes at different levels of radiation has shown that these are usually less shade tolerant than grasses (Watson et al. 1984; Barro et al. 2012), although this is not necessarily a rule (Johnson et al. 2002) and may change under soil nitrogen stress. The large, productive and reproductive success of forage legumes in agroforestry depends on their ability to adapt to decreasing levels of light with tree canopy closure with advancing age (Balocchi and Phillips 1997). Physiologically, the forage legumes operate as C3 plants, and could potentially support shading at intermediate levels. However, quite different responses have occurred with regard to the agronomic performance of forage legumes under shade. There is still inadequate scientific information regarding the performance of legumes under shade in subtropical environments of Brazil (Varella et al. 2009).

## 10.3 Potential of Trees for Silvopastoral Systems in Southern Brazil

The trees most commonly used in silvopastoral systems in cold regions of Brazil are species adapted to climate and soil conditions of the region, with relatively rapid growth rate (about 2 m high per year) and with great value of the wood in the market (Table 10.3). Historically the integration of beef cattle with forests date from the mid-eighteenth century (Chang 1985), known as traditional "faxinais" systems, mainly established in the area of occurrence of Araucaria native forests (*Araucaria angustifolia*). The first studies of silvopastoral systems in the Brazilian subtropics sought to utilize livestock as a secondary component of the system. Cattle were introduced in conventional tree plantations as a strategy to improve cash flow in the early years of

**Table 10.3** Predominant tree species applied in silvopastoral systems in cold zones of Brazil

| Number of frost events | Tree species or hybrids used in silvipastoral systems | Tree species tested in experimental studies |
|---|---|---|
| 0–1 | *Eucalyptus urophylla*; *E. urophylla x E. grandis*; *Corymbia citriodora*; *C. camaldulensis*; | Tropical Pinus (Gutmanis 2004) |
| | | Native species (Melotto et al. 2009; Nicodemo et al. 2010) |
| 0–5 | *E. grandis*; *E. urophyllax E. grandis*;*C. citriodora*; *C. camaldulensis*; *Grevillea robusta*; | Native species (Melotto et al. 2009; Nicodemo et al. 2010) |
| | | *Kaya ivorensis* e *Toona ciliata* (Porfírio-da-Silva et al. unpublished data) |
| | | *Leucaena* spp. (Sampaio et al. 2009) |
| ≥5 | *E. dunnii*; *E. benthamii*; *Acacia mearsii*;*E. grandis*; *E. saligna*; *Grevillea robusta*; *Araucaria angustifolia* | Native species (Radomski and Ribaski 2010; Porfírio-da-Silva et al. 2012) |
| | | *Populus* spp. (Otto et al. 2009) |
| | | *Pinus elliotti* (Ribaski et al. 2005) |
| ≥20 | *E. benthamii*; *Pinnus taeda*; *P. elliottii*; *Araucaria angustifolia* | *Pinus taeda* (Soares et al. 2009) |

forest cultivation, besides getting the benefits of controlling the development of unwanted plants in the understory and reducing the risks of fire inside the forest (Baggio and Schreiner 1988; Schreiner 1994; Varella and Saibro 1999; Silva et al. 2001).

Although there is potential for use of many other tree species, *Eucalyptus* spp. is the most commonly planted one in cold regions of Brazil, followed by *Grevillea robusta* (Martins et al. 2015), *Acacia mearsii*, and subtropical Pinus spp. The genus Eucalyptus is important for Brazil because of the raw material for pulp production, coal, wood, panels, posts, poles, sawn timber, furniture, packaging and other commercial uses. This tree species and its different cultivars are well adapted to southern Brazil conditions and can compose silvopastoral systems in almost all territories. In Zones 1 and 2 (Fig. 10.2), successful plantings have been carried out with hybrids between *Eucalyptus grandis* x *E. urophylla*; especially in the western portions of these zones, where frequent droughts can also occur during the summer and autumn seasons. Because of the rapid growth of these hybrids during the first spring and summer seasons, plants can tolerate the cold temperatures in the first winter period. *E. grandis* has been planted successfully in the southern portions of Zone 2 and north- northwest of Zone 3 (Higa and Wrege 2010), except on the coastal area where the species may be affected by various diseases caused by high relative humidity (Alfenas et al. 1983). The *Corymbia citriodora* and *C. camaldulensis* species has also been observed in silvopastoral systems across Zone 2, except at the coastal area where the tree-pasture systems are rare. The *E. benthamii* and *E. dunnii* are more tolerant to low temperatures and this is the main reason for them to be widely planted in silvopastoral systems of Zone 3. *E. benthamii* is the most tolerant species to frosts among all species of Eucalyptus sp. (Jovanovic and Booth 2002; Paludzyszyn Filho and Santos 2013) and has been planted in Zone 4. *E. saligna* also has been planted in integrated systems in this area (Silva 1998; Varella 1997). In addition to *E. benthamii*, subtropical species of pine (*Pinus elliottii* and *P. taeda*) have also been used for silvopastoral systems in Zone 4. These pines have also been planted in Zone 3 of southern Brazil.

In Brazil, soil and water conservation is a priority when planning the introduction of trees into pastures. This means that the orientation of the tree rows during system establishment must be a priority in order to promote soil and water conservation. Observation of the apparent path of the sun in the sky to orient the arrangement of tree rows is secondary to consideration of soil and water conservation issues. The solar energy that reaches the land in cold zones of Brazil is high, ranging from 4 to 6 $kW.hm^{-2}$ (Ceballos and Bottino 2006). The region where the incidence of solar radiation is lowest (4.5 kW.hm$^{-2}$) partially coincides with Zone 4 in Fig. 10.2, and it is distributed along to the coastline and nearby Itajai River valley, and this fact has been associated with increasing cloudiness in these regions.

The spatial arrangement of trees is critical for the success of a silvopastoral system. Establishment is facilitated by proper spatial distribution of trees in the field and this should aim to promote soil and water conservation, favor the transit of machines and benefit ruminants' thermal comfort. According to several experiences in Brazil (Table 10.4), the most effective arrangement is the alleys, where trees are planted in strips (single or multiple rows) with wide spacing between strips. This basic arrangement can also be adjusted according to the commercial use of wood (Porfirio-da-Silva et al. 2009). For the tree component, there basically are two primary considerations in a silvopastoral system: (i) production of a great volume of fine wood (firewood, charcoal, stanchions, podiums, etc.) in the initial part of the trees rotation (i.e. about 6 years for Eucalyptus sp.). In this case, establishment should use the highest number of trees per area possible without limiting understory growth, usually between 600 and 1000 trees per hectare; and (ii) production of thick wood (for sawmill, rolling, etc.) in the final part of trees rotation (over 10 years old, depending on species and site) by planting few trees per area, i.e. about 250 trees per hectare. The different arrangements may be planted more closely spaced and managed by thinning trees over the development period to produce wood for different purposes (thin wood in the early years of the silvopastoral system and timber in the final years of the rotation).

In cold zones of Brazil, silvopastoral systems are commercially oriented and the predominant tree species used are grown to produce wood products (Table 10.3). The productivity and the quality of tree products in silvopastoral systems depends on the genetics involved (trees, forage crops, and livestock), the quality of the site (climate and soil), the spatial arrangement of trees, and management practices applied on arboreal, forage and animal components. Few studies have analyzed the productivity and quality of tree components in silvopastoral systems compared to conventional plantations in southern Brazil. For example, a 4 year silvopastoral system initiated with the goal of producing wood logs at 424 trees ha$^{-1}$ showed mean annual increment of 33.2 m$^3$ ha$^{-1}$ year$^{-1}$, whereas for the clone *E. urophylla* × *E. grandis* on a monoculture at 1111 trees ha$^{-1}$ was 43.5 m$^3$ ha$^{-1}$ year$^{-1}$. In terms of biomass produced in the system at 1111 trees ha$^{-1}$ was the most productive. However, the volume per tree at 424 trees ha$^{-1}$ was the greatest at 0.313 m$^3$ per tree, whereas at 1111 trees ha$^{-1}$ volume reached only 0.156 m$^3$ per tree (Medeiros, unpublished data).

## 10.4 Animal Performance and Behaviour under Trees

The use of grazing animals in commercial forests must be planned taking into account that these activities require distinct management strategies. These integrated systems aim to improve production per unit area, whereas respecting the principle of sustained yield, maintaining the productive potential of renewable natural resources, and social and economic conditions of the local community (Silva and Saibro 1998). However, within a systemic view, the animal component demands attention because the positive interaction between

**Table 10.4** Regular tree spacing and arrangement applied in silvopastoral systems in southern Brazil

| | Thin wood (charcoal, firewood, fence posts) | | | Logs (timber) | | |
|---|---|---|---|---|---|---|
| Tree arrangement | Tree spacing (m) | Trees per ha[a] | % Area occupied per row | Tree spacing (m) | Trees per ha[a] | % Area occupied per row |
| Single row | 14×2 | 357 | 14 | 14×4 | 179 | 14 |
| Single row | 14×2 | 357 | 14 | 28×4 | 89 | 7 |
| Double row | 14×2×3 | 417 | 25 | 18×3 | 185 | 11 |
| Triple row | 14×3×1.5 | 1000 | 40 | 20×3 | 167 | 10 |

[a]Tree mortality was not included over the time (Adapted from Porfírio-da-Silva et al. 2009)

tree-animal-pasture in a silvopastoral system is driven by correct decisions with regard to stocking rate or grazing intensity and this is related to carrying capacity of understory pastures, as well as grazing behavior.

### 10.4.1 Animal Behavior and Performance under Eucalypt-Pasture Systems

The first study of animal performance in a silvopastoral system with eucalyptus in Rio Grande do Sul State was conducted in an area of 180 ha and tree spatial arrangement of 2×3 m. From June 1992 to March 1993 a test with 87 zebu heifers, up to 2 years old and average liveweight of 221 kg was conducted during which the animals were rotationally grazed. The animals gained 0.567 kg per day with a gain per area of 12.55 kg $ha^{-1}$ (Silva et al. 1993). The total global radiation at the ground level inside the system ranged from 18 % in summer to 22 % in winter, compared to full sunlight (Ferreira et al. 1993). The forest allowed better environment for animal welfare, because, as compared to open pasture, temperature was warmer in winter and cooler in summer (Krause et al. 1993).

In another study, the impact of cattle and sheep grazing during the establishment phase of an *Eucalyptus saligna* forest with different tree densities was evaluated (204, 400 and 816 trees $ha^{-1}$) in the Central area of Rio Grande do Sul (Varella 1997; Varella and Saibro 1999). The authors concluded that the damage caused by cattle and sheep to trees was severe when tree height was less than 182 cm for weaned calves and 154 cm for lambs. Therefore, grazing could start at 6–7 months of eucalypt age in this region. This study also showed that cattle frequently caused damage by chewing leaves and tips of lateral or apical branches, trampling seedlings, and by rubbing or scratching seedlings. The previous experience of the animals to graze under these systems and the availability and quality of understory pasture also affected the extent of damage to seedlings. The authors concluded that animals could work as efficient biological tools to control herbaceous competition to seedlings at the establishment phase and this would be more economical than the use of pre and post emergent herbicides, in addition to reducing the risks of fire inside the forest. In a subsequent study, the impact of grazing on the growth of *Eucalyptus saligna* over the establishment phase was evaluated. Fucks (1999) showed that sheep grazing grassland under trees with an offer of 10 % (i.e. 10 kg of DM per 100 kg LW per day) resulted in significant increase in tree height and diameter at breast height (DBH) and this produced a greater volumetric increment rate and better quality of forest product at the three eucalypt densities. These results showed a benefit of animal presence and positive effects on tree development due to a decrease of understory herbaceous competition.

In the central area of Rio Grande do Sul, Silva et al. (1998) studied the effects of three grazing intensities (forage allowance of 6 %, 9 % and 13 % i.e. 6, 9 and 13 kg of pasture dry matter per 100 kg LW per day) and two eucalypt densities (1666 and 833 trees $ha^{-1}$) on animal behavior and performance (Fig. 10.4). This experimental site consisted of *Eucalyptus saligna* and annual ryegrass (*Lolium multiflorum* L.) + arrowleaf clover (*Trifolium vesiculosum* Savi cv. Yuchi) overseeded on native pasture. Continuous grazing with steers started when trees were at an average of 2.30 m height (9 months old) and stocking rates were adjusted monthly according to forage supply. The author reported that survival and growth of trees (tree height and diameter) were not affected by stocking at all grazing intensities and tree densities. Only 4.4 % of trees were damaged by steers, but without affecting further tree growth. It was shown that grazing management in silvopastoral systems, without limiting forage intake and quality of available forage, resulted in good animal performance and tree development. In addition, the mean daily gain (MDG) of steers was represented by a linear model, whereas the mean gain per area (MGA) was a curvilinear relation under 1666 trees $ha^{-1}$ (Fig. 10.4a). Under the 833 trees $ha^{-1}$, both MDG and MGA were fitted by curvilinear models (Fig. 10.4b). Silva

**Fig. 10.4** Relationship between mean daily gain (*MDG*) and mean gain per area (*MGA*) with pasture allowance (kg DM per 100 kg liveweight) under 1666 (5a) and 833 (5b) eucalypt trees per hectare and respective stocking rates (5c) in a silvopastoral system with eucalypt, annual ryegrass and arrowleaf clover (*Trifolium vesiculosum*). Data are average of three replicates collected from 6th September and 9th November 1995 at UFRGS

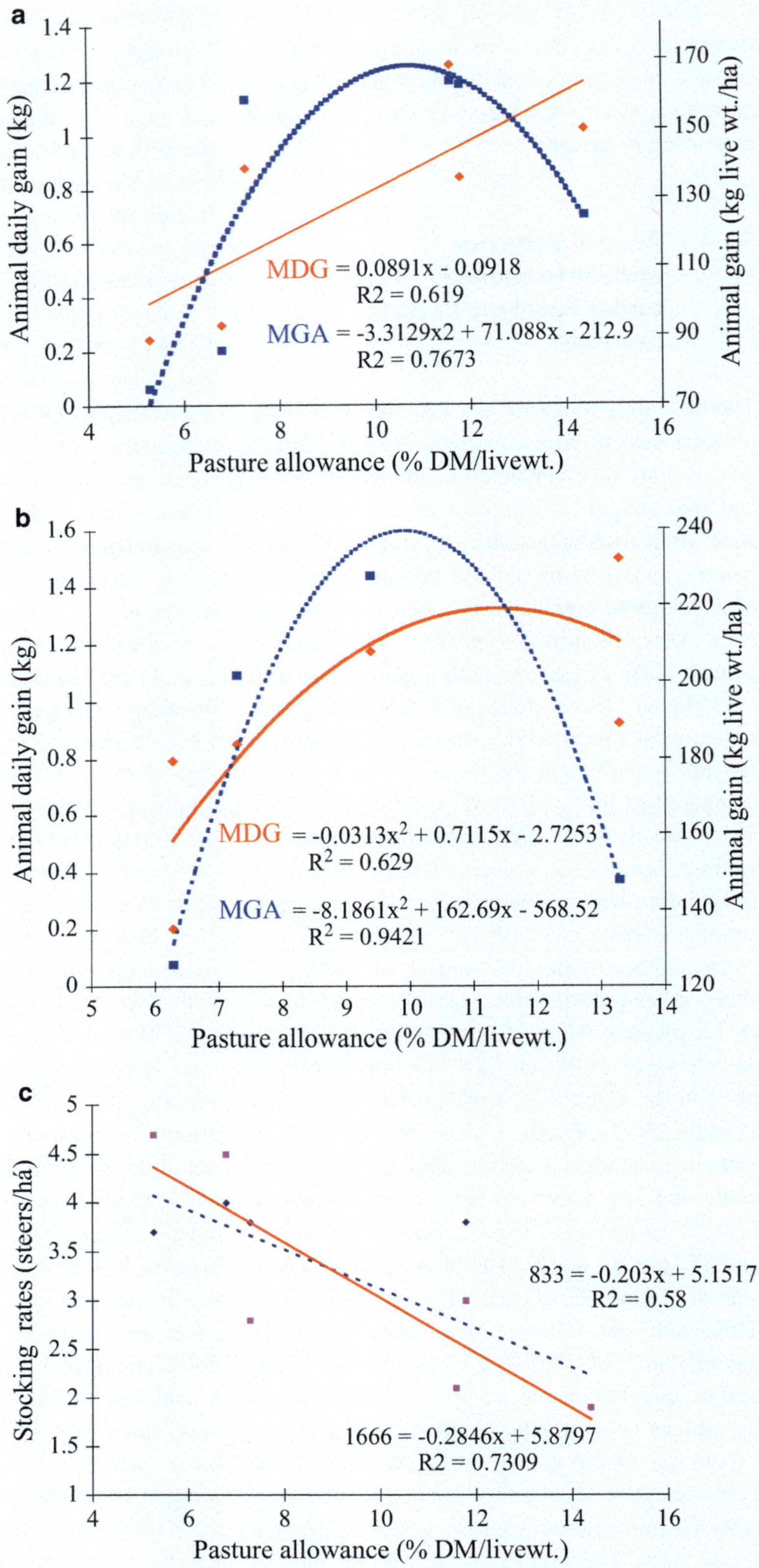

(1998) concluded that gain per animal in a silvopastoral system increased with forage supply. The gain per area decreased at the lowest pasture allowance because of reduced individual performance by animals. Above 11 % of pasture allowance, MGA was also reduced due to low stocking rates (Fig. 10.4c). At intermediate stocking rate the optimum animal LWG per head and per area were obtained. For instance, the animal performance was about 220 kg LWG per ha at 10 % of herbage allowance during the first grazing period (i.e. 64 grazing days) under 833 trees $ha^{-1}$. In contrast, under 1666 trees $ha^{-1}$, animal performance was 161 kg LWG per ha at the same grazing intensity level. The best animal performance obtained in 2 years of experiment time was 455 kg LWG per ha under 833 trees $ha^{-1}$ with stocking rates adjusted to 10 % of pasture allowance. This performance was 108 % higher than under 1666 trees $ha^{-1}$, showing the importance of low densities to get best results in silvopastoral systems. As a reference, average animal production on open native pastures is about 70 kg LWG per ha per year at the same location in southern Brazil (Maraschin 1998). Under the lowest tree population, animal production was 30 % greater than the highest tree population in the first year after tree establishment and this result doubled by the second year of the experiment. In addition, grazing was interrupted under 1666 trees $ha^{-1}$ at tree age of 1.5 years, mainly due to strong shading to understory pasture (Silva et al. 1996). The conclusion of this study was that optimum pasture allowance in a silvopastoral system ranged from 9 to 11.5 % (Silva et al. 2011), whereas in open grasslands of southern Brazil, ranges between 9.5 and 12 % have been reported (Moraes et al. 1995; Maraschin 1998).

### 10.4.2 Animal Behavior and Performance under Black Wattle-Pasture Systems

The first study involving cattle grazing under black wattle was conducted in a commercial forest at 2 years old during winter period in Rio Grande do Sul State in 1992. The conclusions of this work were: (i) beef cattle were well adapted under the Acacia forest and were harmless to trees when stocking rates were controlled at intermediate levels; (ii) the integration of forestry and cattle grazing was profitable, showing a mean gross margin of 6.28 %; (iii) the costs of maintaining the forest and livestock are minimized by sharing infrastructure and human resources.

In 1995, a long-term experiment was established by the Rio Grande do Sul State Foundation for Agricultural Research (FEPAGRO) in collaboration with UFRGS and a private forestry company. This study was located in Tupanciretã (RS) with the aim to evaluate the interactions at soil-plant-animal-microclimate interface in a silvopastoral system with two populations of *Acacia mearnsii* and three understory $C_4$ pastures: *Brachiaria brizantha*, *Panicum maximum* cv. Gatton, and grassland infested by Annoni grass (*Eragrostis plana*), an undesirable weedy species. Beef cattle grazed pastures continuously with stocking adjustments to maintain pasture allowance between 10 and 12 %. The results showed 66 kg LWG per ha grazing on *P. maximum* and 33 kg LWG grazing on *B. brizantha* pastures during 63 days in winter under a 2×3 m *Acacia mearnsii* forest. In contrast, mean gain per ha increased to 75 and 60 kg grazing *P. maximum* and *B. brizantha*, respectively, under a

**Table 10.5** Average daily gain (LWG) per animal and per hectare and stocking rate in silvopastoral systems with black wattle and perennial $C_4$ pastures grazed for 63 days in the winter

| Understory pasture | Tree density | | | | | |
|---|---|---|---|---|---|---|
| | 1667 trees $ha^{-1}$ (2×3 m) | | | 1000 trees $ha^{-1}$ (2×5 m) | | |
| | Kg LWG (per day) | Kg LWG (per ha) | Heads/ha | Kg LWG (per day) | Kg LWG (per ha) | Heads/ha |
| *P. maximum* | 0.536 | 66 | 2.2 | 0.627 | 75 | 2.6 |
| *B. brizantha* | 0.750 | 33 | 1.6 | 0.726 | 60 | 2.8 |

Data are averages of two replicates. FEPAGRO experimental site in Tupanciretã/RS (Castilhos et al. 1999)

2 × 5 m Acacia forest (Table 10.5). These systems proved to be efficient during the winter season when beef cattle usually lose LW in open grasslands of southern Brazil. Over summer, animal performance under trees increased to 169 and 195 kg LWG per ha under the 2 × 5 m Acacia forest grazing *P. maximum* and *B. brizantha*, respectively (Table 10.6). The authors concluded that the silvopastoral systems with black wattle were sustainable, showing potential increase in animal production and income, particularly at low tree densities, for beef farms in Rio Grande do Sul.

At year 6 in this study, both tree densities were thinned to 50 % of the original population and *B. brizantha* and *E. plana* pastures were replaced by *Panicum maximum* cv. Aruana and *Digitaria diversinervis*, respectively, in the experimental plots. At this stage, Castilhos et al. (2009) reported that black wattle showed similar volume production ($m^3$ $ha^{-1}$) with or without grazing at 7 years after planting. For instance, Castilhos et al. (2009) showed that the volume of wood produced at this experimental site was 166, 143, 86 and 51 $m^3ha^{-1}$ under the Acacia populations of 1667, 1000, 833 and 500 trees $ha^{-1}$, respectively. In addition, animal performance on *D. diversinervis* showed the greatest daily liveweight gain of all pastures as a result of higher nutritive value compared to *P. maximum*. However, this $C_4$ pasture provided the lowest grazing-days as a result of a remarkable decrease in pasture yield during the water stress period occurred in summer. According to the authors, this result explained the lowest liveweight gain per area and carrying capacity of *D. diversinervis* of all pastures under trees. In contrast, both *P. maximum* cultivars showed greater carrying capacities and gains per hectare compared to *D. diversinervis* (Table 10.7). The conclusion of this study was that silvopastoral systems using populations of *Acacia mearnsii* lower than 833 trees $ha^{-1}$ compromised wood production, whereas increased animal performance grazing underneath $C_4$ pastures. From this work, it was also shown that Annoni Grass (*E. plana*) was shade intolerant. Several authors report Annoni Grass as the main weed that degrades southern Brazil rangelands and occupy about 20 % of the total area. Therefore, the tree-pasture systems become a sustainable alternative to recover those infested areas by using shade tolerant pastures combined to trees.

Finally, there is still a lack of scientific information on grazing behavior and animal welfare measured under silvopastoral systems. In the cold subtropical area of Brazil, livestock face extreme temperatures in winter and summer. Tree protection combined with accumulated forage may be strategic for sheep, dairy and beef farms which usually lose production at these stages. In the near future, this information may be available because a recent long term experimental site has been initiated at EMBRAPA South Livestock aiming to recover degraded grasslands by Annoni Grass and to evaluate thermal comfort of Brangus (Angus × Nelore cattle breed) at critical climate periods under silvopastoral systems.

### 10.4.3 Animal Behavior and Performance under Mixed Tree Crop-Livestock System

In southern Brazil, integration of trees, crops, and livestock in rotational systems are still unusual. Experiences have been applied by forestry companies and farmers more recently. This usually involves rotational systems between annual crops (maize, soybean, Sorghum spp., winter cereals), annual fruits (watermelon, melon, pumpkin) and pastures under spaced eucalypt plantations (Balbino et al. 2011; De Melo 2012). Likewise, experiments on mixed tree-crop-livestock systems have become frequent in scientific institutions of southern Brazil. For instance, an experiment with crop-livestock-forestry integration, using a mixture of tree species and winter pastures has been conducted since 2006 in Parana State. The tree components were *Eucalyptus dunnii*, *Schinus terebinthifolius* and *Grevillea robusta* and this was implemented in the Model Farm Station of IAPAR (Porfírio-da-Silva 2012). The trees were planted in an alternate arrangement within the row. The spatial arrangement was simple rows, with spacing of 14 m × 3 m, allocated across the predominant direction of the slope due to soil conservation issues. Since tree establish-

**Table 10.6** Pasture residual dry matter (RDM), average daily gain (ADG), average gain per area (AGA), pasture carrying capacity (CC) and stocking rate (STOCK) in silvopastoral systems with two densities of black wattle and three $C_4$ pastures

| Understory pasture | Tree density | | | | | | | | | |
|---|---|---|---|---|---|---|---|---|---|---|
| | 1666 trees/ha (2×3 m) | | | | | 1000 trees/ha (2×5 m) | | | | |
| | RDM (kg/ha) | ADG (kg/hd/day) | AGA (kg/ha) | CC (kg/ha/day) | STOCK (hd/ha) | RDM (kg/ha) | ADG (kg/hd/day) | AGA (kg/ha) | CC (kg/ha/day) | STOCK (hd/ha) |
| *P. maximum* cv. Gatton | 2422A[b] | 0.644A | 104 | 372 | 1.70 | 3200A | 0.696A | 169 | 520 | 2.55 |
| *B. brizantha* cv. Marandu | 1720B | 0.573AB | 105 | 394 | 1.85 | 2995A | 0.690A | 195 | 698 | 1.85 |
| *E. plana* | 1182B | 0.539AB | 95 | 418 | 1.80 | 1417B | 0.417B | 122 | 741 | 3.25 |
| Mean[a] | 1775a | 0.585a | 101a | 395a | 1.78a | 2537b | 0.601a | 162b | 653b | 2.55b |

Pasture allowance was maintained between 10 and 12 % during the period of 13th November 1998 to 18th February 1999. Data are averages of two replications. FEPAGRO experimental site in Tupanciretã/RS (Silva et al. 1999)

[a]Means followed by the same letter in the row, between the tree densities, do not differ at 5 % probability level by Duncan test

[b]Means followed by the same capital letter in the column, for pastures and tree densities interactions, do not differ at 5 % probability level by Duncan test

**Table 10.7** Average daily gain (ADG), Average gain per area (AGA), average daily gain per area (ADGA), grazing-days per hectare (GD) and real pasture allowance (PA) in silvopastoral systems with two populations of *Acacia mearnsii* and three $C_4$ pastures

| Pasture | ADG[a] | AGA[a] | ADGA[a] | GD[a] | PA |
|---|---|---|---|---|---|
| | kg/hd | kg/ha | kg/ha/day | An.day/ha | kg DM/100 kg LW |
| *P. maximum* cv. Gatton | 0.738 a | 337.6 a | 3.13 a | 445.5 a | 13.9 a |
| *P. maximum* cv. Aruana | 0.799 a | 328.2 a | 3.04 a | 406.3 a | 11.9 a |
| *D. diversinervis* | 0.844 a | 289.7 a | 3.11 a | 333.9 a | 9.4 a |
| Mean | 0.794 | 318.5 | 3.09 | 395.2 | 11.7 |

Pasture allowance was maintained at 12 % between 1st December 2003 and 18th March 2004. Data are averages of the two tree densities and replications. FEPAGRO experimental site at Tupanciretã/RS (Lucas 2004)
[a]Means followed by the same letter in the column do not differ significantly by the F test at 5 % probability

ment, the underneath area has been managed with corn and soybeans for grain production in summer and under a no till system. In winter, a mixture of black oats and annual ryegrass pastures (*Avena strigosa* and *Lolium multiflorum*) has been sown in sequence to summer crops. This experiment evaluated two levels nitrogen fertilization (90 kg $ha^{-1}$ and 180 kg $ha^{-1}$ N) on understory pasture. In this agroforestry system, grazing started at 41 months of trees age. At this stage the author reported that all tree species received damage by cattle, but severity was great only on *S. terebinthifolius*, therefore this tree species was considered unsuitable for integration systems (Porfirio-da-Silva et al. 2012). When trees were 29 months old, corn yield was similar under the three systems, with a mean grain production of $4.1 \pm 0.3$ ton $ha^{-1}$ and mean increment of wood production was 1.03 $m^3$ per hectare over the corn cycle. In addition, the productivity of a soybean crop under trees at 56 months of age was 3.7 ton grain $ha^{-1}$ or 19 % less than in crop-livestock integration conducted in an adjacent area, whereas the increase in wood production was 6.6 $m^3$ $ha^{-1}$ over this crop cycle. The mean pasture DM yield was 2210.3 kg $ha^{-1}$ and the mean daily gain of steers was $0.86 \pm 0.31$ kg $ha^{-1}$ $day^{-1}$ or $440.6 \pm 75.9$ kg $ha^{-1}$ over the two grazing cycles (Porfirio-da-Silva 2012). The author observed that fertilization of winter pastures combined with the maintenance of a residual pasture height of 20 cm was essential for cattle production and strategic for no till management of summer crops.

## 10.5 Challenges for Research, Development and Technology Transfer on Silvopastoral Systems

Silvopastoral systems are dynamic and complex, particularly when considering the multiple interactions between trees, pasture and livestock in time and space (Balbino et al. 2011). The experience of implementation of integrated crop-livestock-forest systems in the last 25 years indicates the need for new and adapted models of production, technical assistance and rural extension for assuring the sustainability of Brazilian agriculture. To achieve that, it is important that research institutions conduct long-term experiments to investigate and transfer technologies to farmers and technicians continuously. Therefore, it is imperative that institutions develop joint programs for Research and Development (R&D) and Technology Transfer in silvopastoral systems since trees establishment period to the harvest of forest products. In addition, it is also important to understand that public and private technical advice require more training and qualification to manage these dynamic and sustainable systems.

In southern Brazil, the potential areas for silvopastoral systems are usually the ones currently occupied with extensive beef cattle and sheep extensive systems. To increase interest in this type of integration, it is essential that R&D offer suitable models that match producer interests and needs of the region. In this respect, issues related to animal welfare, the strategic forage supply during periods of extreme weather, and soil conservation should be highlighted in R&D programs. Besides, the self-consumption of wood in rural properties for energy and constructions, as well as for direct sale may also attract livestock producers to adopt these integrated systems. Therefore, the following opportunities and challenges for research and technology transfer in silvopastoral systems in southern Brazil are:

- To study new designs and spacing between trees for silvopastoral systems to allow minimum radiation level of 50 % on understory plants throughout the tree cycle;
- To offer alternatives of perennial and annual forage species and mixtures tolerant to shade, as well as climate and soil conditions in southern Brazil;
- To develop new forage cultivars adapted to shade;
- To quantify the benefits of shading for animal welfare (thermal comfort, performance and grazing behavior);
- To find profitable alternatives for selling or processing of timber and other forestry products from low tree density systems;
- To develop genetic improvement of trees for integration systems, highlighting plant architecture, biological cycle and quality of harvesting product;
- To develop silvopastoral system models involving native trees and forage species to meet requirements of the Brazilian environmental laws (legal reserves) on farms;
- To improve management of farms using silvopastoral systems;
- To develop and implement new training programs for technical and extension staff for silvopastoral systems on farms;
- To quantify environmental benefits provided by silvopastoral systems (carbon balance, mitigation of greenhouse gases, soil and water conservation);
- To develop soil nutrient tables to support annual fertilization decisions for tree-pasture systems;

Besides long term experimental areas and continuous research support from agencies, it is important that institutions invest in multidisciplinary teams, capable of responding to the opportunities and challenges mentioned. Important research institutions in southern Brazil, such as units of EMBRAPA, Federal Universities, and State Research Organizations should cooperate to achieve these goals as quick as possible. A collaborative network of scientists, extension agents, and consultants should develop demonstrative areas on farms for continuous training and technology transfer developed by research institutions. Currently, in southern Brazil, there is a reasonable physical structure available for national and international research institutions able to carry on long term and collaborative works on silvopastoral systems located at different sites on this region, such as:

- The silvopastoral system experimental site located at EMBRAPA South Livestock Systems (CPPSUL): established in 2013 and located in Bagé, Rio Grande do Sul State, border area with Uruguay. It's an area of 34 ha, containing three levels of radiation on the native pasture (full sun; 800 and 400 trees $ha^{-1}$), subdivided into two management systems (an intensive system using improved pasture with cool season forage species (annual ryegrass, red clover and birdsfoot trefoil) and a high fertility soil opposed to a conservative system using selective application of herbicide and livestock rate control) with the aim to recover degraded native pasture infested by the Annoni Grass weed (*Eragrostis plana*). The *Eucalyptus grandis* trees were established in triple rows spaced 2 m between plants and 3 m between rows. The spacing between the triple rows is 34 m or 14 m, resulting in a model of $3 \times 2 \times 34$ m and $14 \times 3 \times 2$ m.
- The experimental Farm Canguiri of the Federal University of Paraná (UFPR): located in Pinhais City, State of Parana. This area uses different integration models in order to obtain results over the complete cycle of the system: (i) integrated crop-livestock-trees system; (ii) crop-livestock system; (iii) Livestock-trees system; (iv) Crop-tree system and (v) integrated tree-pasture system. The trees were planted in single rows with *Eucalyptus*

*benthamii* spaced 14 m between rows and 2 m within the rows at an initial density of about 250 trees ha$^{-1}$. For comparative purposes, there is still a monoculture system (plantation) established with at 3×2.5 m, resulting in an initial density of 1333 trees ha$^{-1}$ and a final density of 800 ha$^{-1}$.

- The Model Farm of the Agronomic Institute of Paraná (IAPAR): located in Ponta Grossa City in the South Central Region of the Paraná State. This experimental site has three major areas of study:(i) an experimental area for crop-livestock-tree system. It was established in 2006 with *E. dunnii*, *Schinus terebinthifolius* and *Grevillea robusta* trees, planted in single rows and spaced at 14×3 m, resulting in 238 trees ha$^{-1}$. In 2013 the *S. terebinthifolius* was removed from experimental area because leaves and branches were noticed to be palatable to grazing animals, therefore barking and chewing became limiting for the integration (Porfírio-da-Silva et al. 2012). The total area of the experiment comprises 10.9 ha and was splitted into six paddocks of crop-livestock integration and other six of tree-crop-livestock system. Over the cool season, the pasture areas between trees rows were sown with a mixture of forage oats and annual ryegrass for grazing, followed by soybeans and maize in summer; (ii) to make experiments in plots with understory plants and between the double rows of *Eucalyptus dunni*. It was established in 2007 in the density of 330 trees ha$^{-1}$ (21×4×3 m). After thinning in 2011, the density decreased to 155 trees ha$^{-1}$ (Pontes et al. 2012).; (iii) to make experiments with different densities and spatial arrangements of trees, *Eucalyptus benthamii* was established in 2008 in single, double and triple rows of trees. The spacing between rows is 21 m, resulting in the following spatial arrangements: 21×2 (single line); 21×2×3 (double line) and 21×2×3 (triple line).

**Acknowledgements** The authors acknowledge the research groups from UFPR, UFRGS, UTFPR Pato Branco, FEPAGRO and IAPAR) for have kindly offered information about the Experimental sites and progressing studies.

We also acknowledge Professor Emeritus Don Ball from Auburn University for English and for technical review of this chapter and valuable comments on the text.

## References

Alfenas AC, Jeng RS, Hubbes M (1983) Virulence of *Cryphonectria cubensis* on Eucalyptus species differing in resistance. Eur J Plant Pathol 13(4):197–205

Amaral MF, Mezzalira JC, Bremm C et al (2013) Sward structure management for maximum short-term intake rate in annual ryegrass. Grass For Sci 68:271–277

Associação Brasileira de Produtores de Florestas Plantadas (ABRAF) Anuario Estatistico da ABRAF 2010 ano base 2009. Brasilia/DF, 2010. 140p. Available in: http://www.ipef.br/estatisticas/relatorios/anuario-ABRAF-2010-BR.pdf. Accessed 3 Nov 2015

Baggio AJ, Schreiner HG (1988) Análise de um sistema silvipastoril com *Pinus elliottii* e gado de corte. In: Boletim de Pesquisa Florestal (16). Available from EMBRAPA. http://ainfo.cnptia.embrapa.br/digital/bitstream/CNPF-2009-09/4854/1/baggio.pdf. Accessed July 2015

Balbino LC, Cordeiro LAM, Porfirio-Da-Silva V et al (2011) Evolução tecnológica e arranjos produtivos de sistemas de integração lavoura-pecuária-floresta no Brasil. Pesq Agrop Brasileira 46(10):1–12

Baldissera TC (2014) O ambiente luminoso: do impacto no crescimento e desenvolvimento em nível de planta forrageira a dosséis em sistemas integrados de produção agropecuária. Thesis, Universidade Federal do Paraná

Baldissera TC, Pontes LS, Barro RS et al (2014) Altura do dossel forrageiro de seis espécies perenes tropicais em sistema arborizado. In: 1st Simpósio Internacional de Arborização de Pastagens em Regioes Subtropicais, Colombo: Embrapa Florestas

Balocchi OA, Phillips CJC (1997) Grazing and fertilizer management for establishment of Lotus uliginosus and Trifolium subterraneum under Pinus radiata in southern Chile. Agrofor Syst 37:1–14

Barro RS, Saibro JC, Silva JLS et al (2006) Rendimento de matéria seca de cinco forrageiras de ciclo hibernal sob duas densidades arbóreas de *Pinus elliotti* e ao sol pleno. In: Proceedings of 43rd Reunião Anual da Sociedade Brasileira de Zootecnia, João Pessoa

Barro RS, Saibro JC et al (2008) Rendimento de forragem e valor nutritivo de gramíneas anuais de estação fria submetidas a sombreamento por Pinus elliottii e ao sol pleno (Forage yield and nutritive value of cool-season annual forage grasses shaded by Pinus elliottii trees and at full-sun). R Bras de Zootec 37:1721–1727

Barro RS, Varella AC, Bangel FV et al (2010) Screening native C4 pasture genotypes for shade tolerance in Southern Brazil. In: Proceedings of 15th Australian

Society of Agronomy Conference, Lincoln, November 2010

Barro RS, Varella AC, Lemaire G et al (2012) Forage yield and nitrogen nutrition dynamics of warm-season native forage genotypes under two shading levels and at full sun. Braz J Ann Sci 41:1589–1597

Barro RS, Saibro JC, et al (2014) Morphological acclimation and canopy structure characteristics of Arachis pintoi under reduced light and in full sun. Trop Grassl 2:15–17

Castilhos ZMS, Silva JLS, Guterres E et al (1999) Desempenho de espécies forrageiras de verão em sistema silvipastoril com acácia negra (*Acacia mearnsii* de Wild). In: Proceedings of 36th Reunião anual da Sociedade Brasileira de Zootecnia, Porto Alegre

Castilhos ZMS, Barro RS, Savian JF et al (2009) Produção arbórea e animal em sistemas silvipastoril com acácia-negra (*Acacia mearnsii*). Braz J Med Biol Res 60:39–47

Ceballos, JC, Bottino, MJ (2006) Solar radiation in South America, period 1998-2004: Some aspects of a satellite-based data base. In: XIV Congresso Brasileiro de Meteorologia, 2006, Florianópolis, SC. Anais do XIV Congresso Brasileiro de Meteorologia. Available in:http://www.cbmet.com/cbm-files/14-06214951bf3c0dff8374177933b83865.pdf. Accessed 3 Nov 2015

Chang MY (1985) Faxinais no Paraná. Instituto Agronômico do Paraná – IAPAR, Londrina

Da Silva SC, Bueno AAO, Carnevalli RA et al (2009) Sward structural characteristics and herbage accumulation of *Panicum maximum* cv. Mombaça subjected to rotational stocking managements. Sci Agric 66(1):8–19

Da Trindade JK, Da Silva SC, Souza SJ Jr et al (2007) Composição morfológica da forragem consumida por bovinos de corte durante o rebaixamento do capim-marandu submetido a estratégias de pastejo rotativo. Pesq Agrop Brasileira 42(6):883–890

Deiss L, Moraes A, et al (2014) Oat tillering and tiller traits under different nitrogen levels in an eucalyptus agroforestry system in subtropical Brazil. Cienc Rural (1):71–78

IBGE Instituto Brasileiro de Geoestatística (2006) Sistema IBGE de Recuperação Automática. Banco de Dados Agregados. Estabelecimentos na agropecuária: unidades Brasil IBGE, Rio de Janeiro. http://www.sidra.ibge.gov.br/bda/pecua/default.asp?z=t&o=24&i=P

Ferreira VP, Schaffer G, Krause CA et al (1993) Microclima de floresta de eucalipto no Rio Grande do Sul: II- distribuição da radiação solar. In: Proceedings of VIII Congresso Brasileiro de Agrometeorologia, Porto Alegre, July 1993

Ferreira ASJF, Valentim CMS, Andrade GML et al (2006) Distribuição da biomassa de raiz de genótipos de Panicum spp. nas condições ambientais do Acre. In: Anais Reunião Anual da Sociedade Brasileira de Zootecnia, 43, SBZ, João Pessoa, PB. 2006. Available in: http://www.sbz.org.br/reuniaoanual/anais/?idiom=en. Accessed 8 Jan 2016

Filho Paludzyszyn E, dos Santos PET (2013) Escolha de cultivares de eucaliptos em função do ambiente e do uso. Comunicado Técnico (316). Available from EMBRAPA. http://ainfo.cnptia.embrapa.br/digital/bitstream/item/97498/1/CT-316-Escolha-de-cultivares.pdf. Accessed 05 July 2015

Fucks LFM (1999) Dinâmica da pastagem nativa, desempenho de ovinos e desenvolvimento arbóreo em sistema silvipastoril com três populações de Eucalyptus saligna. Thesis, Universidade Federal do Rio Grande do Sul

Gutmanis D (2004) Estoque de carbono e dinâmica ecofisiológica em sistemas silvipastoris. Thesis, Universidade Estadual Paulista

Higa RCV, Wrege M (2010) Zoneamento climatico de *Eucalyptus grandis* para a região Sul do Brasil. In: Documentos (209), Available from EMBRAPA. http://www.infoteca.cnptia.embrapa.br/infoteca/bitstream/doc/883452/1/Doc209.pdf. Accessed 05 July 2015

Johnson SE, Sollenberger LE et al (2002) Nutritive value of rhizoma peanut growing under varying levels of artificial shade. Agron J 94:1071–1077

Jovanovic T, Booth T (2002) Improved species climatic profiles: a report for the RIRDC/L&W Australia/FWPRDC/MDBC Joint Venture Agroforestry Program. In: RIRDC Publication (02/095). Available from https://rirdc.infoservices.com.au/items/02-095

Kirchner R, Soares AB, Sartor L et al (2010) Desempenho de forrageiras hibernais sob distintos níveis de luminosidade. Braz J Vet Res Anim Sci 39:2371–2379

Krause CA, Aragones RS, Ferreira VP et al (1993) Microclima de floresta de eucalipto no Rio Grande do Sul: I- temperatura, umidade relativa e velocidade do vento. In: Proceedings of VIII Congresso Brasileiro de Agrometeorologia, Porto Alegre

Lucas NM (2004) Desempenho animal em sistema silvipastoril com acácia-negra (*Acacia mearnsii* de wild.) e rendimento de matéria seca de cultivares de panicum maximum jacq. sob dois regimes de luz solar. Thesis, Universidade Federal do Rio Grande do Sul, Porto Alegre, RS

Maraschin GE (1998) Utilização, manejo e produtividade das pastagens nativas da região Sul do Brasil. In: Proceedings of Ciclo de Palestras em Produção e Manejo de Bovinos de Corte. Ênfase: manejo e utilização sustentável de pastagens, ULBRA, Canoas/RS

Martins EG, Neves EJM, Sturion JA et al (2015) Volume de madeira e ganho genético de progênies de grevílea (*Grevíllea robusta* Cunn.) de segunda geração na região de Londrina, Paraná. Boletim de Pesquisa Florestal (51). Available from EMBRAPA. http://ainfo.cnptia.embrapa.br/digital/bitstream/item/117556/1/boletim51-pag45-55.pdf. Accessed 05 July 2015

Melo IB (2012) Integração lavoura-pecuária-floresta no norte do Rio Grande do Sul: estudo de caso. In: Fontaneli RS, dos Santos HP, Fontaneli RS (eds)

Forrageiras para integração lavoura-pecuária-floresta na região sul-brasileira, 2ed, Brasília, pp 461–487
Melotto A, Nicodemo MLF, Bocchese RA et al (2009) Sobrevivência e crescimento inicial em campo de espécies florestais nativas do brasil central indicadas para sistemas silvipastoris. Rev Árvore 33(3):425–432
Montoya LJV et al (1994) Aspectos de arborização de pastagens e viabilidade técnica e econômica da alternativa silvipastoril. In: Luciano J. Montoya Vilcahuaman; Moacir J. Sales Medrado. (Org.). I Seminário sobre Sistemas Agroflorestais na Região Sul. Curitiba Paraná 5:157–171
Moraes A, Maraschin GE, Nabinger C (1995) Pastagens nos ecossistemas de clima subtropical pesquisas para o desenvolvimento sustentável. In: Proceedings of Simpósio sobre Pastagens nos Ecossistemas Brasileiros: Pesquisa para o Desenvolvimento Sustentável, Brasília
Nicodemo MLF, Porfírio-Da-Silva V, Pezzopane JRM et al (2010) Desenvolvimento de árvores nativas em sistemas silvipastoril e agrossilvipastoril na região Sudeste. In: Proceedings of 47th Reunião Anual da Sociedade Brasileira de Zootecnia, Salvador
Otto GM, Motta ACV, Reissman CB (2009) Adubação nitrogenada em sistema silvipastoril álamo: pastagens de inverno. Rev Árvore 33(3):443–441
Pedreira BC, Pedreira CGS, Da Silva SC et al (2007) Estrutura do dossel e acúmulo de forragem de Brachiaria brizantha cv. Xaraés em resposta a estratégias de pastejo. Pesq Agrop Brasileira 42(2):281–287
Pontes LS, Giostri AF, Baldissera T et al (2012) Interactive tree and N supply effect on six C4 forage grasses productivity. In: II International symposium on integrated crop-livestock systems, Porto Alegre
Pontes LS, Baldissera TC, Barro RS, et al (2014) Interactive N supply and cutting intensity effect on canopy height at 95% light interception. In: General meeting of the European Grassland Federation
Porfirio-Da-Silva V (2012) Produtividade em sistema de integração lavoura-pecuária-floresta no subtrópico brasileiro. Thesis, Universidade Federal do Paraná
Porfirio-Da-Silva V, Medrado MJS, Nicodemo MLF et al (2009) Arborização de pastagens com espécies florestais madeireiras: implantação e manejo. Embrapa Florestas, Colombo
Porfírio-Da-Silva V, Moraes A, Moletta JL et al (2012) Danos causados por bovinos em diferentes espécies arbóreas recomendadas para sistemas silvipastoris. Pesq Florestal Brasileira 3(70):67–76
Radomski MI, Ribaski J (2010) Excentricidade da medula em Grevillea robusta e Corymbia citriodora cultivados em sistema silvipastoril. Embrapa Florestas. Comunicado técnico, 248 Colombo: Embrapa Florestas, 4 pp
Ribaski J, Dedececk RA, Mattei VL et al (2005) Sistemas silvipastoris: estratégias para o desenvolvimento rural sustentável para a metade sul do Estado do Rio Grande do Sul. Embrapa Florestas Comunicado Técnico 150
Saibro JC (1992) Programa de estímulo à integração de grupos e centros de pesquisa com o setor empresarial. Integração Silvipastoril de Eucalipto com Pastagens na Depressão Central no Rio Grande do Sul. (Proc. No 91/1684-3). Annual Technical Report, 1992. Fapergs/Riocell S.A./Ufrgs, Porto Alegre
Saibro JC (2001) Animal production from tree-pasture association systems in Brazil. In: International Grassland Congress, Lecture FEALQ, São Pedro
Saibro JC, Barro RS (2009) Indicadores biofísicos e sustentabilidade em sistemas silvipastoris In: Workshop Integração Lavoura-Pecuária- Floresta No Bioma Pampa, Embrapa Clima Temperado (ed). Pelotas 2009
SBS. Sociedade Brasileira de Silvicultura (2008) Fatos e Números do Brasil Florestal. Dezembro, 2008. Available in: http://www.sbs.org.br/FatoseNumerosdoBrasilFlorestal.pdf. Accessed 14 Sept 2014
Saibro JC, Medeiros R, Silva JL et al (2008) Rendimento de forragem e valor nutritivo de gramíneas anuais de estação fria submetidas a sombreamento por Pinus elliottii e ao sol pleno (Forage yield and nutritive value of cool-season annual forage grasses shaded by Pinus elliottii trees and at full-sun). Braz J Ann Sci 37:1721–1727
Sampaio OB, Schaffrath VR, Pinto ER. (2009) Sistema silvipastoril em pastagens naturais e artificiais. In: Congresso Brasileira de Sistemas Agroflorestais, 7. Proceedings 2009, Luziânia-GO, 5p. Available in: http://www.sct.embrapa.br/cdagro/tema02/02tema07.pdf. Accessed 7 Sept 2014
Sartor LR, Soares AB, Adami P et al (2006) Produção de forragem de espécies de inverno em ambiente sombreado. SynScy 1:13–21
Schreiner HG (1987) Tolerância de quatro forrageiras a diferentes graus de sombreamento. Pesq Agropec Bras 61–72
Schreiner HG (1994) Relatório de consultoria técnica em agrossilvicultura. Centro Nacional de Pesquisa de Florestas/EMBRAPA, Colombo/PR, Brazil, 76
Sentelhas PC, Angelocci LR (2012) Geada: ocorrência, prevenção e controle. Slide class of discipline Agricultural Meteorology LCE306. Available in: http://www.leb.esalq.usp.br/aulas/lce306/Aula12_2012_Geada_modificada_em_pdf.pdf. Accessed at 16 Aug 2014
Silva JLS (1998) Produtividade de componentes de um sistema silvipastoril constituído por Eucalyptus saligna e pastagens cultivada e nativa no Rio Grande do Sul. Thesis, Univesidade Federal de Viçosa
da Silva SC, Carvalho PCF (2005) Foraging behaviour and herbage intake in the favourable tropics/subtropics. In: McGilloway DA (Org.). Grassland: a global resource. Wageningen Academic, Wageningen, pp 81–95
Silva JLS, Garcia R et al (1996) Desempenho de bovinos e seus efeitos sobre as árvores em florestas de eucalipto (Eucalyptus saligna) na região fisiográfica da Depressão Central do RS. In: Simpósio Internacional Sobre Ecossistemas Florestais, Belo Horizonte, 1996

Silva JLS, Saibro JC (1998) Utilização e manejo de sistemas silvipastoris. In: Ciclo De Palestras Em Produção e Manejo de Bovinos de Corte. Ênfase: manejo e utilização sustentável de pastagens. Anais. Canoas. ULBRA, pp 3–28

Silva JLS, Saibro JC, Soares LHB (1993) Desempenho produtivo de bovinos via pastejo do sub-bosque forrageiro em mata de eucalipto. In: Reunião anual da Sociedade Brasileira de Zootecnia, 30. SBZ. Anais, Rio de Janeiro, pp 61

Silva JLS et al (1998) Desempenho animal e de pastagem de inverno em sistema silvipastoril com eucalipto no ano de seu estabelecimento no RS. In: Proceedings Reunião Anual da Sociedade Brasileira de Zootecnia, 1998

Silva JLS, Castilhos ZMS, Savian JF et al (1999) Desempenho animal e forragem residual em sistema silvipastoril com acácia-negra (Acacia mearnsii de Wild) e pastagens de verão no RS. In: XXXVI Reunião anual da Sociedade Brasileira de Zootecnia. SBZ. Anais. Porto Alegre, pp 015

Silva JLS, Saibro JC, Sousa ZMC (2001) Situação da pesquisa e utilização de sistemas silvipastoris no rio grande do sul. In: Sistemas Agroflorestais Pecuários: Opções De Sustentabilidade Para Áreas Tropicais e Subtropicais. Carvalho MM, Alvim MJ, Carneiro JC (Org.) Brasília, pp 258–283

Silva JLS et al (2009) Desenvolvimento de sistemas silvipastoris no Rio Grande do Sul. In: Workshop Integração Lavoura-Pecuária-Floresta Na Embrapa Resumos e palestras apresentados. Brasília: Embrapa, 2009

Silva JLS, Varella AC, Saibro JC, Castilhos ZMS (2011) Manejo de animais e pastagens em sistemas de integração silvipastoril. Embrapa Clima Temperado Documentos 335. Pelotas, 98 pp

Soares AB, Sartor LR, Adami PF et al (2009) Influência da luminosidade no comportamento de onze espécies forrageiras perenes de verão. Braz J Ann Sci 38:443–451

United Nations/FAO (2014) El estado de los Bosques del Mundo-Potenciar los beneficiossocioeconómicos de los bosques. Roma. Available in: http://www.fao.org/3/a-i3710s.pdf. Accessed at 14 Sept 2014

Varella AC (1997) Uso de herbicidas e de pastejo para o controle da vegetação nativa no ano do estabelecimento de três densidades de Eucalyptus saligna Smith. Dissertation, Universidade Federal do Rio Grande do Sul

Varella AC, Saibro JC (1999) Uso de bovinos e ovinos como agentes de controle da vegetação nativa sob três populações de eucalipto. Braz J Ann Sci 28(1):30–34

Varella AC, Ribaski J, Silva VP et al (2008) Recomendações para a escolha e manejo de plantas forrageiras em sistemas silvipastoris no Sul do Brasil. Série EMBRAPA, Documentos 76

Varella AC, Silva VP, Ribaski J et al (2009) In: Fontaneli RS, Santo HP, Fontaneli RS (Org) (eds) Estabelecimento de Plantas Forrageiras em Sistemas de Integração Floresta-Pecuária no Sul do Brasil. Forrageiras para a Integração Lavoura-Pecuária-Floresta na Região Sul-Brasileira Embrapa Trigo, Passo Fundo, pp 283–303

Varella AC, Moot DJ, Pollock KM, Peri PL, Lucas RJ (2011) Do light and alfalfa responses to cloth slatted shade represent those measured under an agroforestry system? Agrofor Syst 81:157–173

Watson VH, Hagedorn C et al (1984) Shade tolerance of grass and legume germplasm for use in the southern forest range. J Range Manage 37(3):229–232

Zanini GD, Santos GT, Sbrissia AF (2012) Frequencies and intensities of defoliation in Aruana Guinea grass swards: accumulation and morphological composition of forage. Rev Bras Zootech 41:905–913

# Opportunities and Challenges for Silvopastoral Systems in the Subtropical and Temperate Zones of South America

11

Pablo Luis Peri, Francis Dube, and Alexandre Costa Varella

**Abstract**

The aim of this chapter is to synthesize the features of silvopastoral system research and practices presented in the previous chapters. Also, based on the analysis of the various systems presented so far, there are recommendations on future research and policy to promote silvopastoral systems in the region, and thus capitalize on the various opportunities and benefits that these productive systems provide.

**Keywords**

Animal component • Institutional structure • Policy guidelines • Productive silvopastoral systems • Tree component • Research • Undestorey component

## 11.1 Overview

The current importance of silvopastoral systems in different regions of South America can be determined by the extent of silvopastoral practices in the field and the amount of research that is carried out. This book contains information on different aspects (pasture, animal and tree production, carbon sequestration, conservation) of silvopastoral systems in native forests and tree plantations in southern South America (Argentina, Chile and Southern Brazil). Based on original research articles, case studies, and regional overviews, the book summarizes, for the first time, the current state of knowledge on:

1. Highly productive systems in the humid Mesopotamia region (Misiones and Corrientes provinces in northeastern Argentina) (Chap. 2)
2. Lower Delta of the Paraná River with production of poplar and willow for multiple uses (Chap. 3)
3. Indigenous forest of the western Chaco (Chap. 4)

P.L. Peri (✉)
Universidad Nacional de la Patagonia Austral (UNPA), Instituto Nacional de Tecnología Agropecuaria (INTA), CONICET, CC 332, 9400 Río Gallegos, Argentina
e-mail: peri.pablo@inta.gob.ar

F. Dube
Department of Silviculture, Faculty of Forest Sciences, University of Concepción, Victoria 631, Casilla 160-C, VIII - Concepción, Chile

A.C. Varella
Brazilian Agricultural Research Corporation (EMBRAPA), South Livestock Center (CPPSUL), BR 153, Km 603, 242, 96401-970 Bage, RS, Brazil



P.L. Peri et al. (eds.), *Silvopastoral Systems in Southern South America*, Advances in Agroforestry 11, DOI 10.1007/978-3-319-24109-8_11

4. Development of silvopastoral systems based on ponderosa pine plantations established on natural grasslands in northwestern Patagonia (Chap. 5)
5. Southern Patagonia experiences with silvopastoral systems in native *Nothofagus antarctica* forests (Chap. 6)
6. Silvopastoral systems with *Acacia*, *Atriplex* and *Prosopis* in arid and semiarid zones of Chile (Chap. 7)
7. *Nothofagus obliqua*- and *Pinus radiata*-based silvopastoral systems in semiarid to humid zones of the Mediterranean region and silvopastoral systems with exotic species such as Cherry, Poplar, Stone pine and Walnut in Chile (Chap. 8)
8. Silvopastoral systems with introduced species such as *Pinus contorta*, *P. ponderosa* and *Pseudotsuga menziesii*, and native *Nothofagus pumilio* and *N. antarctica* forests in the Aysén and Magallanes regions of Patagonia (Chap. 9)
9. Silvopastoral systems in southern Brazil that include a variety of $C_3$ and $C_4$ pastures (native and exotic), beef cattle, performance sheep and fast-growing trees species (Chap. 10).

There is a wide range of subtropical and temperate silvopastoral systems that have been established for a variety of purposes, using several species in different spatial and temporal configurations. The main types of silvopastoral systems chosen by the authors to illustrate traditional practice and current research are summarized in Table 11.1. This range of productive systems is a clear illustration of how farmers have integrated tree and pasture/grassland species in their land use systems to reach higher production per unit of land area, risk avoidance, product diversification, and sustainability. Thus, silvopastoral systems practiced in the region provide multiple products (e.g., food, wood, fodder) and services (e.g., maintenance of soil fertility, control of erosion, microclimate improvement, biodiversity enhancement, watershed protection, and carbon sequestration). Also, the range of conditions in southern South America (geography, climate, culture, and markets) has provided the basis for the different silvopastoral systems implemented in the region. The distribution of natural vegetation (rain forests, savannas, dry land ecosystems and cold temperate forests), which is largely determined by climate (mean annual temperatures ranging from 5 to 20 °C, and rainfall ranging from 200 to 2,000 mm), also influences the type of silvopastoral systems.

In the Argentinean provinces of Corrientes and Misiones (Mesopotamia region), silvopastoral systems with highly productive pine trees and $C_4$ grasses have mainly been adopted by cattle farmers as an alternative to diversity production and increase the profitability as compared with traditional farming systems. Farmers have requested technologies for the establishment and management of silvopastoral systems in order to obtain a better timber quality, suitable forage resources and higher stocking rates. Scientists and land managers have studied these tree-pasture-animal synergies specifically on the effect of planting arrangements, planting densities, and tree combinations on the productivity of the forage and animal components; these research results have helped in enhancing the adoption of silvopastoral systems from small farmers to large-scale companies.

The high demand for beef production in the Delta region (Argentina) during the recent past has led to the need to move from the traditional plantation systems to more intensive silvopastoral systems that combine production of wood and cattle. Current research work on these riparian systems is very much concerned with the development of technologies for the design of a forest-forage system that should lead to sustainable cattle-raising. The acquired knowledge now used to focus on the establishment of poplars and willow-based silvopastoral systems aiming at production of high quality sawn wood, veneers and wood pulp, as well as biomass for bio-energy.

In the western Chaco region (Argentina), silvopastoral systems in mixed native forests and degraded savannas are being designed to specifically address the problem associated to dense shrub thickets and overstocked secondary forests caused by livestock overgrazing, over logging, changes in the fire regime and fencing (Kunst et al. 2006). This process known as "woody plant encroachment" has occurred in many parts of the world, especially in African savannas (Sankaran et al. 2005). Woody plant communities reduce

**Table 11.1** The importance of the silvopastoral systems reported for the temperate and subtropical zones of South America

| Region country | Systems described | Main tree species component | Main understory component | Main livestock component | Extent, comments |
|---|---|---|---|---|---|
| Mesopotamia, Argentina | Silvopastoral | Mainly evergreen *Pinus taeda*, *P. elliotti*, hybrid pines | *Brachiaria brizantha*, *Axonopus* sp. | Cattle (Bradford, Brangus) | Mainly small and medium-sized farms as an alternative to diversify production, cattle raising as their main activity |
| | Intercropping | *Grevillea robusta*, *Eucalyptus* sp., pine | Yerba mate grasses, cassava, tobacco | – | Small commercial farm and subject of further research |
| Delta, Argentina | Silvopastoral | Deciduous *Populus* sp., *Salix* sp | Natural grasslands, *Lolium multiflorum*, *Brommus catharticus* | Cattle (Aberdeen Angus, Hereford) | From small and medium-sized farmers fir diversification alternatives to forest companies aiming at quality wood production |
| Chaco, Argentina | Silvopastoral, forest and degraded savanna grazing | Mixed and secondary native forest of *Schinopsis* l*orentzii*, *Aspidosperma quebracho-blanco Prosopis* sp. | Native grasslands, *Cenchrus ciliaris cv. Texas* (Buffel grass); *Panicum maximum* (Gatton panic) | Large farmers: cow-calf operations (criollo, British and half Zebu breeds) | Low intensity roller-chopping' (RBI) to manage mainly shrub communities (secondary forests and shrub thickets) is proposed at production level |
| | | | | Small farmers: mixed, cattle and goats | |
| Patagonia, Argentina | Silvopastoral | *Pinus ponderosa* | Natural grasslands, mainly *Festuca pallescens* and wetlands | Criollo goat, cattle (Hereford), sheep (Merino) | Transhumance livestock based on goat. Low adaptation of silvopastoral systems compared with detailed research information and numbers of productions units |
| | Silvopastoral, Forest grazing | Deciduous *Nothofagus antarctica* native forest | Natural grasslands, naturalized species such as *Dactylis glomerata*, *Trifolium repens* | Cattle (Hereford), sheep (Corriedale) and mixed (cattle + sheep) | Extensive production |

(continued)

**Table 11.1** (continued)

| Region country | Systems described | Main tree species component | Main understory component | Main livestock component | Extent, comments |
|---|---|---|---|---|---|
| Arid and semiarid zones, Chile | Alley-cropping, riparian strip shelterbelt | *Acacia* sp., *Atriplex* sp., *Prosopis* sp. | Alfalfa, fodder trees | Small farmers: mixed cattle, goats and sheep | Small research plots |
| | | | | | Extensive semi natural |
| | | | | | Timber belts containing quality timber species, charcoal production, erosion control, protein bank, fodder, live fences |
| Temperate Mediterranean zones, Chile | Alley-cropping, forest grazing, woodlot | *Pinus radiata*, *P. pinea*, *Nothofagus obliqua*, *Juglans regia*, *Prunus avium*, *Populus* sp. | Natural grasslands, *Avena strigosa*, *A. sativa*, *Dactylis glomerata*, *Festuca* sp., *Lolium perenne L. multiflorum*, *Medicago sativa*, *Phalaris acuatica*, *Trifolium repens*, *T. subterraneum* | Small farmers: mixed cattle and sheep, beef cattle (Hereford, red Angus) | Applied by forest companies and partnership programs, Extensive production, Small and medium-sized farms to diversify production, Weed control, wrosion control, fuelwood, sawn wood, charcoal, high-value timber species, fodder, fruit and nut production, small and medium research plots |
| Patagonia, Chile | Forest grazing, woodlot | *Pinus contorta*, *P. ponderosa*, *Pseudotsuga menziesii*, *Nothofagus antarctica*, *N. pumilio* | *Dactylis glomerata*, *Festuca* sp., *Medicago sativa*, *Trifolium* sp., *Lolium perenne*, *Holcus lanatus*, *Poa pratensis* | Beef cattle (black Angus, Hereford), sheep, dairy cows (Holstein) | Mainly medium-sized farms as an alternative to diversify production, extensive cattle farming, animal shelter, windbreaks for pasture and forages, woodlots |

| | | | | | |
|---|---|---|---|---|---|
| Cool zone, Southern Brazil | Silvopastoral | *Eucalyptus* sp.; *Pinus* sp.; *Acacia mearnsii*; *Grevillea robusta* | Annual and perennial temperate and tropical forages | Beef cattle and sheep | Increasing wood production in traditional beef and sheep cattle farming |
| | Intercropping with rotational understory plants | *Eucalyptus* sp; *Acacia mearnsii*; Yerba Mate (*Ilex paraguariensis*); *Mimosa scabrella* | Maize; sorghum; soybean; sunflower; watermelon (1st and 2nd years), annual and perennial temperate forages (3rd and following years) | Beef cattle from 3rd and following years | Applied by forestry companies and their partnership programs |
| | Silvopastoral for animal shelter and soil amelioration | *Eucalyptus* sp.; *Corymbia* sp.; *Grevillea robusta*; *Araucaria angustifolia*; *various native trees* | Grasslands; *Brachiaria* sp.; *Paspalum* sp.; ryegrass | Beef and dairy cattle | Wide spaced trees planting for animal protection from extreme climate; Trees planting contour for soil and water conservation |
| | Forest grazing | *Araucaria angustifolia forest*; *other native forests* | South Brazilian grasslands | Beef cattle | Practiced in extensive beef cattle farming. Animal taken to forest grazing in autumn and winter seasons |
| | Fruits and nut trees | Temperate fruits and nut trees | Ryegrass; oats; white clover | Weaned calves or heifers; sheep | Traditionally planted for shelter and soil amelioration |

forage availability due to competition for resources (sunlight and water) and hinder livestock and personnel movements due to their high stem density and thorns. In this context, it is interesting to note that silvopastoral system in western Chaco are established using the mechanical treatment called "low intensity roller-chopping" (RBI) based on a mechanical disturbance that improves the system productivity while biodiversity, soil properties and tree advance regeneration are preserved.

In northern Patagonia (Argentina) 82,000 ha have been afforested with exotic coniferous species. In northern Neuquén, the province with the greatest forest development, smallholders who practice the traditional transhumance livestock based on "local criollo goat" use as much as 90 % of the land. However, in the south-central region, 60 % of high forestry potential lands are found within large private landowners who are dedicated to extensive cattle. Although pine forest plantations have shown to be a profitable and competitive activity compared to extensive livestock raising after 40 years of state promotion, only 10 % of the regional potential has been achieved. It comprises the cattle culture of the region, which in the last 10 years began to promote silvopastoral systems based on the conversion of plantations initially established on natural grasslands in the forest-steppe ecotone. Currently, approximately 15,000 ha of silvopastoral systems with cattle have been implemented and five forestry subsidy projects for smallholders to implement agroforestry systems with "local criollo goat" have been presented to authorities. However, these advances contrast with the great knowledge reached from research trials on *Pinus ponderosa*-based silvopastoral systems that the authors reviewed.

Another extensive system in Patagonia (Argentina) includes silvopastoral systems in native *Nothofagus antarctica* forest that become an economically, ecologically and socially productive alternative. There is valuable information from the Patagonian experience (practice and research) in this type of silvopastoral system that was gained in the last 15 years. As a result, the productivity and nutritive value of understory grassland interacting with environmental and silvicultural practices, the livestock production, the adaptive silvopastoral management (strategic separation of homogenous areas, stocking rate adjustments, protection of regeneration from herbivores browsing) and the existence of ecophysiological data (litter decomposition, nutrients dynamic and carbon storage) is well known. The economic and biodiversity impact of native forest grazing is now giving way to integrated land-use systems. And as in other temperate regions of the world, silvopastoral systems show potential for the alleviation of environmental problems from overgrazing.

In Chile, the Forestry Institute (INFOR) has developed the National Agroforestry Program (NAP) during 2002–2014 with the objective of generating the required information needed for the establishment of suitable agroforestry systems in the country. This includes the study of the species having the highest potential for these systems, and the establishment of demonstration agroforestry units to be used as extension tools and to encourage its usage, preferably by smallholder farmers. Sources of funding included those provided directly by the Ministry of Agriculture and the Institute for Agricultural and Livestock Development (INDAP). Between 2006 and 2013, INFOR has established 63.5 ha of alley cropping systems on small farms, 495.1 ha of silvopastoral systems, 486.8 ha of windbreaks, 15.3 ha of riparian buffer strips, 31.9 ha of bioenergy trials and 21.1 ha of beekeeping systems, totaling nearly 1114 ha in 1600 properties. As such, silvopasture represents 44.4 % of the overall extent of agroforestry systems implemented by INFOR across Chile. According to several scientific studies that were performed during that period, alley cropping systems have led to reduction in soil loss by 1,700 %; additionally, the establishment of trees in pastures has reduced wind-speed by 200 %, and increased prairie productivity by 41 % (Sotomayor 2015).

In arid and semiarid Chile, reforestation with exotic and indigenous species has permitted to recover highly degraded soils and produce firewood and fodder for livestock. These tree species, which are often leguminous and nitrogen fixers, are well adapted to areas with minimum amount of water throughout the year. They also grow well on low fertility soils, and under conditions of high sun radiation and extreme diurnal temperature variations. They are also an important resource for the local peasants and indigenous communities as they provide them with food and shelter for their livestock.

In the temperate region of central to south-central Chile (Mediterranean zone), a silvopastoral system that was recently established in an over mature *Nothofagus obliqua* forest in the foothills of the Andes is described in details (Dube et al. 2015). Historically, an important portion of *Nothofagus* forests in the Biobío Region have been over exploited either for cattle grazing or fire/sawn wood production. Non-systemic cattle grazing impede the establishment of natural regeneration and cause soil compaction from trampling and overstocking. Such a trial is relevant in the prevailing situation whereby the National Forest Service does not have any regulation regarding the use of forests for pastures, knowledge on adequate stocking rates for cattle is insufficient. The objectives of the study are to: (i) assess the limits of compatibility (mid and long-term) between silvopastoral and conventional management of forest properties with several stands of over-mature *Nothofagus obliqua*; (ii) optimize silvopastoralism through the establishment of a system that includes summer and winter operations, by respecting the principle of sustainability in the use of the forest ecosystem by preventing soil degradation, excluding livestock from riparian areas and allowing the gradual renewal of forest resources, leaving open the possibility of a future transition to a traditional use of the forest, if productivity so permits, or as socio-economic or environmental considerations should so require; and (iii) provide greater stability to the small farm or campesino families, by training them to manage a technically feasible system that is sustainable under the given conditions, in both the mid and long terms.

Chapter 8 also discusses the technical and economic aspects of *Pinus radiata*-based silvopastoral systems. The results of the study showed that radiata pine had promising potential for use in silvopastoral systems in diverse environments, from semiarid to humid zones, being the most employed by small agroforestry producers. Forest companies through partnership programs also encourage the introduction of cattle in their radiata pine plantations at specific development stages in order to reduce the amount of fuel load (through grazing of natural pasture). Finally, small and medium-sized farmers are gradually adopting high-quality hardwood species to diversify the overall production and obtain added-value products from their property. The experimental trials that were established in the last 20 years, which include exotic species such as Cherry (*Prunus avium*) and Walnut (*Juglans regia*) have shown to be successful, and therefore promising for the development of the region as they will help improve the quality of life of small farm owners through the regular income from the sale of diversified woody and non-woody forest products.

In Chilean Patagonia (Aysén and Magallanes regions), the principal uses of agricultural lands are as natural prairies for extensive cattle farming. The limited rates of afforestation are due to the low acceptance by cattle farmers for traditional forestry activities. Pronounced slopes that increase the risk of erosion, low-fertility soils, limited road connectivity, and large distances from the main economic and production centers undoubtedly affect the capacity for agricultural production (SERPLAC 2005). Adverse weather conditions, such as low winter temperatures and strong spring and summer wind gusts are largely responsible for the extensive low-productivity cattle farming in the region. However, windbreaks for pastures and forages and other established silvopastoral systems (e.g., shelterbelts, woodlots) have shown to help solve the problems associated with soil erosion and low animal productivity. For instance, conifer crown coverage reduced the wind speed by 200 % in silvopastoral systems compared to open pastures. Over 12 years of research have demonstrated that silvopastoral systems with introduced cold-tolerant conifers are feasible and beneficial, and that silvopastoralism in *Nothofagus pumilio* and *N. antarctica forests* can be used provided forest sustainability requirements are emphasized. As such, a systemic approach consisting of the forest, prairie and cattle components should always be considered when designing such silvopastoral systems.

In southern Brazil, agroforestry practices probably dated from the Portuguese and Spanish colonization period. However, the first reported agroforestry experience was "Faxinal System" with Yerba Mate (*Ilex paraguariensis*) and Bracatinga tree (*Mimosa scabrella*) in the 1970s (Ribaski and Ribaski 2015). Since then, several other agroforestry systems have been developed and applied by farmers and forest companies, and scientists have published a number of studies, in particular on silvopastoral systems. Chapter 10 initially emphasizes the history and importance of integrated livestock-forest systems in the cold subtropical zone of Brazil and focuses on related experiences and main scientific data that were collected between latitudes 24° and 33° S, comprising the states of Rio Grande do Sul, Santa Catarina and the south-central region of Paraná. This chapter also characterizes the current land uses in southern Brazil, emphasizing the predominance of rangelands (about 50 % of total arable land area), whereas commercial forests only occupy 12.5 %. In contrast, the increasing demand and insufficient wood supply as the Brazilian market opened, economy grew and middle class consumers increased is highlighted. This scenario, along with the prices and market instability in agriculture and livestock sectors, and the increase of concerns on environmental and animal welfare have brought opportunities for integrating forestry with livestock systems from the 1990s. As a result, the scientific community performed a number of researches, which results are reported in Chap. 10. The main focus areas are: Potential of forage plants and trees for silvopastoral systems; Pasture and tree management; and Animal performance, behavior and management under trees. In addition, examples of successes and experiences of tree-pasture-agriculture integration from the private sector are also detailed in the chapter. Finally, Chap. 10 addresses the main challenges and opportunities for research, development and technology transfer on silvopastoral systems in southern Brazil.

## 11.2 Key Research Needs

Analyzing the literature (1997–2012) that provided technical support to the implementation and management of silvopastoral systems in Argentina, it was deduced that most of the information was generated in the pasture/grassland component (Fig. 11.1). The knowledge generated in regard to the production and management of the tree component (24 %) represented almost twice the animal component. However, in most of the country the animal production represents the main annual income of silvopastoral systems. This highlights the lack of information related to animal production, animal welfare and livestock management. A shortcoming was also detected in the knowledge regarding environmental services of silvopastoral systems, mainly for those developed in native forest. Thus, only 8 % of the published works in the country provided information on environmental services such as biodiversity conservation, carbon sequestration, its protective soils and watersheds functions (including aspects of water quality). Similarly, information related to social (such as employment, working conditions, land tenure) and economic aspects (profitability, added value of products, markets, and so on) of silvopastoral systems is scarce.

Another aspect is the importance of spatial and temporal scales of studies that are performed in silvopastoral systems in Argentina, since the patterns and processes that occur in these production systems are dependent on the scale on which they have been observed. Most studies corresponded to a spatial scale study (plots, sample units) less than 500 $m^2$, decreasing the proportion of works as the spatial scale increases (Fig. 11.1). From this, 57 % of the information in silvopastoral systems has been generated at a spatial scale less than 1 ha. In many cases the findings from research have been extrapolated to larger scales. Therefore, studies at higher scales (landscape level) are still needed to answer important aspects of these production systems such as the strategic

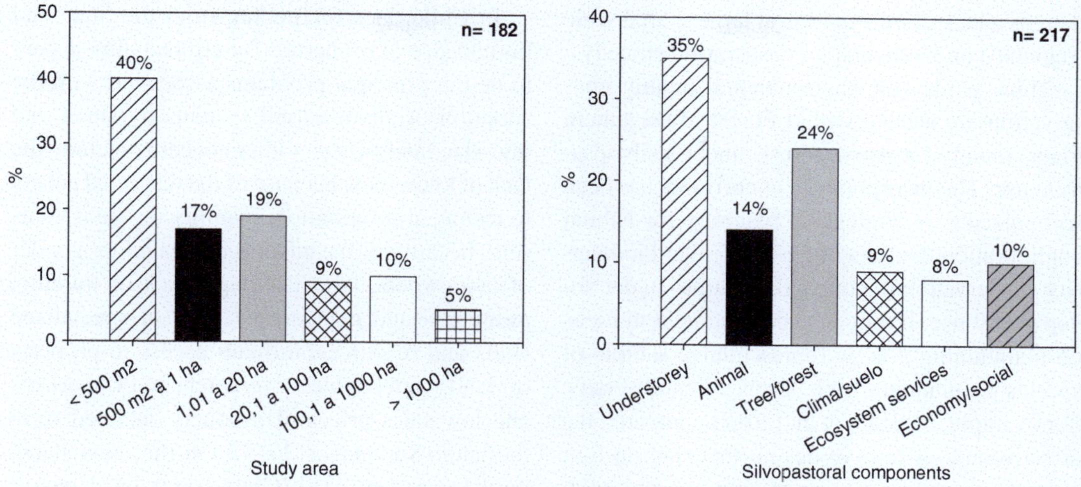


**Fig. 11.1** Percentage distribution (%) of the information related to the different components and according to the spatial scale (plot size or area of study) of silvopastoral systems developed in Argentina (Source: Peri 2012)

use of the environment at farm level, adjustment of stocking rates, more realistic lambing values, personnel management, water availability for animals, and predator effects. Reviewing the temporal scale of the silvopastoral research in Argentina, it was determined that 42 % of the studies were evaluated for at least 1 year, 36 % from 1 to 2 years, and the remaining 22 % of the works had results of more than 2 years (Peri 2012). However, there are processes occurring in silvopastoral systems over a longer time scale, such as the impact of livestock on plant communities, assessing the growth of the tree component under silvicultural treatments interacting with the understory, and the effect of grazing on soil physicochemical processes or on wildlife habitat. Short-term studies do not reveal slow changes, which occur over years and decades, or allow the interpretation of the cause and effect relations of these slow changes. Therefore, it is important to conduct ecological research at temporal scale broader than it has been possible with conventional research funding, especially in silvopastoral systems carried out in native forests (Chaco and Patagonia). Understanding the importance of the spatial-temporal aspects in silvopastoral systems allow us to achieve a better strategic planning for future studies and facilitate the interpretation of the derived results.

Biodiversity and ecosystem dynamics are complex, and environmental drivers, both of natural and anthropogenic origin, induce changes that are commonly translated into changes in ecosystem services and human wellbeing. In South America, silvopastoral systems became an economical, ecological and social productive alternative being in some areas more productive, profitable and sustainable than forest monocultures or sole animal production systems. Usually, silvopastoral systems show higher conservation value compared with other productive land uses (intensive forestry or agriculture). However, livestock grazing is one of the most widespread land uses in South America and is arguably the land use that has had the greatest impact on regional biodiversity. Lands modified by human activities usually have less habitat diversity and complexity than natural systems, and offer poorer food and shelter sources for many native species (Brockerhoff et al. 2008). There are multiple factors to consider, including species diversity, spatial extent, and future climate changes. In this book, information on biodiversity in silvopastoral systems and the use of bioindicators at stand level have been reported. This may provide early warning indicators of environmental changes, monitor a specific ecosystem stress or indicate levels of taxonomic diversity at a site. However, there is a lack of information at large spatial scale (regional) on biodiversity conservation strategy.

Thus, guides for environmental quality conservation are needed within Forest Management Plans, mainly for native forests under a silvopastoral use. The incorporation of corridors has been recognized as a strategy in planning for habitat configuration, and may be used to plan biodiversity conservation in consolidated areas under silvopastoral use. It may also be beneficial to focus on maintaining a large representative sample of species to maintain high levels of biodiversity. For example, Mokany et al. (2013) compared the effectiveness of reserve design focusing on four strategies: connectivity, aggregation, representativeness and a balance of these three. Their results indicated that the best design depends strongly on the conservation goals of the designers.

The concept of ecosystem services has recently received increasing attention in scientific and policy contexts because of its capacity to bridge the connections between ecosystems and social systems (Reyers et al. 2013), as well as to integrate ecological, socio-cultural and economic approaches to knowledge building and policy development (Chan et al. 2012). The Millennium Ecosystem Assessment provided the most comprehensive assessment of the state of the global environment and classified ecosystem services in (i) supporting services (the services that are necessary for the production of all other ecosystem services including soil formation, photosynthesis, primary production, nutrient cycling and water cycling), (ii) provisioning services (products such as food, fiber, fuel, genetic resources, biochemicals, natural medicines, ornamental resources and fresh water), (iii) regulating services (climate, water and erosion regulation, disease regulation, pollination), and (iv) cultural services (the non-material benefits that people obtain from ecosystems through spiritual enrichment, recreation and aesthetic). Among these recognized categories of ecosystem services, cultural and regulating ecosystem services are those that have received less scientific attention in silvopastoral systems in South America, although their human demand will increase in the next years in industrialized and rural societies.

In Chile, lack of funding from the State and limited size of properties for afforestation appear to be the principal problems associated with the adoption of silvopastoral systems by small and mid-size landowners. Other concerns include the lack of know-how because of the restricted access to technical information, inability to "do it yourself" because of the advanced age and, or gender of landowners, labor shortage, lack of training, awareness and competence, lack of specialized skills and resources, difficult access to property or land, indebtedness, remoteness of markets, and low sales prices of products obtained from the land (Sotomayor 2015). On the other hand, until December 2012, Chile had a legal instrument that promoted afforestation (Decree Law No. 701 from 1974 and amendments), which contemplated incentives to establish silvopastoral systems and windbreaks in private properties. Unfortunately, the ending of Decree Law 701 aggravated the situation of unavailable financing as indicated by farmers, making even more difficult the promotion of these systems among the community (Sotomayor 2015).

Rural communities depend mostly on what can be grown and raised on their land, which rarely exceed 20 ha. Recent socioeconomic studies (e.g., AMBAR 2010) have shown that several rural households are classified as being poor, a condition that is often related to the inefficient use of local and natural resources and inadequate planning in the mid and long terms. Given the magnitude of the problem, the Faculties of Forest Sciences, Agronomy and Social Sciences at the University of Concepción decided to create the Center for Investigations in Agroforestry (CIAF). The Center will (i) conduct and promote interactive research (with true replicates, as opposed to Demonstration Units which usually consist of pseudo replicates) agroforestry practices that improve the production and environmental protection functions (related to water, soil, vegetation) of farm and forestland; (ii) conduct and promote inter and trans disciplinary research on the social, economic, market and policy dimensions of agroforestry systems; (iii) implement and conduct a technology transfer and extension program in order to increase awareness of the various stakeholders and increase the adoption of sustainable agroforestry practices; and (iv) develop and implement a collaborative agroforestry program with national (e.g., INFOR, INIA, INDAP) and international investigation centers (e.g., INTA, GIRAF-Laval University, UMCA) in the areas of research, teaching and technology transfer.

Research priorities in applied sciences to be addressed by CIAF shall include: (i) role of silvopastoralism in climate change adaptation and mitigation (carbon sequestration, GHG emissions, soil conservation and improvement); (ii) production of biomass for bioenergy; (iii) use of riparian buffer strips to protect watercourses; (iv) establishment and compatibility of silvopastoral systems with native species (in managed indigenous forest, and since the establishment of native species plantations); (v) maintenance of water quality and quantity near silvopastoral systems, protection of non-woody forest products, maintenance of biodiversity. Additionally, the lines of research in social and rural development are: (i) development of silvopastoral schemes tailored to the needs of First Nations (Lafkenches and Pehuenches) and other local communities; (ii) global sustainability of production systems (economic sustainability and product diversification, and environmental sustainability and biodiversity conservation); and (iii) optimization of the overall productivity of forestry, agricultural and livestock components.

In South Brazil, traditional agriculture, livestock and forestry systems rather than integrated systems are still predominant. However, pressures from environment, market (prices and consumer preferences), labor availability, land-use efficiency and government (taxes, social function of land) on farms have brought opportunities to the integrated systems, such as silvopastoral systems. Therefore, the proposition of feasible models for forest-livestock systems in contrast to traditional cropping systems, the extensive cattle and sheep systems and the specialization of plantations still continue to be the main target for research and extension services. In this context, long-term scientific data is necessary to recommend silvopastoral models for the different zones in South Brazil and encourage farmers to invest on them. As shown in Chap. 10, there are different cold zones that would demand a number of silvopastoral models in South Brazil. Additionally, research needs on silvopastoral system could be listed based on technological and socioeconomic approaches. From the technological aspects, priorities are: (i) development of forage (shade and water stress tolerant) and trees (adapted plant architecture, growth and development) cultivars; (ii) quantification of the benefits of tree shading for animal welfare and performance (thermal comfort, live weight gain and grazing behavior); (iii) development of soil fertilization protocols (establishment and maintenance) specific to silvopastoral systems and rotational understory programs; and (iv) setting strategies of animal feeding using silvopastoral systems (protein bank; pasture deferral for feeding shortage periods; rotational and continuous grazing). From socioeconomic aspects, research needs and priorities are: (i) development of silvopastoral models using native trees and grasslands to meet requirements of the Brazilian environmental laws (The Brazilian Forest Code); (ii) valuing environmental services from silvopastoral systems (carbon budgets, mitigation of greenhouse gases, soil and water conservation, biological diversity); (iii) developing silvopastoral models for recovery of degraded land (erosion, weed occupation and burned areas); (iv) implementation of public policies that support a strategy of territorial management using geospatial tools (satellite imagery and geo-referenced information system) capable of supporting decisions and rules for forestry and agroforestry systems and reducing environmental concerns; and (v) building of new training and capacity programs from research institutions to consultants and rural extension staff on silvopastoral systems. The skills to manage integration systems, such as tree-pasture systems require specific knowledge and a systemic approach in contrast to the specialized and intensive systems. Therefore, collaboration between scientific institutions, official extension services and private sector are needed to enable adequate transfer of technologies and increase the land use area with silvopastoral systems in southern Brazil.

## 11.3 Decision Making: Policy and Socioeconomic Aspects

Many challenges and opportunities arise for scientific institutions, rural extension agencies and public policies aiming to promote more sustainable and diversified farming systems. There are differences in the motivation between landowners in the region that determined the contrasting design of the silvopastoral systems, such as improve their economic profits, animal welfare, self-consumption of wood in rural properties, soil conservation and land use efficiency. Limits to wider adoption are perhaps the most complex area of silvopastoral development. Perceptions of land-use specialists, extension agents and the farmers impact on the development of silvopastoral systems in the region. There are also institutional problems linked to adoption by farmers, which are often underestimated by agronomic and forest researchers. The problems associated with forest product harvesting, processing and marketing, together with the strategy of producing added-value wood and animal products also are key factors for silvopastoral system development.

Therefore, legal framework, policy and planning appear to be key areas for silvopastoral system development in the region. In Argentina, National Law No. 26,331 for the Environmental Protection of Native Forests promotes the conservation of indigenous forests through land planning, sustainable management and tightening the regulations associated with land-use change. This requires all the provinces to develop a Land Use Planning Process (LUPP) with respect to native forests in a participatory fashion. In the last few years (2010–2014 period), more than 70 % of the budget has been destined to silvopastoral system plans in native forest (1,423,194 ha) that includes forestry inventories, silvicultural practices, adjustment of stocking rates and fencing for strategic separation in homogenous areas (Peri et al. 2015). Also, the Argentine government enacted National Law 25,080 and 26,432 to provide a federal framework for forest plantation by promoting investments in new forest enterprises through reimbursement of planting costs and tax benefits. This could support silvopastoral systems by financing plantations at low densities (6×6 m) that represent approximately 10 % of the area afforested under the Act (Peri et al. 2015). However, many farmers (as it was highlighted for Patagonia native forest) have been slow to adopt an integrated silvopastoral system management plan, possibly because of the lack of convincing evidence of positive economic returns or long term benefits from ecosystem service values. In this context, an integral economic analysis is required, quantifying timber products, annual animal production, soil conservation benefits, as well as the landscape values associated with silvopastoral systems.

To expand silvopastoral land use systems and farmer adoption, a multi-agency, interdisciplinary and participatory strategy is required. In this context, it is critical to involve landowners who practice silvopastoral systems to share their experiences related to adequate management and land use. For this, it is also important to increase awareness of the interdependence among stakeholders and key actors to address issues in common in silvopastoral practices. In the region, extension advice is provided mainly by National and Provincial institutions (e.g., in Argentina the National Agricultural Institute – INTA). However, in many situations the advisory function is fragmented among several agencies. In any extension program, the perceptions of farmers in relation to the practice of silvopastoral systems on their farms are important. The use of silvopastoral research sites for demonstration developed in different regions of South America by the university and agency-based research is an important additional resource for the learning process and may contribute to farmer adoption. Joint venture schemes with the scientific sector, industry, government, or other financial sources can also be useful to increase silvopastoral adoption. The challenges associated with the adoption of sustainable silvopastoral systems in the region provide the opportunity for production diversification and profitability for farmers and society.

In Chile, a recurring question among national and local government authorities is whether it is possible to prove the contribution of agroforestry to the rural economy and its sustainability, in order to promote its use. Several studies from Chile and other countries value the contribution of trees and shrubs in agroforestry systems in protecting the natural environment, where reduction of erosion and modification of microclimate for the benefit of agriculture and livestock have been documented, especially if they are used with adequate spatial arrangement (Sotomayor 2010; Dube et al. 2012). Some of the benefits that have been identified after traditional cropping, livestock or forest systems were converted into silvopastoral systems include (i) diversification of productive activity of small and medium-size agricultural businesses, making efficient and sustainable use of available resources of the property (Snaydon and Harris 1979); (ii) enhancement of the scenic beauty and the value of the property (Sotomayor 2009); and (iii) soil conservation, protection and improvement. These benefits have undoubtedly helped to promote the use of agroforestry systems in Chile.

## 11.4 Conclusion

Silvopastoral systems in South America are widespread and therefore ecologically and culturally important. In the region, sustainable silvopastoral systems are suggested as a key solution to the conflict between expanding agricultural production and conserving natural ecosystems. Strategic planning with an appropriate combination of pastures/grassland and animals with trees, at real production spatial space and time is needed. Silvopastoral systems interventions and planning at landscape level should combine production with protection functions to meet some of the society expectations. To supplement the knowledge-base on advances in silvopastoral systems in the region covered in this book, further research is required in order to understand the biological basis of any yield advantage, provide information at larger temporal-spatial scales and quantify aspects related to ecosystem services. This should be carried out in a systematic way that would allow the development of guidelines for improving the ecological and economic combining ability of certain trees, pasture/grassland and animals. Finally, improvement of the policy guidelines and institutional structure for extension services that favor silvopastoral systems in the region is required.

## References

AMBAR – Consultoría e Ingeniería Ambiental (2010) Confección y elaboración de una guía de manejo en terreno fiscal Ranchillo Alto en la Región del Biobío, 189p (in Spanish)

Brockerhoff E, Jactel H, Parrotta J, Quine C, Sayer J (2008) Plantation forests and biodiversity: oxymoron or opportunity? Biodivers Conserv 17:925–951

Chan KMA, Satterfield T, Goldstein J (2012) Rethinking ecosystem services to better address and navigate cultural values. Ecol Econ 74:8–18

Dube F, Espinosa M, Stolpe N, Zagal E, Thevathasan NV, Gordon AM (2012) Productivity and carbon storage in silvopastoral systems with *Pinus ponderosa* and *Trifolium* spp., plantations and pasture on an Andisol in Patagonia, Chile. Agrofor Syst 86(2):113–128

Dube F, Müller-Using B, Stolpe N, Zagal E, Doussoulin M (2015) Silvopastoreo en robledales viejos con distintos grados de cobertura como una opción de manejo sustentable en la Región del Bíobío, Chile. In: Proceedings of the 3rd National Congress of Silvopastoral Systems – 8th International Congress of Agroforestry Systems, Iguazú, Misiones, Argentina. INTA Editions, p 490 (in Spanish)

Kunst C, Monti E, Perez H, Godoy J (2006) Assessment of rangelands of southwestern Santiago del Estero for management and research. J Environ Manag 80:248–265

Mokany K, Harwood TD, Ferrier S (2013) Comparing habitat configuration strategies for retaining biodiversity under climate change. J Appl Ecol 50:519–527

Peri PL (2012) Implementación, manejo y producción en SSP: enfoque de escalas en la aplicación del conocimiento aplicado. In: Proceedings of the 2nd National Congress of Silvopastoral Systems, Santiago del Estero, Argentina. INTA Editions, pp 8–21 (in Spanish)

Peri PL, Paez JA, Marcovecchio J, Carranza C, Laclau P, Schlichter T (2015) Política forestal en apoyo a la implementación de sistemas silvopastoriles en Argentina. In: Proceedings of the 3rd National Congress of Silvopastoral Systems – 8th International Congress of Agroforestry Systems, Iguazú, Misiones, Argentina. INTA Editions, pp 391–396 (in Spanish)

Reyers B, Biggs R, Cumming GS, Elmqvist T, Hejnowicz AP, Polasky S (2013) Getting the measure of ecosystem services: a social-ecological approach. Front Ecol Environ 11:268–273

Ribaski J, Ribaski SAG (2015) Sistemas agroflorestais na região sul do Brasil. In: Alves FV, Laura VA, Almeida

RG. Sistemas Agroflorestais. A agropecuária sustentável. Brasília, Embrapa Editions, pp 46–68 (in Portuguese)

Sankaran M, Hanan NP, Scholes RJ, Ratnam J, Augustine DJ, Cade BS, Gignoux J, Higgins SI, Le Roux X, Ludwig F, Ardo J, Banykwa F, Bronn A, Bucini G, Caylor KK, Coughenour MB, Diouf A, Ekaya W, Freal CJ, February EC, Frost PGH, Hiernaux P, Hrabar H, Metzeger KL, Prins HHT, Ringrose S, Sea W, Tews J, Worden J, Zambatis N (2005) Determinants of woody cover in African savannas. Nature 438:846–849

Serplac/Gobierno Regional de Aysén (2005) Atlas Región de Aysén, pp 11–13 (in Spanish)

Snaydon RW, Harris PM (1979) Interactions belowground: the use of nutrients and water. In: Willey RW (ed) Proceedings of the international workshop on intercropping. ICRISAT, Hyderabad, pp 188–201

Sotomayor A (2009) Sistemas silvopastorales, alternativa productiva para un desarrollo sustentable de la agricultura en Chile. In: Proceedings of the 1st National Congress of Silvopastoral Systems, Posadas, Misiones, Argentina. INTA Ediciones, pp 26–47

Sotomayor A (2010) Sistemas silvopastorales, alternativa de producción integrada para un desarrollo sustentable de la agricultura en Chile. Cienc Investig For 16(1):19–51 (in Spanish)

Sotomayor A (2015) Sistemas agroforestales y su contribución a un desarrollo silvoagropecuario sustentable en Chile. In: Proceedings of the 3rd National Congress of Silvopastoral Systems – 8th International Congress of Agroforestry Systems, Iguazú, Misiones, Argentina. INTA Editions, pp 444–452 (in Spanish)

The manufacturer's authorised representative in the EU is Springer Nature Customer Service Centre GmbH, Europaplatz 3, 69115 Heidelberg, Germany. If you have any concerns regarding our products, please contact ProductSafety@springernature.com

Printed and bound by CPI Group (UK) Ltd, Croydon, CR0 4YY
05/10/2026
02228133-0012